W9-BQM-613

PHYSICS

Prisms split white light into the colors of the rainbow.

THE WORLD BOOK ENCYCLOPEDIA OF SCIENCE

VOLUME 2

PHYSICS

WORLD BOOK, INC.
a Scott Fetzer company

CHICAGO

Staff

World Book, Inc.
233 N. Michigan Ave.
Chicago, IL 60601

For information on sales to schools and libraries, call 1-800-WORLDBK (967-5325), or visit our Web site at http://www.worldbook.com

Library of Congress Catalog Card No. 00-109505
ISBN: 0-7166-3399-X (set)
ISBN: 0-7166-3351-5 (vol. 2)
Printed in the United States of America

4 5 6 7 8 9 06 05 04 03 02 01 00

Contents

Preface

Physics, like the other volumes in the *Encyclopedia of Science,* deals with a specific subject, in this case, physics. The subject is introduced through an explanation of the nature and states of matter. Then forces and their effects on static and moving objects are described, followed by an account of the different forms of energy and their various interactions with matter. Finally, this volume contains a section on time and relativity, the latter being probably the most important single advance in modern physical science.

The editorial approach

The object of the *Encyclopedia of Science* is to explain for adults and children alike the many aspects of science that are not only fascinating in themselves but are also vitally important for an understanding of the world today. To achieve this, the books in this series are straightforward and concise, accurate in content, and are clearly and attractively presented.

The often forbidding appearance of traditional science publications has been completely avoided in the *Encyclopedia of Science.* Approximately equal proportions of illustrations and text make even the most unfamiliar subjects interesting and attractive. Even more important, all of the drawings have been created specially to complement the text, each explaining a topic that can be difficult to understand through the printed word alone.

The thorough application of these principles has created a publication that covers its subject in an interesting and stimulating way, and that will prove to be an invaluable work of reference and education for many years to come.

The advance of science

One of the most exciting and challenging aspects of science is that its frontiers are constantly being revised and extended, and new developments are occurring all the time. Its advance depends largely on observation, experimentation, and debate, which generate theories that have to be tested and even then stand only until they are replaced by better concepts. For this reason, it is difficult for any science publication to be completely comprehensive. It is possible, however, to provide a thorough foundation that ensures that any such advances can be comprehended. It is the purpose of each book in this series to create such a foundation, by providing all the basic knowledge in the particular area of science it describes.

How to use this book

This book can be used in two basic ways.

The first, and more conventional, way is to start at the beginning and to read through to the end, which gives a coherent and thorough picture of the subject and opens a resource of basic information that can be returned to for re-reading and reference.

The second allows the book to be used as a library of information presented subject by subject, which the reader can consult piece by piece as required.

All articles are prepared and presented so that the subject is equally accessible by either method. Topics are arranged in a logical sequence, outlined in the contents list. The index allows access to more specific points.

Within an article, scientific terms are explained in the main text where an understanding of them is central to the understanding of the subject as a whole. There is also an alphabetical glossary of terms at the end of the book, so that the reader's memory can be refreshed and so that the book can be used for quick reference whenever necessary.

Each volume also contains a section on the various careers that pertain to the volume's subject.

The sample two-page article *(right)* shows the important elements of this editorial plan and illustrates the way in which this organization permits maximum flexibility of use.

(A) **Article title** gives the reader an immediate reference point.

(B) **Section title** quickly shows the reader how information is arranged within the article.

(C) **Main text** consists of narrative information set out in a logical manner, avoiding biographical and technical details that might tend to interrupt the story line and hamper the reader's progress.

(D) **Illustrations** include specially commissioned drawings and diagrams and carefully selected photographs, which expand, clarify, and add to the main text.

(E) **Captions** explain the illustrations and make the connection between the textual and the visual elements of the article.

(F) **Labels** help the reader to identify the parts of the illustrations that are referred to in the captions.

(G) **Theme images,** where appropriate, are included in the top left-hand corner of the left-hand page, to emphasize a central element of information or to create a visual link between different but related articles.

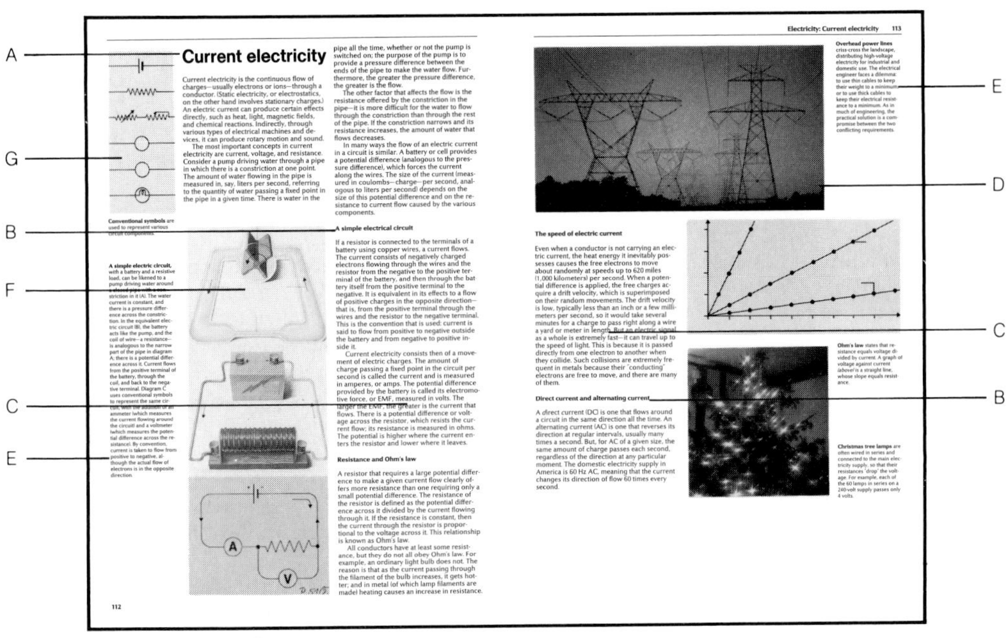

Electricity: Current electricity 113

Current electricity

Current electricity is the continuous flow of charges—usually electrons or ions—through a conductor. (Static electricity, or electrostatics, on the other hand involves stationary charges.) An electric current can produce certain effects directly, such as heat, light, magnetic fields, and chemical reactions. Indirectly, through various types of electrical machines and devices, it can produce rotary motion and sound.

The most important concepts in current electricity are current, voltage, and resistance. Consider a pump driving water through a pipe in which there is a constriction at one point. The amount of water flowing in the pipe is measured in, say, liters per second, referring to the quantity of water passing a fixed point in the pipe in a given time. There is water in the pipe all the time, whether or not the pump is switched on; the purpose of the pump is to provide a pressure difference between the ends of the pipe to make the water flow. Furthermore, the greater the pressure difference, the greater is the flow.

The other factor that affects the flow is the resistance offered by the constriction in the pipe—it is more difficult for the water to flow through the constriction than through the rest of the pipe. If the constriction narrows and its resistance increases, the amount of water that flows decreases.

In many ways the flow of an electric current in a circuit is similar. A battery or cell provides a potential difference (analogous to the pressure difference), which forces the current along the wires. The size of the current (measured in coulombs—charge—per second, analogous to liters per second) depends on the size of this potential difference and on the resistance to current flow caused by the various components.

Conventional symbols are used to represent various circuit components.

A simple electrical circuit

If a resistor is connected to the terminals of a battery using copper wires, a current flows. The current consists of negatively charged electrons flowing through the wires and the resistor from the negative to the positive terminal of the battery, and then through the battery itself from the positive terminal to the negative. It is equivalent in its effects to a flow of positive charges in the opposite direction—that is, from the positive terminal through the wires and the resistor to the negative terminal. This is the convention that is used: current is said to flow from positive to negative outside the battery and from negative to positive inside it.

Current electricity consists then of a movement of electric charges. The amount of charge passing a fixed point in the circuit per second is called the current and is measured in amperes, or amps. The potential difference provided by the battery is called its electromotive force, or EMF, measured in volts. The larger the EMF, the greater is the current that flows. There is a potential difference or voltage across the resistor, which resists the current flow; its resistance is measured in ohms. The potential is higher where the current enters the resistor and lower where it leaves.

A simple electric circuit, with a battery and a resistive load, can be likened to a pump driving water around a closed pipe with a constriction in it (A). The water current is constant, and there is a pressure difference across the constriction. In the equivalent electric circuit (B), the battery acts like the pump, and the coil of wire—a resistance—is analogous to the narrow part of the pipe in diagram A; there is a potential difference across it. Current flows from the positive terminal of the battery, through the coil, and back to the negative terminal. Diagram C uses conventional symbols to represent the same circuit, with the addition of an ammeter (which measures the current flowing around the circuit) and a voltmeter (which measures the potential difference across the resistance). By convention, current is taken to flow from positive to negative, although the actual flow of electrons is in the opposite direction.

Resistance and Ohm's law

A resistor that requires a large potential difference to make a given current flow clearly offers more resistance than one requiring only a small potential difference. The resistance of the resistor is defined as the potential difference across it divided by the current flowing through it. If the resistance is constant, then the current through the resistor is proportional to the voltage across it. This relationship is known as Ohm's law.

All conductors have at least some resistance, but they do not all obey Ohm's law. For example, an ordinary light bulb does not. The reason is that as the current passing through the filament of the bulb increases, it gets hotter; and in metal (of which lamp filaments are made) heating causes an increase in resistance.

112

Overhead power lines criss-cross the landscape, distributing high-voltage electricity for industrial and domestic use. The electrical engineer faces a dilemma: to use thin cables to keep their weight to a minimum, or to use thick cables to keep their electrical resistance to a minimum. As in much of engineering, the practical solution is a compromise between the two conflicting requirements.

The speed of electric current

Even when a conductor is not carrying an electric current, the heat energy it inevitably possesses causes the free electrons to move about randomly at speeds up to 620 miles (1,000 kilometers) per second. When a potential difference is applied, the free charges acquire a drift velocity, which is superimposed on their random movements. The drift velocity is low, typically less than an inch or a few millimeters per second, so it would take several minutes for a charge to pass right along a wire a yard or meter in length. But an electric signal as a whole is extremely fast—it can travel up to the speed of light. This is because it is passed directly from one electron to another when they collide. Such collisions are extremely frequent in metals because their "conducting" electrons are free to move, and there are many of them.

Direct current and alternating current

A direct current (DC) is one that flows around a circuit in the same direction all the time. An alternating current (AC) is one that reverses its direction at regular intervals, usually many times a second. But, for AC of a given size, the same amount of charge passes each second, regardless of the direction at any particular moment. The domestic electricity supply in America is 60 Hz AC, meaning that the current changes its direction of flow 60 times every second.

Ohm's law states that resistance equals voltage divided by current. A graph of voltage against current *(above)* is a straight line, whose slope equals resistance.

Christmas tree lamps are often wired in series and connected to the main electricity supply, so that their resistances "drop" the voltage. For example, each of the 60 lamps in series on a 240-volt supply passes only 4 volts.

C A R E E R S

Electronics engineers design a wide variety of electronic equipment. Here, an electronics engineer works on a robot.

Biophysicists apply the tools and techniques of physics to the study of life processes. Biophysics is considered an *interdisciplinary science* because it combines the theories and principles of biology, chemistry, and physics to explore the atomic details of organic structure and function and the mechanisms of heredity.

Much biophysical research is focused on the relationship between the structure of a molecule and its biological function. Biophysicists use many sophisticated techniques in their work, including electron microscopy, X-ray diffraction, magnetic resonance spectroscopy, and electrophoresis. These techniques allow biophysicists to obtain detailed information about the molecular structure of proteins, nucleic acids, viruses, and parts of cells, such as chromosomes and ribosomes.

Electrical engineers use their knowledge of electrons, magnetic fields, and electric fields to produce and distribute electricity for many different purposes. Electrical engineering is the largest branch of engineering.

Many electrical engineers work in research, development, and design to find new and improved ways of using electric power. Most work for power companies, electrical equipment manufacturers, the construction industry, military and space programs, and government agencies. Others teach at colleges and universities, or work as sales representatives or consultants.

Some electrical engineers design generators that transform water power, coal, oil, and nuclear fuels into electricity. Others design and develop electrical machinery used in factories to manufacture products. Electrical engineers also design the ignition systems used in automobile, airplane, and boat engines.

Electronics engineers are electrical engineers who specialize in electronic equipment. Electronics engineers deal with relatively small amounts of electricity compared with other electrical engineers. Some of these specialists develop master plans for the parts and connections of *integrated circuits,* which control the electric signals in most electronic devices. Others design industrial robots, medical and scientific instruments, communications satellites, and missile-control systems.

Electronics engineers also design such consumer items as television sets and stereo equipment. They often specialize in the design, construction, and programming of complex computer systems. Some apply their skills in telecommunications, where they design systems for transmitting and receiving messages over long distances.

Electro-optical engineers specialize in finding new ways to use laser and fiber-optics technology. They also design, modify, and test laser equipment and systems.

Electro-optical engineers work in many industries, including medicine, computers, robotics, manufacturing, materials processing, and construction. Many electro-optical engineers are employed in the telecommunications industry, where optical fiber cables are fast replacing metal wiring. The defense industry also employs electro-optical engineers, who use laser technology to improve navigation systems and to provide range information for weaponry and missile targets.

Some electro-optical engineers focus their research specifically on optical fibers. They may devise new ways to remove impurities in optical fibers or design optical fibers to exact dimensions and chemical compositions.

Nuclear engineers use their knowledge of nuclear energy to solve engineering problems and find practical applications for the discoveries of nuclear physicists and other scientists. These highly trained professionals are involved in all aspects of the production, use, and maintenance of nuclear energy. They work in many different fields, including medicine, food processing, water-supply management, agriculture, and the power industry.

Most nuclear engineers specialize in a particular area, such as research and development, fuel management, safety analysis, operation and testing, sales, or education. Those who focus on design engineering may design the nuclear reactors used to make nuclear power and their many parts, such as core supports, thermal shields, and safety and control systems. Or they may be involved in the design of equipment used to generate nuclear power, process nuclear fuels, or dispose of radioactive waste. Nuclear engineers also work for public electric utility companies, where they are responsible for supervising operations, monitoring the environmental impact of the plant, and ensuring the proper storage and disposal of nuclear waste. In the field of medicine, nuclear engineers design and build equipment used in the diagnosis and treatment of diseases.

Some of these specialists work for government regulatory agencies, where they set the standards for facilities using nuclear energy. They also conduct inspections of these facilities to ensure that nuclear materials are being used in ways that protect public health and the environment. Nuclear engineers concerned specifically with protecting the health and safety of workers employed at facilities using nuclear energy are known as nuclear health physicists, nuclear criticality safety engineers, and radiation protection engineers. These engineers conduct research on radiation hazards and develop job-training programs that help eliminate the risk of radiation exposure for workers.

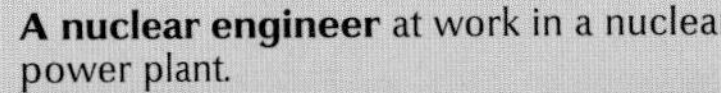

A nuclear engineer at work in a nuclear power plant.

Nuclear medicine technologists use diagnostic imaging equipment to perform tests on patients in hospital and clinics, under the supervision of a physician. Before conducting a test, these technologists prepare and administer one or more radioactive drugs to the patient. Then they use a gamma scintillation camera called a *scanner* to take pictures of the drug as it travels throughout the patient's body or accumulates in certain organs. Throughout the test, the technologist follows these images on a computer screen.

When the test is completed, the technologist performs laboratory tests on the patient's blood volume and fat absorption. The radiologist uses this information to make a diagnosis and recommend a course of treatment.

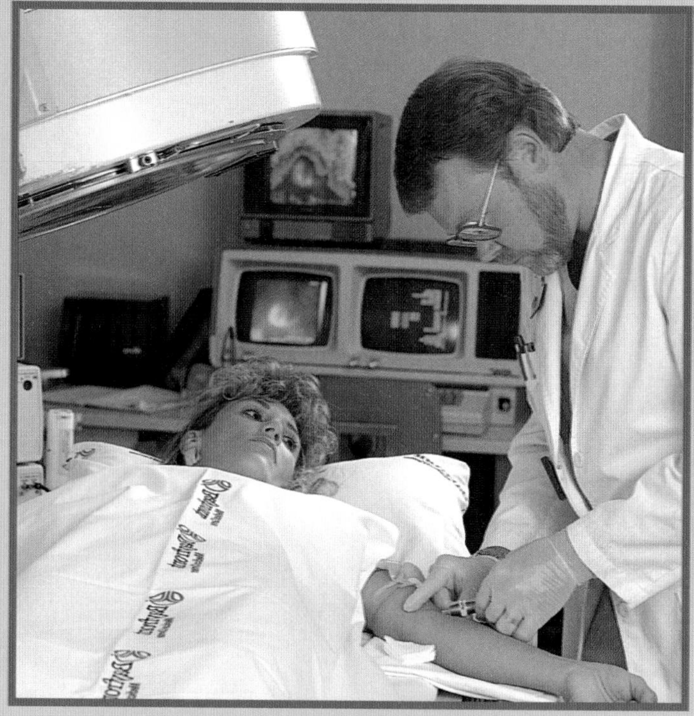

A nuclear medicine technologist prepares a patient for a radiological scan.

Nuclear physicists study the properties, structure, and reaction of the atomic nucleus, focusing their research on radioactivity, fusion, and fission. These physicists gain much of their information from studying nuclear reactions. They use a complex device called a *particle accelerator* to create tiny, high-velocity beams of protons, electrons, and other particles. When one of these particles strikes a nucleus, which results in a nuclear reaction, these researchers use high-precision tools to analyze the radiation emitted during the reaction.

Nuclear physicists use their knowledge of nuclear reactions for peaceful purposes, as well as in the design of nuclear weapons. For example, their research has made many contributions to the field of medicine, such as computerized tomography (CT) and other imaging techniques for diagnosing and treating disease. Research by nuclear physicists has also led to improved methods of sterilizing and preserving food and exploring for oil.

Nuclear physicists at Fermilab in Batavia, Illinois, observe events from collisions between high-energy protons and antiprotons.

Physicists are scientists who study how matter and energy are related to each other and how they affect each other over time and space. More than half of the physicists in the United States are engaged in research and development. These physicists explore the fundamental laws of nature through observations and experiments, and investigate how these laws may be applied to practical problems.

Theoretical physicists formulate laws of physical phenomena and express these laws in mathematical form. Experimental physicists conduct experiments of physical phenomena and compare their results to what was predicted according to the laws and theories developed by theoretical physicists.

Some physicists work in applied research. They use the theories and principles of physics to improve manufacturing products and processes. Others work at colleges and universities, teaching and conducting research. Because physics is such a broad field, most physicists specialize in a particular branch, such as acoustics, the study of sound; fluid physics, the study of liquids and gases; molecular physics, the study of molecules; plasma physics, the study of highly ionized gases; and thermodynamics, the study of heat and other forms of energy.

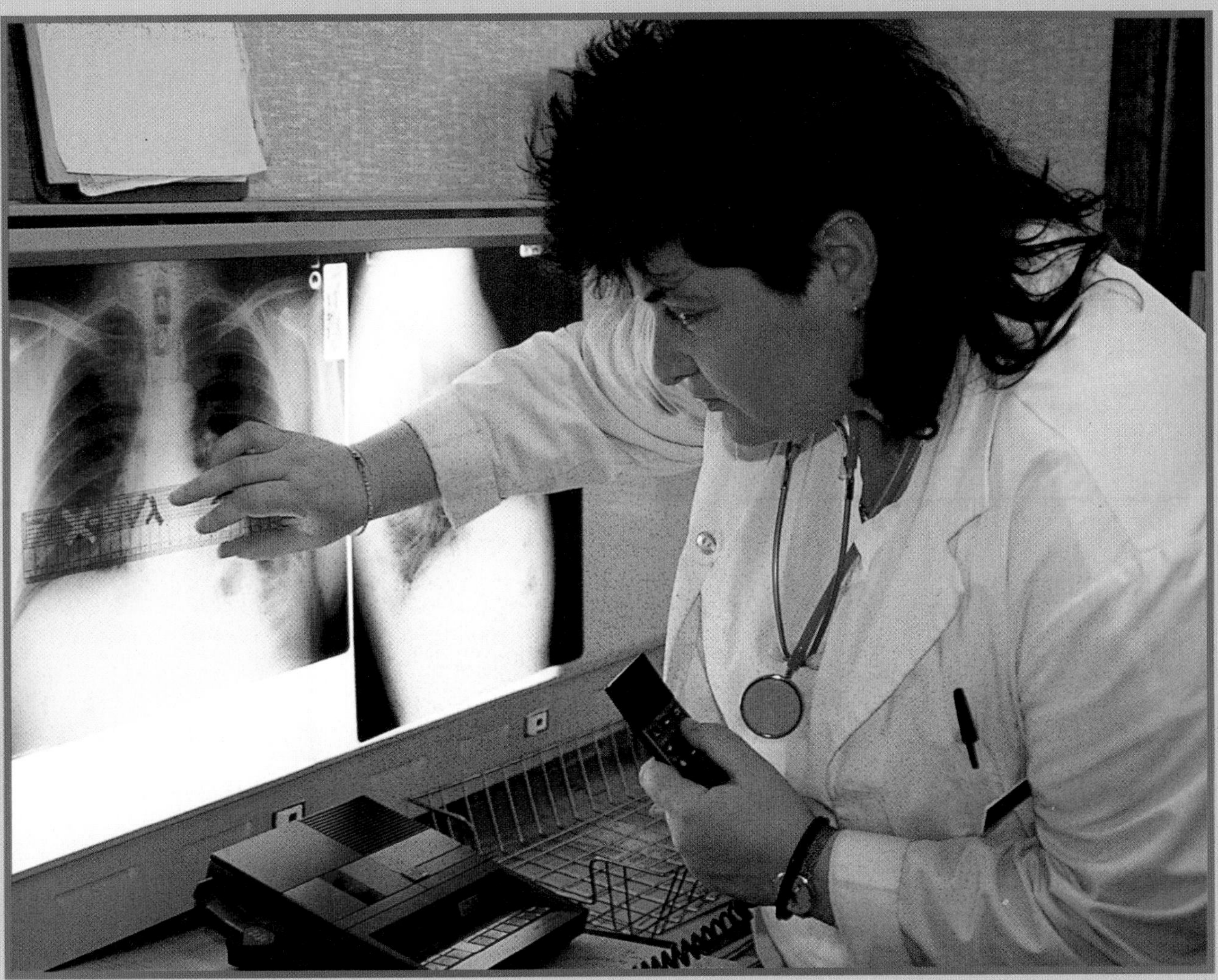

Radiologists use a variety of techniques, such as X rays, ultrasound, and magnetic resonance imagine (MRI), to diagnose and treat disease. In this photograph, a radiologist reads an X ray.

Radiologists are medical doctors who use imaging techniques and radioactive substances to diagnose and treat illness and disease. Imaging techniques, such as X rays, computerized tomography (CT), fluoroscopy, magnetic resonance imaging (MRI), positron emission tomography (PET), and ultrasound, provide a detailed view of the patient's bones, organs, and other internal structures without surgery or other invasive procedures.

Diagnostic radiologists study the images generated by these techniques to assist physicians in diagnosing illness. They also treat internal and external tumors with radiation.

Semiconductor-development technicians work closely with electronics engineers in research laboratories to design and test new kinds of semiconductor devices for use in electronic equipment. They also test production samples to ensure that the new devices are operating according to the designer's specifications and production requirements.

Semiconductor-development technicians use computer-aided design (CAD) to develop complex electronic circuits for new semiconductor devices and test them with oscilloscopes, calibrating devices, and logic and test probes. These highly skilled workers also assist electronics engineers in designing new testing methods and improving testing equipment. In addition, they are responsible for the routine maintenance and calibration of such equipment.

The nature of matter

Physicists study the properties of matter in the universe and how matter interacts through forces. One way to make these studies is to consider the individual components of matter itself. By studying the microscopic and sub-microscopic constituents of matter, physicists can often deduce how large bits of matter will interact.

There can, however, be conceptual difficulties in explaining physics at this sub-microscopic level. The minute particles that make up matter can be "seen" only by the effects they have on visible things, or by the gross phenomena that result when millions of them act together.

Molecules: the basic components of matter

Experimental evidence supports the idea that matter is composed of tiny particles, called molecules, which are continuously in motion. Molecules range in size from about 250 million to about 25,000 to the inch (10^{-10} meters to 10^{-6} meters across).

Despite the minuteness of molecules, evidence of their existence is provided by various phenomena, such as Brownian motion. For example, if smoke particles are viewed with a microscope they are seen to be in a state of rapid, random motion. The comparatively large, visible smoke particles are continually struck by the much smaller, randomly moving air molecules, which cannot be seen.

Every molecule in a substance exerts forces on other molecules nearby. That is why it is difficult to crush a solid (which involves pushing its molecules together against the intermolecular forces), and it is also difficult to stretch a solid (which involves separating its molecules against the force). There is, therefore, both an attractive and a repulsive component of the intermolecular force. These forces are basically electrostatic in nature.

Molecules consist of groups of atoms, which themselves consist of electrons and nuclei. As molecules are forced together, the electrons belonging to the various atoms making up the molecules interact, repelling each other. This repulsion is a very short-range force and predominates when the distance between the molecules is only about 4 thousand-millionths of an inch (10^{-10} meters). The attraction between molecules is of longer range. Its origin is complicated, but it also depends on the electrical interaction between molecules.

In solids and liquids, the molecules move relatively slowly (have relatively low kinetic energies), are close together, and they therefore interact fairly strongly. In gases the molecules are, on average, widely separated and interact relatively briefly. For this reason, a simple analysis of the behavior of gases is much easier than that of solids and liquids.

Atoms: the building blocks of molecules

Just as substances are comprised of molecules, molecules themselves are comprised of atoms. Atoms in molecules are bound together in various ways, although all of these

Inside an atom is a region of unceasing movement as subatomic particles respond to the various short-range forces that act on them. This model conveys the blur of electrons as they rapidly orbit around the nucleus.

interatomic forces arise basically from interactions between electrons in the atoms.

There are three types of particles that can, in a simple description, be considered as making up a typical atom. The central nucleus, with a diameter 10,000 times smaller that that of the whole atom, is comprised of neutrons and protons. (The hydrogen nucleus is unique in having no neutron, only a single proton.) The neutron is a particle with no electrical charge, whereas the proton has a single positive charge. Both have roughly the same mass.

Moving about this central region, held in orbit by the influence of the protons' positive charges, are the electrons. These are subatomic particles, each with a single negative charge and an extremely small mass: $\frac{1}{1836}$ that of a proton.

An atom is the smallest particle that can represent a particular chemical element. Each chemical element is characterized—and identified—by its atomic number, Z, which is the number of protons in the nucleus. And because an electrically neutral atom must contain an equal number of protons and electrons, Z also equals the number of electrons orbiting the nucleus. The mass number A of an atom is the sum of the number of protons and neutrons in the nucleus (thus $A-Z$ is the number of neutrons). As an example, the element helium, which has an atomic number (Z) of 2 and a mass number (A) of 4, (written $^4_2$He) has two protons (and hence two electrons) and two neutrons.

Electrons play the major role in determining the properties of the various elements. At the turn of the century, physicists spent much effort in trying to derive all the observed phenomena relating to elements from a mathematical model of the atom. One key phenomenon was the spectral lines produced by atoms when heated: each element has its own unique set of lines.

In trying to explain the generation and

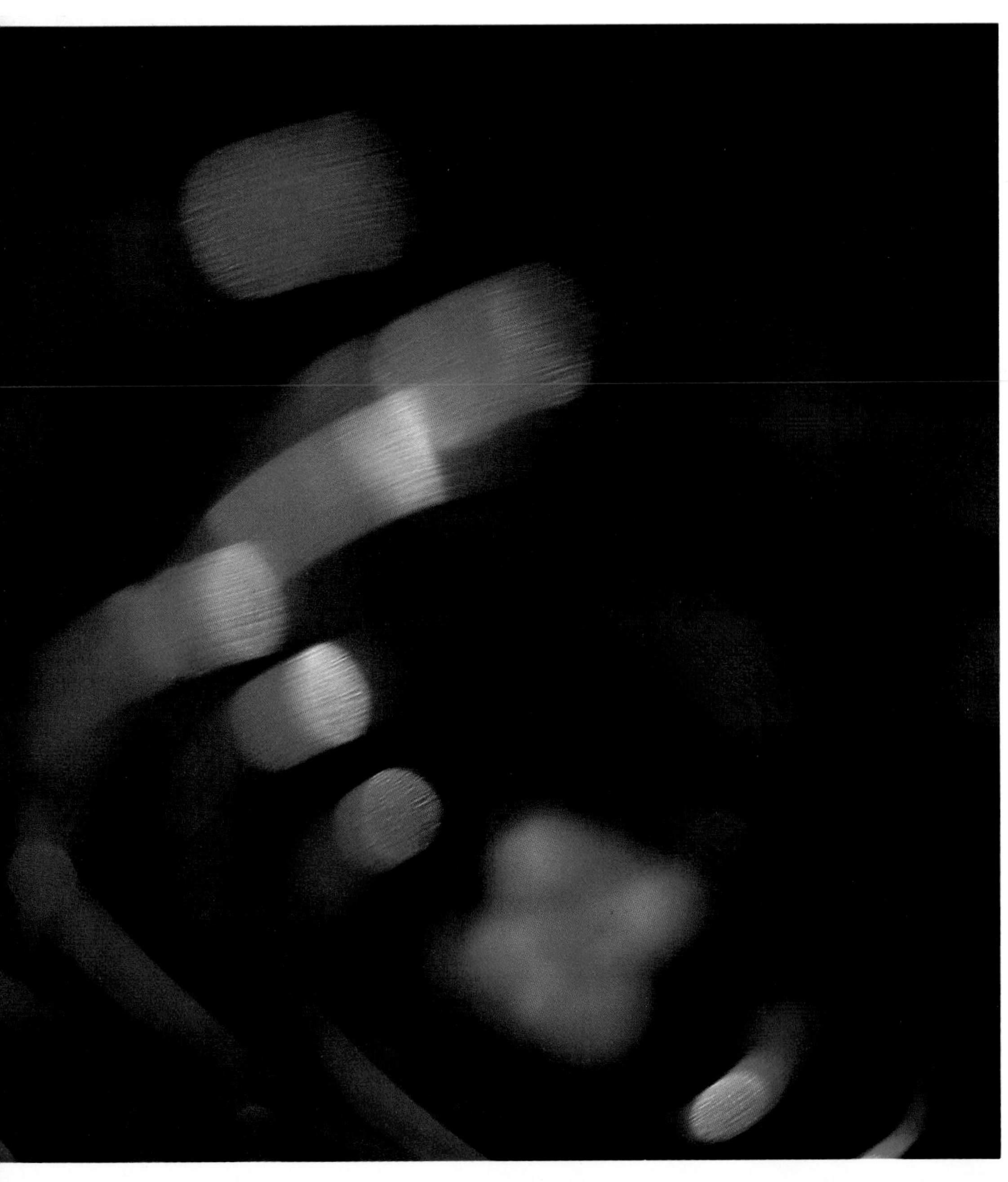

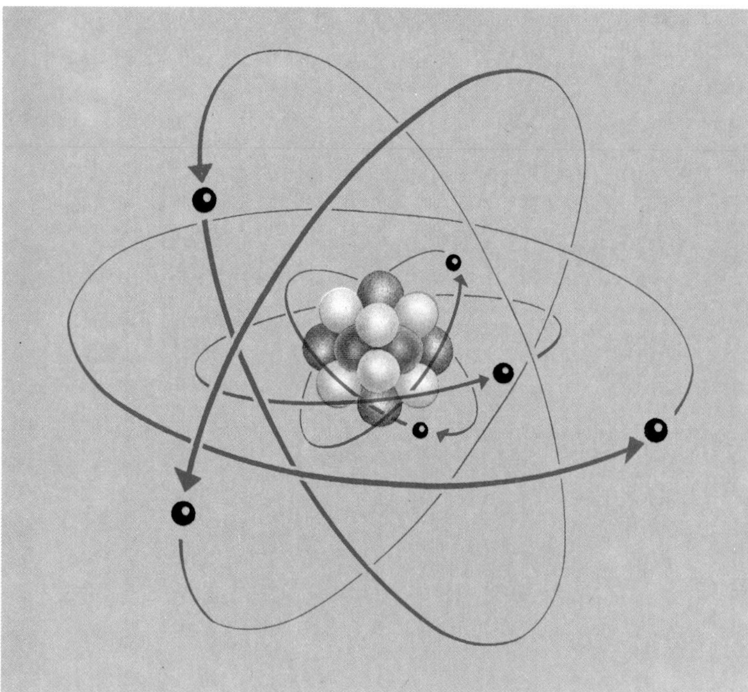

In a simple model of the atom, a central nucleus made up of protons (pink) and neutrons (brown) is orbited by numerous electrons (black) in fixed paths. The atom in this example is carbon, with six protons, six neutrons, and six electrons.

appearance of atomic spectra, physicists found that they had to introduce one of the basic concepts of modern physics: the quantum.

Atomic structure and quantum theory

Close scrutiny of a black-and-white newspaper photograph reveals that the apparent differences in light and dark tones are not the result of gradual shading but rather are created by subtly different sizes of dots, or "packets" of ink on the plain paper. Physicists have discovered that, in dealing with matter on small scales, many phenomena, including energy, must be treated as existing in "packets" or quanta, and not as being continuous. The quantum concept provides the key to understanding how atoms can absorb and emit energy, and how the various atomic spectra are produced.

In a simple quantum-based visualization of the atom, the electrons orbit the central nucleus under the electrostatic attraction of the proton charges, like planets in a miniature solar system. The electrons can move only in certain orbits about the nucleus. But by receiving energy they can become "excited" and move into orbits farther out from the nucleus. By giving up energy, they drop back into lower orbits. With each energy change, radiation of a particular frequency is absorbed or emitted.

From this analysis, the origin of atomic spectra can be explained. The individual spectral lines (of certain frequencies) represent electrons falling from high-energy excited states down to lower states. As they do so, they emit quanta of radiation (photons) of a frequency corresponding to the energy difference.

The electrons in an atom can orbit only in paths in which their orbital angular momentum has a fixed value (actually a multiple n of Planck's constant divided by 2π). The sizes of the various allowable orbits—each labeled by the principal quantum number n, beginning at 1 for the innermost orbit—can therefore be calculated. The observed spectral wavelengths

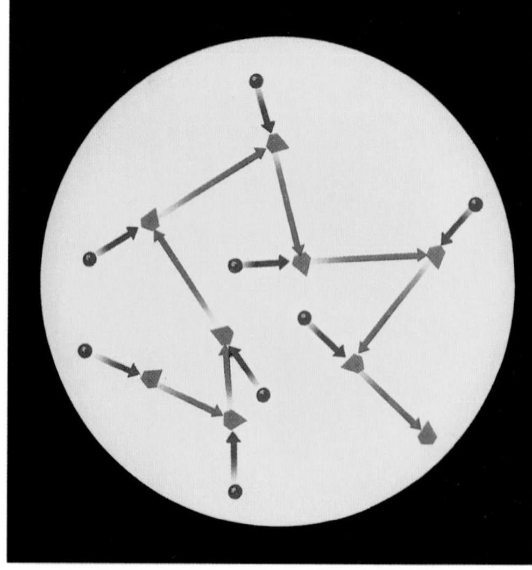

Shafts of sunlight *(left)* stream through the trees and are made visible as the light is scattered by fine particles of dust and smoke in the air. The particles continually dance around, as explained above.

Brownian motion *(above)* is the rapid, random motion of the suspended particles in a suspension. It results when the molecules of a liquid or gas in a suspension move rapidly and collide with the suspended particles.

result from electron "jumps" between orbits of different quantum numbers, and the calculations for the simplest atom, hydrogen, agree fairly well with observation. Discrepancies do occur, however, in the more complicated aspects of the spectra of hydrogen and other elements, and more complicated quantum theory is needed for a more complete description.

The nature of the electron

So far, electrons have been considered simply as point electric charges. Sometimes, however, they behave as if they were waves.

This wave-particle duality of electrons is recognized in de Broglie's relationship, which assigns a wavelength to any particle of known mass and velocity. It lies at the heart of the more sophisticated theory of the atom known as wave mechanics.

According to the theory of wave mechanics, electrons in atoms are not particles moving in orbits, but waves that can be represented mathematically by what is called a wave function (which measures the probability of an electron being at a particular point in space). Peak values of this function can be taken to represent the orbits of the electrons. There is only a high probability—not a certainty—that the electrons will be found in the orbits. The certainty of the old theory has been replaced by a statistical probability measured by the wave function.

According to Heisenberg's uncertainty principle, which arises from wave-particle duality, it is impossible to measure simultaneously both the position and momentum of a particle within certain limits. It can also be shown that it is similarly impossible to measure the total energy and lifetime of a particle simultaneously and with limitless accuracy.

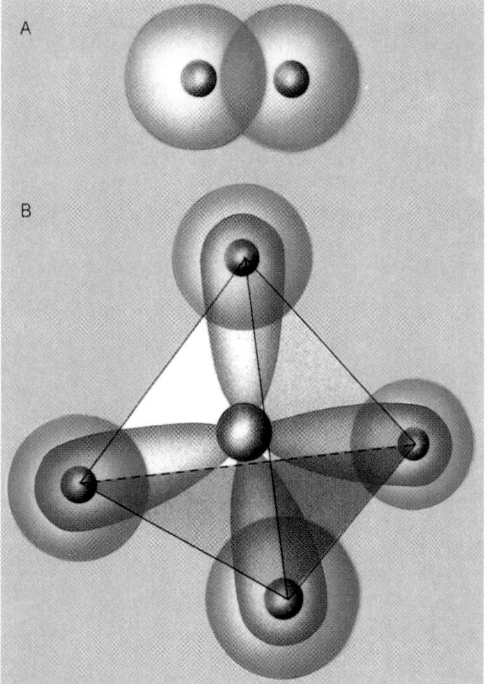

In the wave-mechanical model of the atom, electrons occupy orbitals—regions in space where there is a high statistical probability that an electron will be found. Molecules can be formed by the overlap of atomic orbitals to form covalent chemical bonds as in (A) the hydrogen molecule (consisting of two hydrogen atoms) and (B) methane (one carbon atom and four hydrogen atoms).

A photograph taken with a powerful electron microscope can reveal evidence of individual atoms in a substance. This photograph shows the regular ranks of atoms in a thin crystal of gold, magnified about 25 million times. Although the atoms appear to be solid spheres, they actually consist largely of empty space.

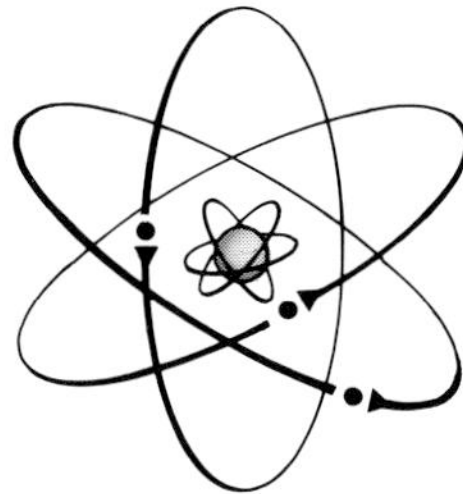

A simple model of the atom visualizes it as having a central nucleus surrounded by orbiting electrons.

Into the nucleus

Near the beginning of the 20th century, the New Zealand scientist Ernest Rutherford carried out one of the most significant experiments in physics. He directed a beam of alpha particles (helium nuclei) at a thin sheet of gold foil. Most of the particles passed straight through, some were deflected slightly from a straight-line path, and a few were scattered through quite large angles. Rutherford deduced that the positively-charged alpha particles were scattered by near collisions with the positively-charged nuclei at the center of the gold atoms. Such scattering techniques are now central to the study of atomic and subatomic structures.

It is now known that every atom contains a nucleus, whose size is about 10,000 times less than that of the atom itself. Most of an atom is empty space. With the exception of hydrogen, every nucleus contains neutrons and protons (collectively called nucleons). It is the so-called strong nuclear force that holds the nucleus together; if the nucleons could not exert this force, the nucleus would fly apart because of the electrical repulsion between the closely packed protons. It turns out that this force is the strongest in nature, being 100 times more intense than the electrostatic force and 10^{45} times stronger than gravity.

The electrons found surrounding the nuclei of atoms appear to be point-like particles. That is, there is no experimental evidence which indicates that electrons have any size at all. Because of this, and because electrons are stable particles, never decaying into any other particles, it is generally believed that electrons are fundamental particles, not being made out of any other smaller particles.

The atomic nucleus was discovered by Rutherford when he observed that gold foil scattered some alpha particles through large angles. He reasoned that the scattering resulted from near-collisions between the positively-charged particles and the positively-charged nuclei of the gold atoms.

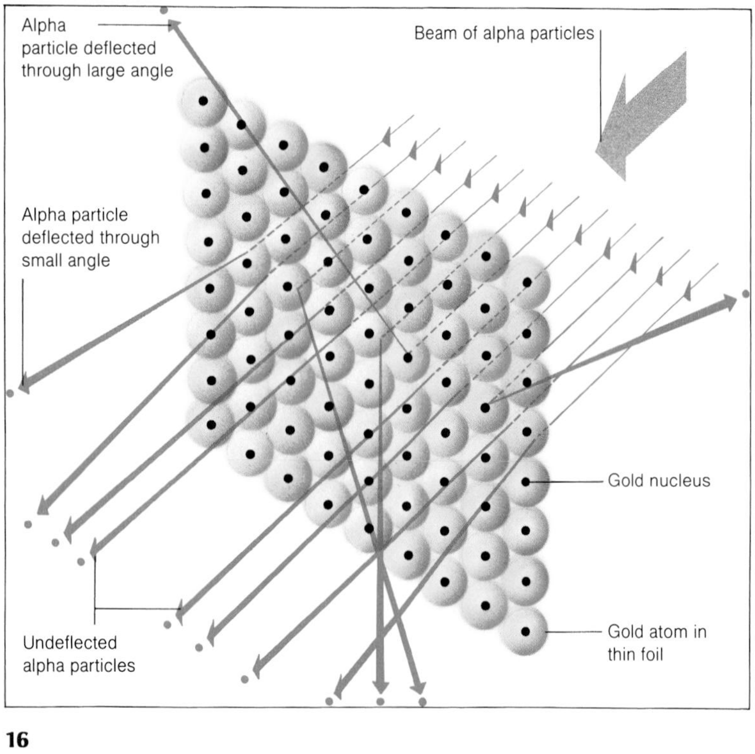

Alpha particle deflected through large angle

Beam of alpha particles

Alpha particle deflected through small angle

Gold nucleus

Undeflected alpha particles

Gold atom in thin foil

Electrons are grouped into the family of fundamental particles called leptons. There are six different leptons in all: electrons, muons, taus, electron neutrinos, muon neutrinos, and tau neutrinos. Of these leptons, only electrons are found in ordinary matter. The other members of the family are usually produced in the radioactive decay of other particles. Electrons, muons, and taus all have an electric charge of -1. The neutrinos have no electric charge.

The protons and neutrons found in the nuclei of atoms are not fundamental particles. They are made from more basic elements called quarks. A total of six kinds of quarks are known to exist, the so-called up, down, charm, strange, top, and bottom quarks. Protons are made out of two up quarks and one down quark. Neutrons are made out of two down quarks and one up quark. Thus, all ordinary matter is made out of only up and down quarks and electrons.

Quarks appear to be point-like particles, as are the leptons. It is therefore likely, but not definite, that quarks make up a family of fundamental particles and are not themselves made out of anything else. Quarks can combine together in many combinations to make up larger particles, and over a hundred of these have been detected in experiments. However, the only combination of quarks which is stable is the one which produces a proton. All other combinations decay rapidly into protons, electrons, neutrinos, and gamma rays. (The quark combination making up a neutron is stable only when it is associated with protons in the nucleus of an atom.) In addition, quarks cannot exist individually. They must always be associated with at least one other quark.

Quarks have a fractional electric charge of plus or minus $\frac{1}{3}$ or $\frac{2}{3}$. However, it is not this electric charge which binds quark combinations together, but rather it is through the exchange of another set of particles called gluons. Gluons can exist in three different states or "colors." Gluons moving between quarks give rise to the so-called strong force which binds quarks and nucleons together.

Probing the nucleus

Experiments to probe the inner structures of both atomic and subatomic particles are frequently carried out using accelerators. These machines, which can be either circular or linear in form, use magnetic and electric fields to accelerate the particles in them to extremely high energies and to cause them to collide. The collisions often create new particles, and the higher the energy of the collision, the more massive are the particles produced.

Consequently, the incentive in this branch of physics is always for higher and higher impact energies, and some accelerators achieve this by smashing together two beams of particles head-on. Others use a storage ring, in which electric and magnetic fields around a circular path give successive "kicks" to the particles until a high energy is attained. The particles are then released into the main collision path, where they collide with the target particles.

Physicists recognize particles produced in collisions by their electronic signatures, shown graphically by computers. The circle shows a computer-generated view of a potential top quark signature, with particle tracks emerging from the center of a collision.

A super neutrino detector built in Japan.

Nuclear stability and radioactivity

For an atomic nucleus to remain stable, the forces at work within it—the electrostatic and strong nuclear forces—must remain in some sort of balance. If a nucleus contains too many neutrons, it is unstable because the strong nuclear force favors pairs of nucleons, and pairs of pairs. A nucleus with too many protons, on the other hand, is unstable because of the relatively strong resultant electrical repulsion between the positively-charged protons.

Unstable nuclei try to become more stable essentially by ejecting material. This process is called radioactivity. It is accompanied by the emission from the nucleus of mass in the form of alpha particles or beta particles, sometimes accompanied by the emission of energy in the form of gamma rays.

An alpha particle consists of two protons and two neutrons (and hence is a helium nucleus, with a mass of 4). Beta decay, on the other hand, involves the emission of an electron or a positron (a particle having an opposite charge to but a similarly negligible mass as an electron). Gamma rays are a form of short-wavelength electromagnetic radiation, resembling high-energy X rays.

When a "parent" nucleus decays by alpha-emission the resulting "daughter" nucleus has a mass lower by 4 (and an atomic number lower by 2); the daughter is a different, lighter element. Thus the metal radium ($^{226}_{88}Ra$) decays by the alpha-process to give the gas radon ($^{222}_{86}Rn$); radium has transmuted into radon.

In beta decay, the mass of the nucleus remains unchanged but the atomic number increases or decreases by one, depending on whether an electron (β^-) or positron (β^+) is emitted. For example, a radioactive isotope of nitrogen ($^{12}_{7}N$) decays by the β^+ process to give carbon ($^{12}_{6}C$).

In the fission of uranium a neutron (brown) strikes a uranium atom (orange), which splits into two roughly equal fragments with the release of usually three more neutrons. Two of these go on to split two more uranium atoms, with the release of heat and more neutrons, and so on. If there is sufficient uranium, a self-sustaining chain reaction results.

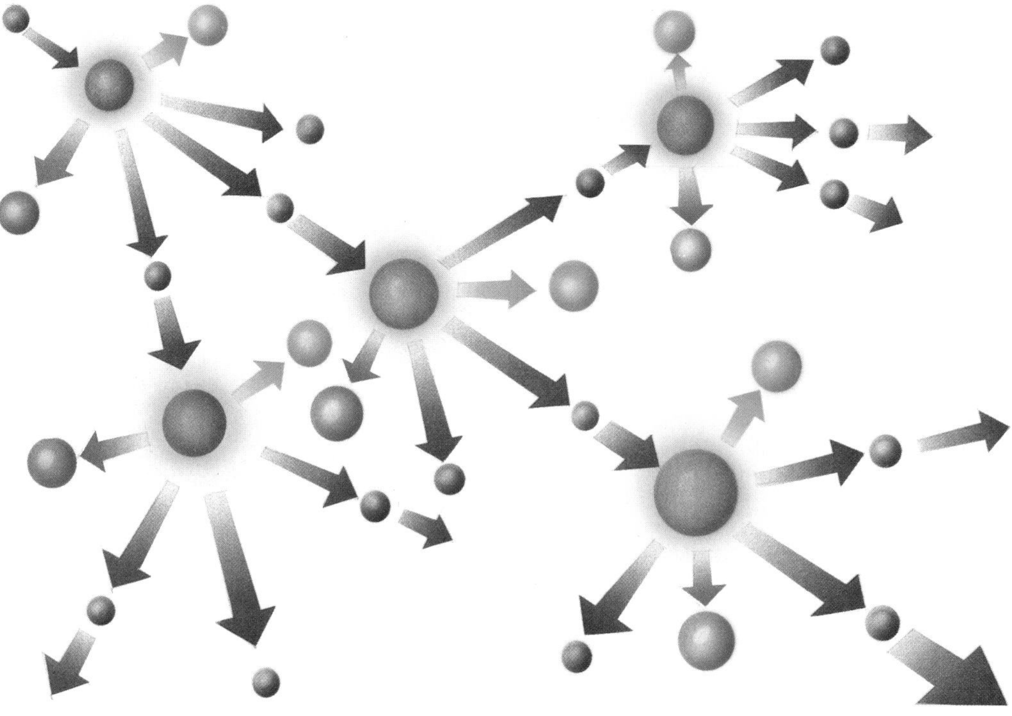

The number of radioactive atoms in a substance decreases exponentially in time. Therefore, for any particular radioactive element, there is a characteristic time called the half-life—the time after which the number of radioactive nuclei in the sample has decreased to half the number originally present. The common, nonfissile isotope of uranium, $^{238}_{92}U$, decays by the alpha process and has a half-life of 4.5×10^9 years (it takes 4,500 million years for half of it to decay).

There are three common radioactive series in nature, called the uranium, thorium, and actinium series; the parent elements are $^{238}_{92}U$, $^{232}_{90}Th$, and $^{227}_{89}Ac$, respectively. By a sequence of alpha and beta decays these radioisotopes pass through a series of others until they finally become stable isotopes of lead. Thus in the uranium series, U-238 decays by emitting alpha particles to Th-234, which itself decays to Pa-234 (protactinium) by the beta process; the series continues until lead (Pb-206) is reached.

Nuclear power stations generate electricity by using the heat of nuclear fission to produce steam to drive turbine generators. Viewed with suspicion by many as potential local hazards and because of the problems of disposing of nuclear waste, the building of new nuclear power stations is now a focus for considerable political comment.

Nuclear energy

It is also possible to cause artificial radioactivity by bombarding a nucleus with high-velocity neutrons. In the best-known example, uranium is struck by neutrons, and each uranium nucleus splits into two roughly equal parts, a process accompanied by the release of two or three neutrons and a great amount of energy.

Above a certain critical mass of such a fissile material (such as U-235), it is possible to ensure that there is always at least one neutron produced by each transformation capable of disrupting another nucleus, and so on in a chain reaction. If this is allowed to run away, out of control, the result is a nuclear explosion of vast power; this is the basis of the atomic bomb. But if the chain reaction is moderated and controlled, the enormous energy produced by the fission process can, for example, be used to heat water and produce steam to drive turbines and then generate electricity. This is the way in which a nuclear power station works.

Light elements can also be used to produce energy. By fusing together the nuclei of two such elements, it is possible to form a new nucleus whose mass is slightly less than the original ones. The "lost" mass appears as energy according to Einstein's relation $E = mc^2$. To be able to fuse, the two nuclei must have sufficient energy to overcome the electrostatic repulsion resulting from their charges. Such energy can be provided only by temperatures of hundreds of millions of degrees. In the stars, however, such temperatures are commonplace, and the fusing of hydrogen to produce helium keeps them shining for thousands of millions of years.

Scientists on earth have already demonstrated the uncontrolled power of nuclear fusion in the hydrogen bomb. By contrast, one area of active research in practical nuclear physics is concerned with finding a method of controlling fusion, so that it can be used productively to generate electricity.

Fusion reactors are the subject of experiments in Europe and the United States. The assembly *(right)* is part of the Joint European Torus (JET) project at Oxford England. The ring of magnets *(left)* shown under construction at Princeton University provide a powerful magnetic field used to contain a plasma of hydrogen used in fusion experiments.

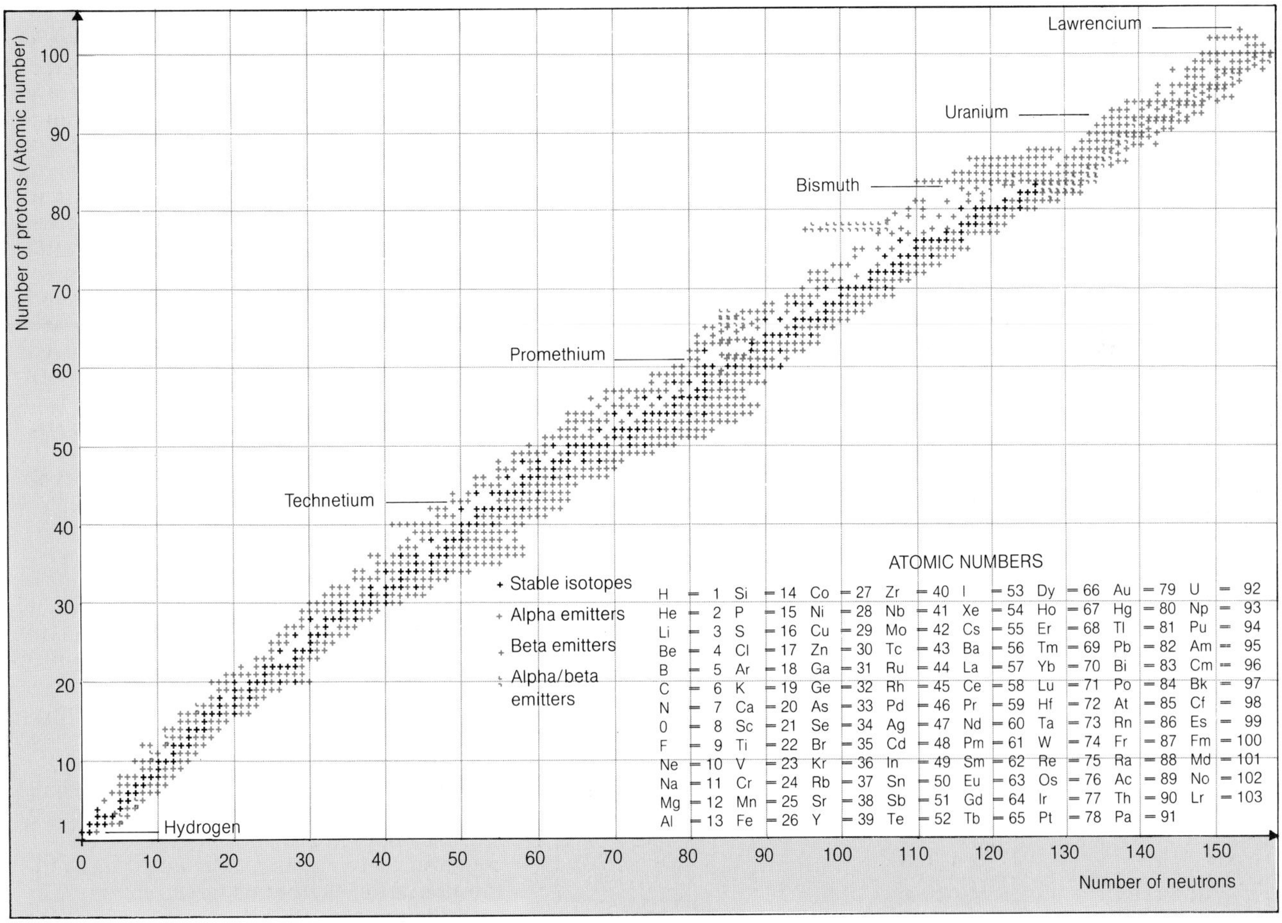

Number of protons (Atomic number)

Number of neutrons

Lawrencium

Uranium

Bismuth

Promethium

Technetium

Hydrogen

+ Stable isotopes
+ Alpha emitters
+ Beta emitters
+ Alpha/beta emitters

ATOMIC NUMBERS

H	1	Si	14	Co	27	Zr	40	I	53	Dy	66	Au	79	U	92
He	2	P	15	Ni	28	Nb	41	Xe	54	Ho	67	Hg	80	Np	93
Li	3	S	16	Cu	29	Mo	42	Cs	55	Er	68	Tl	81	Pu	94
Be	4	Cl	17	Zn	30	Tc	43	Ba	56	Tm	69	Pb	82	Am	95
B	5	Ar	18	Ga	31	Ru	44	La	57	Yb	70	Bi	83	Cm	96
C	6	K	19	Ge	32	Rh	45	Ce	58	Lu	71	Po	84	Bk	97
N	7	Ca	20	As	33	Pd	46	Pr	59	Hf	72	At	85	Cf	98
O	8	Sc	21	Se	34	Ag	47	Nd	60	Ta	73	Rn	86	Es	99
F	9	Ti	22	Br	35	Cd	48	Pm	61	W	74	Fr	87	Fm	100
Ne	10	V	23	Kr	36	In	49	Sm	62	Re	75	Ra	88	Md	101
Na	11	Cr	24	Rb	37	Sn	50	Eu	63	Os	76	Ac	89	No	102
Mg	12	Mn	25	Sr	38	Sb	51	Gd	64	Ir	77	Th	90	Lr	103
Al	13	Fe	26	Y	39	Te	52	Tb	65	Pt	78	Pa	91		

Atomic stability is the key to nuclear physics and to much of chemistry. This chart plots the number of protons in a nucleus (equal to the atomic number) against the number of neutrons for isotopes of all the chemical elements from hydrogen (element number 1) to lawrencium (number 103). Black crosses represent stable isotopes. Unstable isotopes that achieve stability by emitting alpha particles are shown by blue crosses; beta emitters have red crosses. Apart from elements 43 (technetium) and 61 (promethium), all elements up to number 83 (bismuth) have at least one naturally occurring stable isotope. Among the light elements, isotopes with an equal number of protons and neutrons are particularly stable. Beyond bismuth all isotopes are unstable and many are radioactive. Element 92 (uranium) is the heaviest element that occurs naturally on earth; elements 93 to 109 are all man-made. Some of these heavy isotopes can be made to undergo nuclear fission.

States of matter

All single substances can, under the right conditions, exist as a solid, liquid, or gas; these are the three common physical states (or phases) of matter.

Solids, liquids, and gases

Solids possess the property called cohesion; that is, their component particles (atoms or molecules) are held together by attractive forces. As a result, solid substances are rigid and retain their shape unless deformed by external forces.

There are two principal types of solids: "true" and amorphous. True solids have definite crystalline structures (that is, their constituent atoms and molecules are arranged in a regular three-dimensional lattice—like eggs in a stack of egg trays). Most also melt at specific temperatures to become liquids. Examples include metals, ice, and many plastics, in addition to obviously crystalline substances such as common salt and diamond.

In contrast to true solids, amorphous solids do not have crystalline structures; neither do they have specific melting points. Glass and many resins are amorphous solids.

Liquids represent the intermediate stage between solid and gas. A liquid's atoms or molecules have some degree of cohesion and so tend to remain together, but they are not rigidly linked and can therefore move in relation to each other—which is not possible in solids. For this reason liquids flow and in a gravitational field take on the shape of the vessels in which they are contained.

In a gas the constituent particles have negligible cohesion and can therefore move almost completely independently of each other. Like liquids, gases flow and assume the shape of their containers; unlike liquids, however, gases always fill the entire space in their containers—and the container needs to be closed if the gas is not to escape.

Atoms in a solid (A) maintain their regular positions in a crystal lattice and vibrate because of their thermal energy; on being heated (B), they vibrate more vigorously, causing the solid to expand. In a liquid (C), the atoms move about randomly; a liquid has no regular form and takes on the shape of its container. In a gas (D) the atoms move rapidly, collisions with the walls of the vessel resulting in gas pressure; a gas totally fills its container.

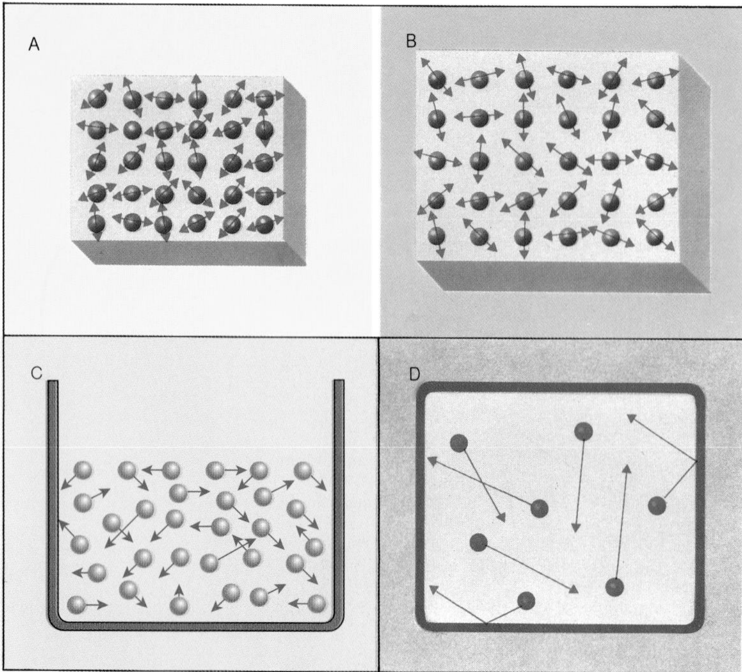

Atomic and molecular forces

The principal attractive forces between atoms comprise ionic bonds and covalent bonds, both of which are electrostatic in nature. Intermolecular forces, which are generally weaker than interatomic attractive forces, include hydrogen bonds and Van der Waals forces; again, both of these are basically electrostatic.

The attractive forces between atoms or molecules in a solid are balanced by repulsive forces. As two of these particles approach each other, the outer electrons of one repel the outer electrons of the other, and the atoms or molecules move apart. But because of the attractive force between them, they move back toward each other again. The overall result of the attractive and repulsive forces is to cause each of the atoms or molecules in a solid to vibrate continually about the same position in a lattice.

The atoms or molecules of a liquid are also affected by attractive and repulsive forces. But a liquid is hotter than the same substance in solid form and its vibrating particles therefore have greater kinetic energy—that is, they vibrate more violently. As a result, the attractive forces cannot hold them in a lattice, and they are relatively free to move.

In a gas the atoms or molecules have so much energy that they have largely broken free of the influences of the attractive and repulsive forces and, therefore, have almost complete freedom of movement, interacting only when they pass near another atom or molecule.

Changes of phase

Increasing the temperature of a substance causes an increase in the kinetic energy of its atoms or molecules, as a result of which they vibrate more violently (in a solid) or move more quickly (in a liquid or gas). At a certain temperature the atoms or molecules of a solid are sufficiently energetic to partly overcome the attractive forces that hold them together; thus the particles become free to move and the solid melts.

Conversely, if the temperature of a liquid is lowered to the point at which the attractive forces overcome the kinetic energy of the atoms or molecules, the liquid freezes to become a solid.

Some solids do not melt when heated under normal conditions; instead they change directly from the solid to the gaseous state. This phenomenon, known as sublimation, occurs in solids such as iodine and frozen carbon dioxide ("dry ice").

The melting point of some substances can be lowered by increasing or decreasing the pressure; for example, if ice at 0° C is subjected to pressure, it melts—despite the fact that it is still at the "freezing point" of water. If the pressure is then reduced, the ice refreezes—a process known as regelation.

A liquid becomes a gas or vapor by evaporation. Unlike melting, evaporation occurs at all temperatures above absolute zero (approximately −459° F (−273° C), the lowest defined temperature). At any temperature higher than this, there are some atoms or molecules at the surface of a liquid that have enough kinetic energy to break free of the surface and form a gas or vapor. Solids also evaporate, although only very slowly unless they are close to their melting points.

Like any gas, the vapor above a liquid has a pressure (if it is enclosed); this is called the saturated vapor pressure. As the liquid is heated, more molecules leave the surface and the vapor pressure increases. When the saturated vapor pressure equals the external pressure, the liquid boils. (This is why water—or any other liquid—boils at a lower temperature in the "thin" atmosphere at high altitudes than it does at sea level, where the atmospheric pressure is greater.) Neglecting the small hydrostatic pressure of the liquid, the vapor pressure is the same as the pressure in the liquid (both equal to external pressure) at its boiling point. Vapor forms not only at the surface but also within the liquid itself, giving rise to bubbles.

A substance that is kept above its boiling point stays in the form of a gas or vapor. Cooling it to below its boiling point results in condensation—the gas or vapor changes to a liquid.

Metal smelting involves all three states of matter. Solid ores and scrap are heated until they melt, and the liquid metal poured into crucibles, or molds. Gases and vapor are given off during the smelting and pouring processes.

Crystals

Most simple chemical compounds consist of crystals. These cannot always be seen clearly because they are often grouped together in a mass that has no particular shape. But if a lump of crystalline material is examined closely, tiny individual crystals can be seen.

All crystals have a definite geometric shape, determined by the way in which the atoms of the substance are linked together. For example, in a crystal of common salt (sodium chloride) the atoms of sodium and chlorine are arranged so that they lie at the corners of a series of (imaginary) cubes; the result is a cubic crystal.

Many substances occur naturally as crystals. Here the yellow crystals of sulfur contrast with the shiny white aggregations of calcite (calcium carbonate).

Crystal systems

Mineralogists recognize 32 different classes of crystals, which are grouped into 7 crystal systems. Each system includes all the crystals that are based on a particular geometric shape, and the ability to recognize crystal shapes is a valuable aid in identifying crystalline substances, such as minerals.

Crystal systems are described by their axes—imaginary lines that join the centers of opposing faces of a crystal. For example, a cubic crystal has three sets of opposing faces and hence three axes; they are of equal length and are all at right angles to each other. Cubic crystals are described as being isometric. But not all isometric crystals are plain cubes. If the corners of a cube are cut off, the result is a polyhedron with six octagonal faces and eight triangular ones, for example. An octahedron, with eight triangular faces, is another common isometric crystal shape.

In addition to the isometric (or cubic) system there are the tetragonal, orthorhombic, monoclinic, triclinic, and hexagonal systems. Each is divided into classes according to how the basic geometric shape of the system has been altered. Some mineralogists recognize a trigonal system, which others maintain is a class belonging to the hexagonal system.

Formation of crystals

A crystal may form when a solution of a substance evaporates. Crystal formation begins as the ions of the substance begin to come out of solution and become arranged in their usual solid framework. More ions deposited on the outside of the crystal merely increase its size—the basic shape does not change. Crystals of common salt and alum can be formed in this way.

Crystals may also form when a vapor or a molten substance solidifies. For example, water vapor may freeze in the air to form hexagonal ice crystals. Crystals of quartz, mica, and pyrite are often found in rocks that were once molten. Crystals can also be formed rapidly in a supercooled liquid—one that has been cooled below its freezing point without actually solidifying.

Extremely pure crystals can be grown atom by atom or molecule by molecule using mod-

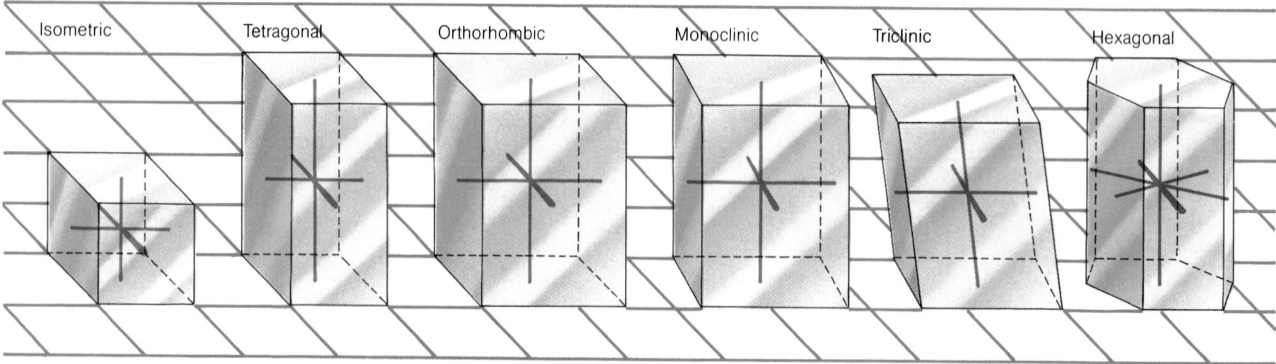

The crystal systems Isometric (cubic) system: three axes of equal length, all at right angles to each other (examples: salt, fluorite, and magnetite). Tetragonal system: three axes, two of equal length, one shorter or longer than the others, all at right angles to each other (examples: rutile and zircon). Orthorhombic system: three axes, all of different lengths, all at right angles to each other (examples: stibnite and topaz). Monoclinic system: three axes, all of different lengths; one axis is at right angles to the other two, which are not at right angles to each other (examples: gypsum, serpentine, and talc). Triclinic system: three axes of unequal length, not at right angles to each other (examples: rhodonite and albite). Hexagonal system: four axes, three of equal length, which meet at angles of 120°; the fourth axis is longer or shorter than the other three and meets them at right angles (examples: beryl, dolomite, and quartz).

ern techniques such as molecular beam epitaxy or sputtering. In these methods a specially prepared surface (substrate) is placed in a very high vaccuum chamber and the atoms or molecules making up the crystal are deposited virtually one at a time upon the substrate. The atoms used in this crystal growth process are obtained by vaporizing a material containing the atoms or by ejecting them with a particle beam.

Allotropes and polymorphism

The atoms of most substances can be arranged in only one way. In some substances, however, there are two or more possible arrangements of the component atoms. A substance of this kind has two or more forms that have the same chemical properties but different physical properties.

Different forms of the same element are known as allotropes. Oxygen, for example, has two allotropes—normal oxygen (O_2), and ozone (O_3), which has molecules consisting of three atoms. Oxygen can be converted into ozone by supplying it with energy. This process occurs naturally in the outer atmosphere, where radiation from the sun provides the energy. Ozone is normally unstable and tends to break down into oxygen.

Pure carbon also has two allotropes—diamond and graphite. A crystal of diamond is in fact a single giant molecule in which every carbon atom is linked to four others by four equal, strong bonds. The bonds are arranged tetrahedrally around each atom, and there are no planes along which the giant molecule can easily be split. This quality is what gives diamond its tremendous hardness—it is the hardest substance that occurs naturally.

Graphite, on the other hand, consists of flat sheets of carbon atoms arranged in adjoining rings. Each carbon atom is linked to three others by fairly strong bonds that all lie in the same plane. The sheets are held together only by weak Van der Waals forces. These can easily be overcome, allowing the sheets to slide over each other. As a result, graphite is a soft, slippery substance that can be used as a lubricant.

Other elements that exist as allotropes include sulfur and phosphorus. Sulfur has two common crystalline allotropes—rhombic sulfur, which is stable above 204.1° F. (95.6° C), and monoclinic sulfur, which is stable below that temperature. Both allotropes contain sulfur atoms arranged in groups of eight, but the groups are joined together in different ways to give two different crystal structures. Phosphorus also has two main allotropes—red and white phosphorus.

Some compounds can also exist in more than one crystalline form, a phenomenon known as polymorphism. Calcium carbonate, for example, is commonly found as calcite, or Iceland spar, which is the main constituent of chalk, limestone, and marble. But it also occurs as aragonite, which is found in some sedimentary rocks and in coral skeletons. Titanium dioxide occurs naturally in three different crystalline forms, as the minerals anatase, brookite, and rutile—which all have the same chemical composition.

Frost on a windowpane reveals the feathery geometrical forms of slowly growing ice crystals. Despite their gross differences, each pattern is based on the same fundamental crystal shape.

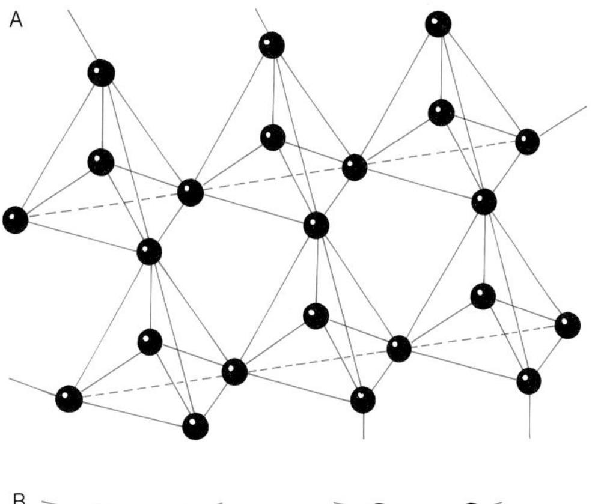

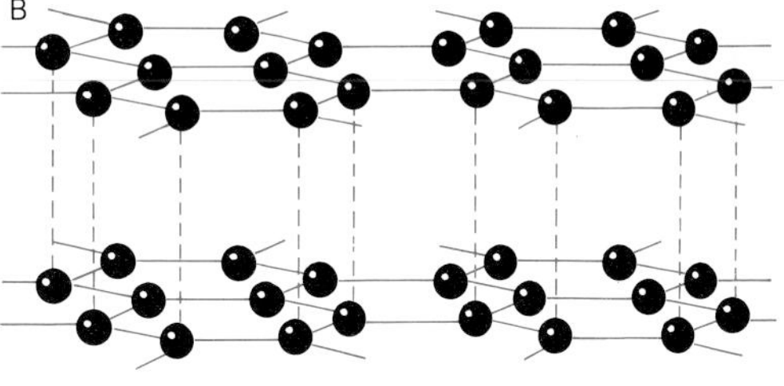

Carbon crystallizes in two allotropic forms, which are related to their atomic structures. Diamond (A) consists of an interlocked network of tetrahedrally bonded atoms, which form a large hard crystal. In graphite (B) the atoms are located in parallel planes of conjoined hexagonal rings, forming a soft substance.

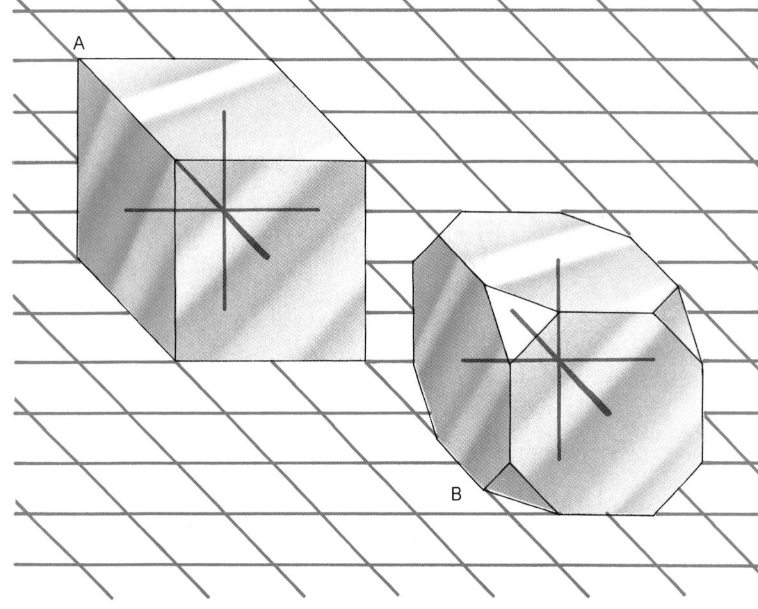

Some crystal shapes can be thought of as derived from the six basic forms by "slicing" off corners. Cutting the corners off the cube (A)—the basic form of the isometric system with six square faces—results in the shape (B), which has six octagonal faces and eight additional triangular ones at the original cube's corners. But the crystal axes remain the same.

Properties of solids

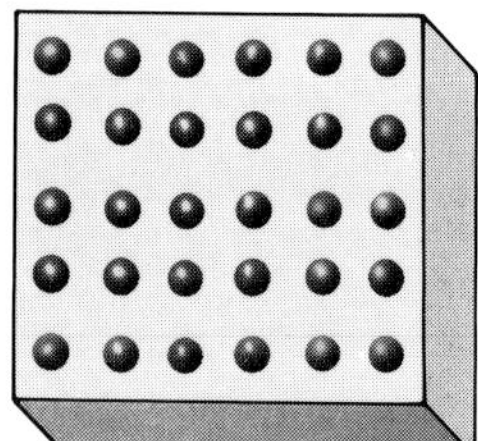

In a pure solid, the atoms are held by electrostatic forces in a regular array in a crystal lattice. This arrangement gives a solid its characteristic properties of strength and hardness.

The forces that hold atoms and molecules in place give solids their strength. Interatomic forces are generally stronger than intermolecular forces, so that solids composed of collections of single atoms, such as most metals and diamond, are the strongest. Molecular solids, such as plastics, iodine, dry ice (frozen carbon dioxide), and some metals and alloys, are softer and melt more easily.

Stress and strain

The strength of a material is determined by measuring its elasticity—that is, how stiff it is and how it behaves when it is stretched or compressed. Elasticity is measured as the ratio of stress to strain.

Stress is defined as the force acting on a material divided by the area over which the force is applied. Stress may be applied in three different ways: tensile, shear, and compressive. Tensile stress causes a material to become elongated (stretched), and shear stress causes it to be twisted out of shape. In both cases the component atoms or molecules are pulled apart as they move from their original positions. When the stress is removed, the attractive forces between the atoms or molecules pull them back together again, and the material is restored to its original shape. Compressive stress, on the other hand, forces the atoms or molecules together, and it is the repulsive forces that restore the shape of the material when the stress is removed.

Strain is measured as the change in size of a material when it is subjected to stress. According to Hooke's law, the strain is proportional to the stress producing it and hence the ratio of stress to strain is constant for any given material. For tensile stress this constant is known as Young's modulus of elasticity. For shear stress the constant is the shear modulus, and for compressive stress it is the bulk modulus of elasticity.

When an elastic solid is stretched, initially the elongation is proportional to the stress producing it—the material obeys Hooke's law. Beyond its elastic limit there is a permanent stretch, and beyond the yield point a small increase in stress produces a large stretch until the material breaks. A knowledge of such properties is essential to engineers in choosing materials for any practical application—from the design of a safety pin to the complex girders of a bridge.

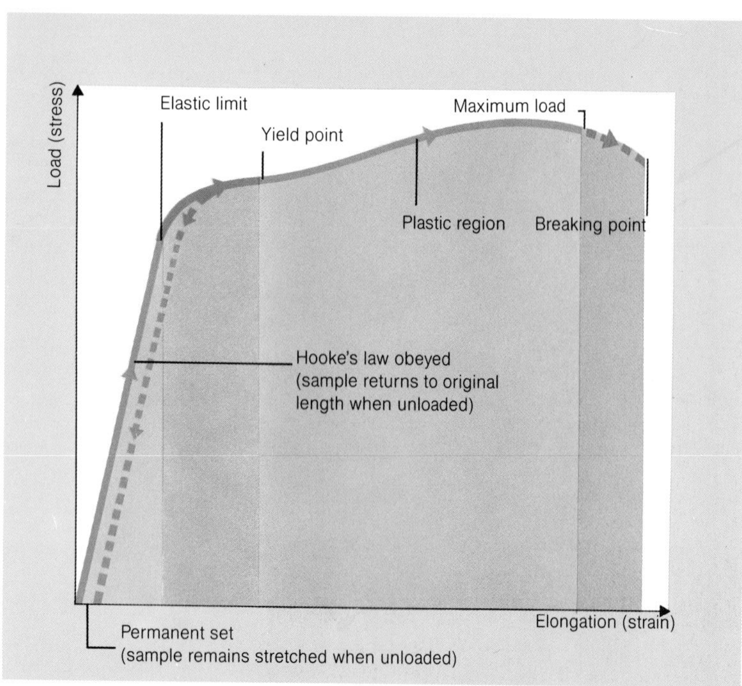

Elastic limit and the yield point

If stress is applied to a material and slowly increased, the material accordingly changes shape. But at a certain point it reaches its elastic limit. If stressed beyond this point, it does not recover its original shape, because in reality the crystalline structures of materials are not perfect. There are weak points caused by tiny cracks and impurities in the material. As a result, layers of atoms or molecules become dislocated (misaligned from their normal locations in the crystal lattice) and start to slide over each other when subjected to great stress. After the material has reached its elastic limit, a small amount of additional stress causes it to reach its yield point, at which new dislocations are formed. The material now becomes plastic and continues to stretch even if the stress is reduced slightly. Finally, a weak point appears where two or more dislocations become jammed. The material becomes thin and brittle at this point and soon fractures.

Knowledge of the elastic limits of materials is very important in all forms of construction. Buildings, bridges, ships, and aircraft have to be designed so that their materials are never subjected to stresses that could take them beyond their elastic limits. Other factors also have to be taken into account. For example, constant flexing of a metal structure may cause it to develop dislocations and fracture—a phenomenon known as metal fatigue. Some brittle materials, such as cast iron and certain plastics, fracture without first becoming plastic if subjected to sudden stress.

Hardness

Another property of solid materials that depends on the strength of the bonds between atoms or molecules is hardness. Hard materials wear away softer ones and so engineers need to know the relative hardness of the materials they are using.

Hardness is usually measured on a scale of 1 to 10 known as Mohs' scale. In 1822, the German mineralogist Friedrich Mohs drew up a list of 10 common minerals in which each one is harder than (and can therefore scratch) those with lower numbers. The minerals are: diamond (hardness 10), corundum (9), topaz (8), quartz (7), feldspar (6), apatite (5), fluorite (4), calcite (3), gypsum (2), and talc (1). Diamond is the hardest naturally-occurring substance and is widely used in industry for grinding and drilling. Talc, the softest material on Mohs' scale, can be scratched with a fingernail.

Working metals

Hardness is not always a desirable quality. For many industrial purposes a metal may be required to be ductile; that is, it should contain enough evenly distributed dislocations to enable it to be stretched beyond its yield point and be drawn out into a wire without break-

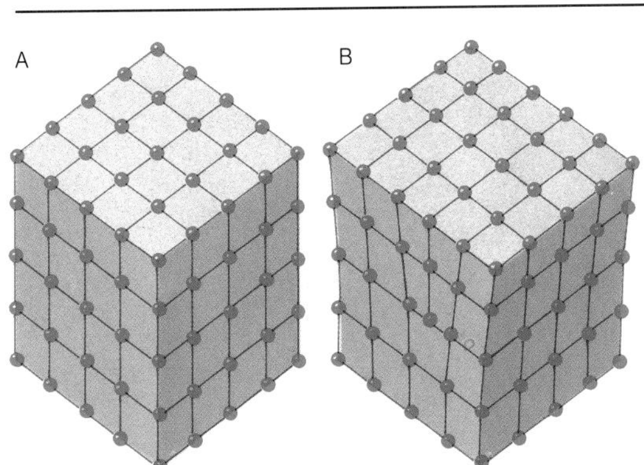

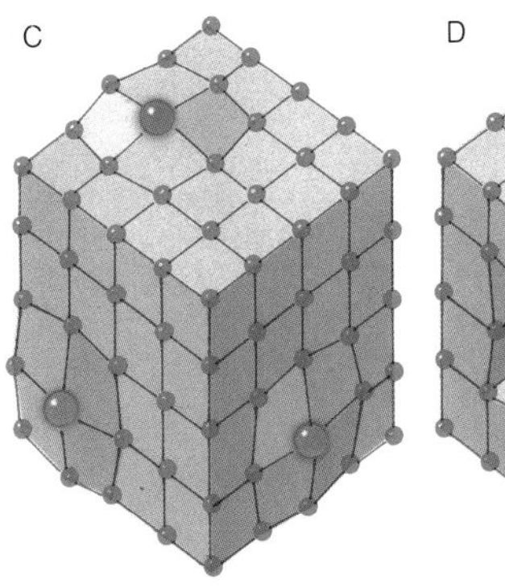

A material's strength is often related to its crystal structure. A normal crystal (A) may become flawed by a defect such as an edge displacement (B), which can move through a solid under stress. "Foreign" atoms (C) or vacancies (D) prevent the movement of such defects.

ing. Or a metal may need to be malleable—that is, able to be hammered or rolled into a thin sheet.

The ductility or brittleness of a metal can be altered to some extent by heat treatment. Annealing—heating a metal to just below its melting point and then cooling it slowly—relieves internal stresses, removes some of the dislocations, and produces a softer, more ductile metal. Heating a metal to a high temperature and then cooling it quickly by quenching (plunging it into cold oil or water) makes it hard and brittle. To reduce the brittleness the metal can then be tempered, which involves heating it to a lower temperature and cooling it rapidly.

Composite solids

Another way of altering the strength of a material is to mix it with another one. Alloys, for example, are made by melting a metal with one or more other substances (which are usually, but not always, also metals) to produce a material whose properties are superior in some required respect to those of its components. Alternatively, a bulk material can be reinforced with strands or fibers of another. For example, plastic resin reinforced with glass fiber or carbon fibers is strong enough to be used for constructing car bodies and boat hulls.

Laboratory testing of the strengths of materials is an important aspect of engineering design and quality control. This machine automatically increases the load on a sample of metal until it breaks and provides data on its strength and elastic properties.

A collapsed bridge is a spectacular example of materials that have failed under stress—in this case (the bridge over the River Seine in Paris, opposite the Palais des Arts) as a result of having been hit by a barge.

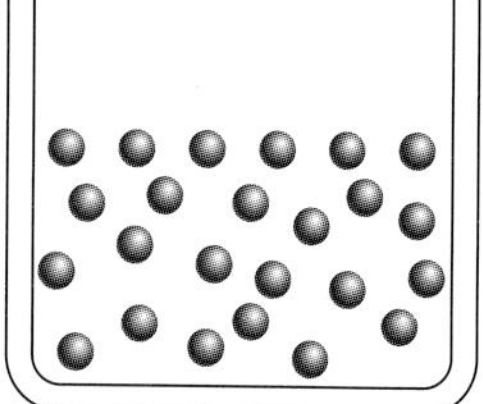

In a liquid, the atoms or molecules are free to move randomly within the constraints of the container and the liquid's surface.

Properties of liquids

As in a solid, the atoms or molecules of a liquid are held together by attractive forces. But these forces are not great enough to hold the atoms or molecules in a fixed pattern; instead, they move about at random. As a result, a liquid can flow and it cannot be stretched or distorted. Like a solid, however, it can be compressed slightly and shows the same sort of elasticity when subjected to compressive stress. Unlike a solid, but like a gas, a liquid exerts pressure, which at any point depends on the depth and the density of the liquid.

Cohesion and adhesion

A liquid does not expand to fill the whole of the volume available to it. Instead, the cohesion between its molecules forces it to maintain a fixed volume (at a given temperature). Generally, of course, it takes on the shape of all or part of its container—but this is only because the force of gravity makes it do so. In zero gravity conditions a liquid takes on the shape with the minimum possible surface area—that is, a sphere. A falling drop of water also tends to take on this shape.

The fact that a liquid tends to take on a spherical shape is due to a phenomenon known as surface tension. Inside a volume of liquid all the atoms or molecules attract each other equally. Each molecule experiences attractive forces in all directions and they cancel out each other. But the atoms or molecules on the surface experience few, if any, attractive forces from the outside. As a result, the forces between them and the inner molecules tend to pull them inwards and toward each other. They therefore act like a skin, holding the surface of the liquid together. Surface tension between soap molecules, for example, holds a soap bubble together. And the soap bubble assumes a spherical shape because the effect of gravity is less than that of surface tension.

Surface tension can be easily demonstrated. A film of soap is formed in a loop of wire. Then a loop of cotton thread is placed carefully on the film. If the film of soap inside the cotton loop is broken, the cotton loop forms a perfect circle—regardless of the shape of the wire loop or of where the cotton loop is placed inside it. This shows that the film of soap contracts to occupy the minimum possible area and that surface tension acts equally in all directions.

Surface tension varies from liquid to liquid. For example, mercury has a very high surface tension. As a result, small drops of mercury on a glass plate are almost spherical and the mercury does not "wet" the glass. Water, on the other hand, does wet glass. This results from another property of liquids called adhesion, which is the attraction of the atoms or molecules of one substance to the atoms or molecules of another substance. When the adhesion of a liquid to a neighboring substance is greater than its cohesion, it overcomes the liquid's surface tension. The adhesion of water molecules to glass molecules, for example, is greater than the cohesion between the water molecules themselves. So water is attracted to glass and tends to spread out.

On the other hand, if the cohesion of a liquid is greater than its adhesion to a neighboring substance, the liquid tries to adopt a shape as near spherical as gravity and the weak adhesion will allow. Thus water droplets in the air are spherical because the cohesion of water molecules is greater than the adhesion of water molecules to air molecules. And the cohesion of mercury atoms is greater than the adhesion of mercury atoms to molecules of glass.

Cohesion and adhesion can be seen clearly when water and mercury are contained in

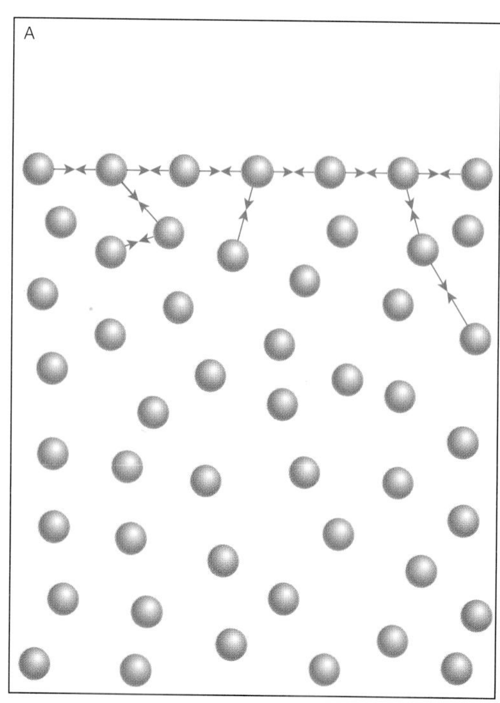

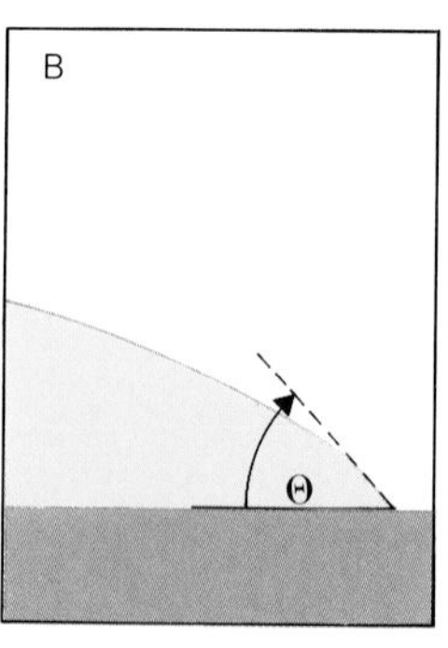

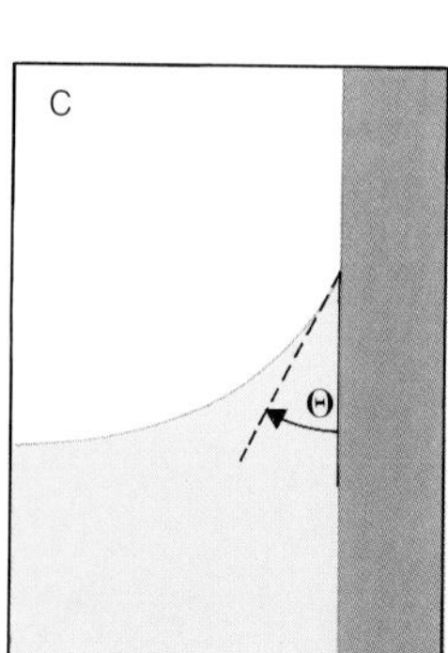

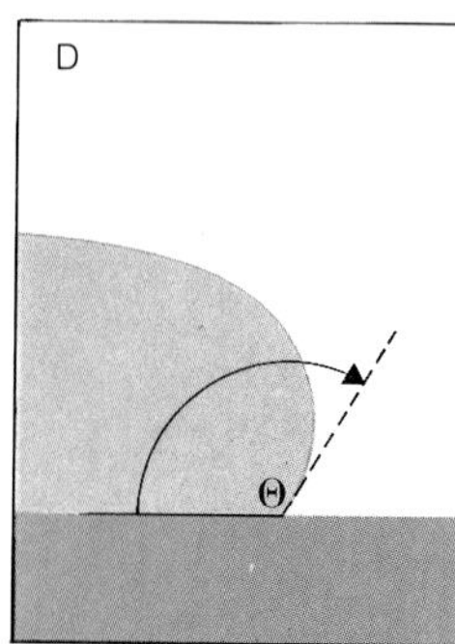

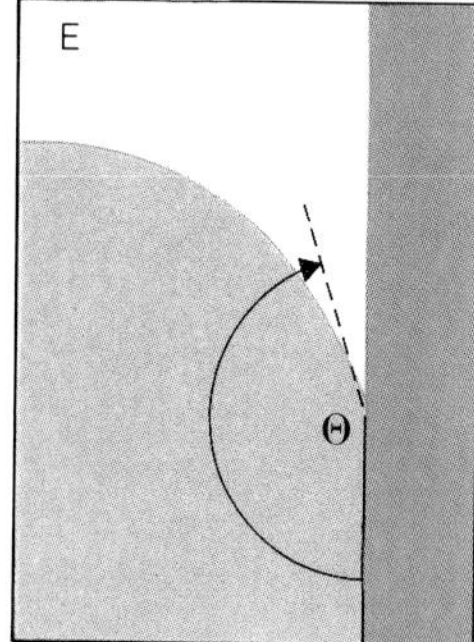

The surface of a liquid behaves as if it has a "skin"—a result of surface tension caused by attractive forces between the atoms or molecules in the surface (A). Adhesive forces between a liquid and a solid surface determine the angle of contact and whether or not the solid is wetted by the liquid. A liquid with a low surface tension (B) has a small angle of contact (Θ) and forms a concave meniscus (C). A liquid with a high surface tension (D) has a large angle of contact and forms a convex meniscus (E).

glass tubes. A tube of water has a concave meniscus at the top; that is, the surface of the water curves upwards toward the glass because of the adhesion of water and glass. A tube of mercury, however, has a convex meniscus, because cohesive forces dominate.

Capillary action

In very narrow tubes, known as capillary tubes, water and mercury behave even more remarkably. If a capillary tube is placed in a bowl of water, the water rises up the tube to a point well above the level of the water in the bowl. This phenomenon is known as capillary action, or capillarity. The principle of capillary action is used in paper tissues, which have tiny pores to draw up fluids.

A bubble of air in water maintains its shape because the air inside it is at greater pressure than the water outside, due to the fact that the inner pressure is aided by surface tension. The pressure difference, or the excess pressure inside the bubble, is given by the formula $2\gamma/r$, where γ is the surface tension and r is the radius of curvature of the bubble. This formula applies to any curved liquid/gas or liquid/liquid interface, including the meniscus formed in a tube.

When a capillary tube is placed in a bowl of water, a meniscus forms immediately and the pressure on the convex (water) side of the meniscus falls below the pressure of the air on the concave side, which is at atmospheric pressure. The radius of curvature of the meniscus is very small, so (from the formula above) the pressure difference is high. As a result, atmospheric pressure on the water in the bowl forces water up the capillary tube until the pressure in the tube at the level of the water in the bowl is equal to atmospheric pressure.

In a capillary tube the radius of curvature of the meniscus and the radius of curvature of the tube are the same. So it follows that the water rises to an even greater height if a narrower tube is used.

The opposite happens when a capillary tube is placed in a bowl of mercury; that is, the level of mercury in the tube falls below the level in the bowl. As before, the pressure is initially higher on the concave side of the meniscus—but this is now the lower (mercury) side, and so the mercury in the tube is forced down by atmospheric pressure.

Altering surface tension

A dry needle can be made to float on water; its weight is not sufficient to overcome the tension of the water's "skin." A pond skater insect uses the same effect to walk on water. But if a drop of detergent is added to the water in which a needle is floating, it sinks rapidly. The detergent lowers the surface tension of water, probably by interspersing its molecules among the water molecules, which therefore lose some of their cohesion.

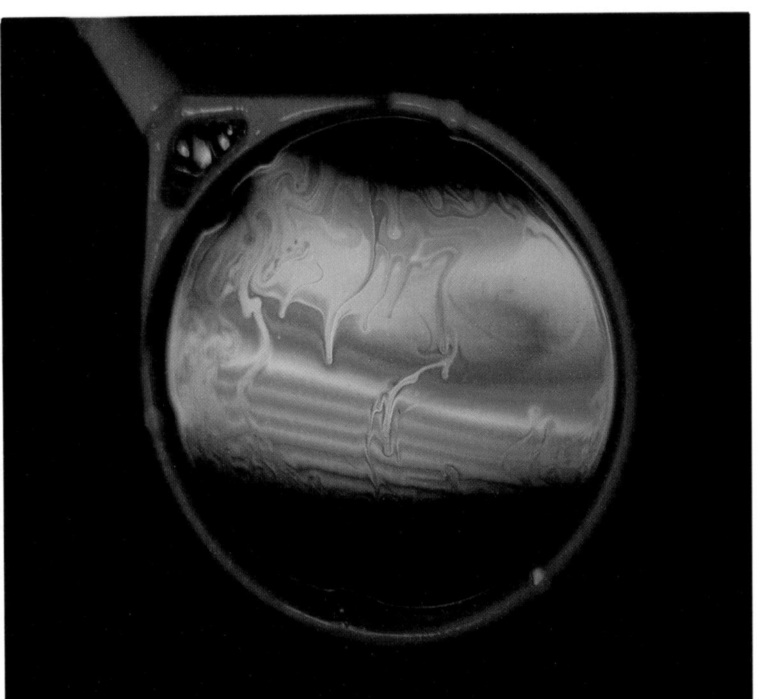

Surface tension causes soapy water to form a thin film across a plastic ring *(above)*. The spectral colors are an optical effect resulting from interference of light reflected from the film.

Mercury *(far left)* has a convex meniscus because it does not "wet" the glass tube, whereas water *(dyed red, right)* wets the glass surface and has a concave meniscus.

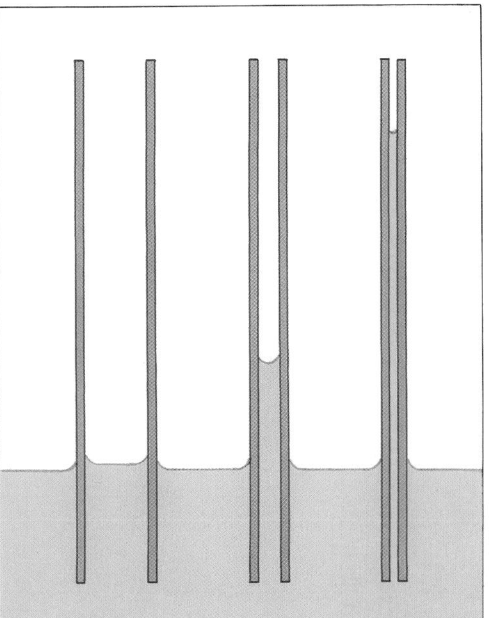

Capillarity causes a liquid with a low surface tension to climb a capillary tube or rise up between a pair of parallel glass plates. The narrower the tube, the greater is the capillary rise. Porous materials such as paper tissues and unglazed pottery absorb water by capillary action.

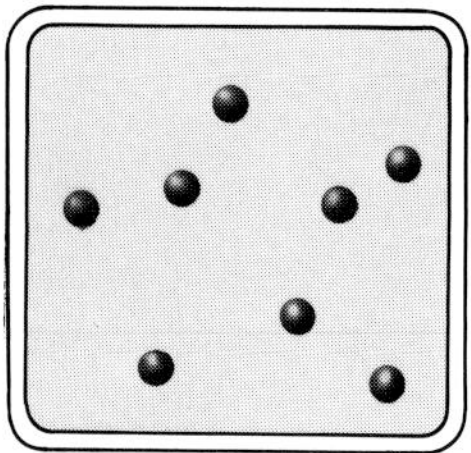

In a gas, the atoms or molecules are free to move about rapidly. Collisions with the walls of the container give rise to gas pressure.

Properties of gases

Because the atoms or molecules of a gas are far apart and move independently of each other, a gas can be expanded or compressed to a much greater degree than can a solid or a liquid. Changes in volume are accompanied by changes in pressure and/or temperature. These changes can be explained in terms of the kinetic theory of gases and the gas laws.

The kinetic theory of gases

The kinetic theory of gases is based on the idea that gases consist of atoms or molecules moving rapidly in all directions. Each of these gas particles has a kinetic energy that depends upon its velocity.

Gas molecules are considered theoretically to be perfectly elastic particles. That is, when two gas molecules collide, they bounce off each other in such a way that their combined energies remain the same. The attraction between the molecules of a gas is almost zero. The volume of the individual molecules is also small compared to the volume of the whole gas. And the time taken for a collision to occur is short compared to the time a molecule spends between collisions.

As well as colliding with each other, the molecules of a gas collide with the walls of their container. As each molecule strikes the wall, it pushes against it with a particular force. In one second the molecule may collide many times with the walls of the container. After each collision, its speed remains unchanged, but its momentum at right angles to the wall is reversed. The force exerted by a single molecule on the walls of the container is the rate at which its momentum changes. Hence the pressure exerted by all the molecules of a gas is their average rate of change of momentum (that is, the average force they exert) per unit area.

Combustible gases are convenient fuels for domestic and industrial heating. But stored gas occupies a large volume, even under pressure, and large gas holders are a common sight in many towns and cities. Some gas holders are rigid structures, whereas others consist of movable telescopic sections that allow for large volume changes.

The gas laws

The pressure, temperature, and volume of a gas are related to each other. This is stated by the three gas laws—Boyle's law, Charles' law, and the pressure law—all of which can be shown to be entirely consistent with the kinetic theory of gases.

In 1662, Robert Boyle experimented with the relationship between pressure and volume. From his work he showed that the pressure of a given mass of gas, at constant temperature, is inversely proportional to its volume; that is, pressure (p) multiplied by volume (V) is constant at constant temperature (T). Kinetic theory agrees with this conclusion, which is known as Boyle's law. If the volume of a gas is reduced and the temperature remains constant, the molecules collide with the walls more frequently and hence the pressure on the walls rises.

The relationship between volume and temperatures is given by Charles' law. This states that the volume of a given mass of any gas at constant pressure increases by $\frac{1}{459}$ of its value at 32° F for every degree Fahrenheit rise in temperature. In metric units, this is $\frac{1}{273}$ of its value at 0° C for every degree Celsius. In other words, at constant pressure, the volume of a gas is proportional to its absolute (Kelvin) temperature; that is, V/T is constant at constant pressure.

Again this is consistent with kinetic theory. If the temperature of a gas is raised, the kinetic energy of its molecules increases and they collide with the walls of the container more frequently and with greater force. So, in order to maintain the existing gas pressure, the volume must be allowed to increase, thus reducing the frequency with which the molecules collide with the walls.

The third gas law, known as the pressure law, can be deduced from Boyle's and Charles' laws. It states that, at constant volume, the pressure of a gas is proportional to its absolute temperature; that is, p/T is constant at constant volume. And from the three gas laws there follows the overall equation of state. This is $pV = nRT$, where R is known as the universal, or molar, gas constant.

Real gases

The gas laws apply to ideal gases—those for which the assumptions of the kinetic theory are true at all times. In practice, however, no such gas exists. Real gases behave somewhat differently and for them the gas laws apply only at moderate to low pressures and moderate temperatures.

Under high pressure the molecules of a gas move closer together, and the assumption that there is little or no attraction between them therefore becomes less and less true. They are attracted to each other by Van der Waals forces. At the same time the combined volume of the individual molecules becomes significantly nearer to the volume of the whole gas, so the volume of the molecules can no longer be ignored. Similarly, at very low temperatures, the molecules of a gas have a greatly reduced kinetic energy and the Van der Waals forces between them begin to have a noticeable effect.

The critical temperature differentiates a gas from a vapor. A gas above its critical temperature cannot be liquefied by pressure alone. Below the critical temperature a gas becomes a vapor, which can be liquefied by applying pressure. For example, the critical temperature of carbon dioxide is 88.7° F (31.5° C), so this gas can be liquefied (under great pressure) at room temperature. Water vapor has a critical temperature of 705.4° F (374.1° C), which is why water is a liquid at normal temperatures and pressures. The critical temperature of nitrogen, on the other hand, is −233° F (−147° C). Nitrogen has to be cooled below this temperature before it can be liquefied.

Gas mixtures

A mixture of gases can be regarded as a solution of one gas in another. As with most mixtures, its properties are a combination of those of the constituents. However, the gases exert pressures independently of each other;

The gas laws are illustrated by the hypothetical experiments *(below)*. In A, a volume of gas is allowed to expand at constant temperature. Doubling the volume halves the pressure, as predicted by Boyle's law. In B, a gas is heated and allowed to expand at constant pressure, demonstrating Charles' law. Diagram C shows Dalton's law of partial pressures: in a mixture of gases, each component contributes the pressure it would exert if it alone occupied the total volume (at the same temperature). No real gas obeys these laws exactly.

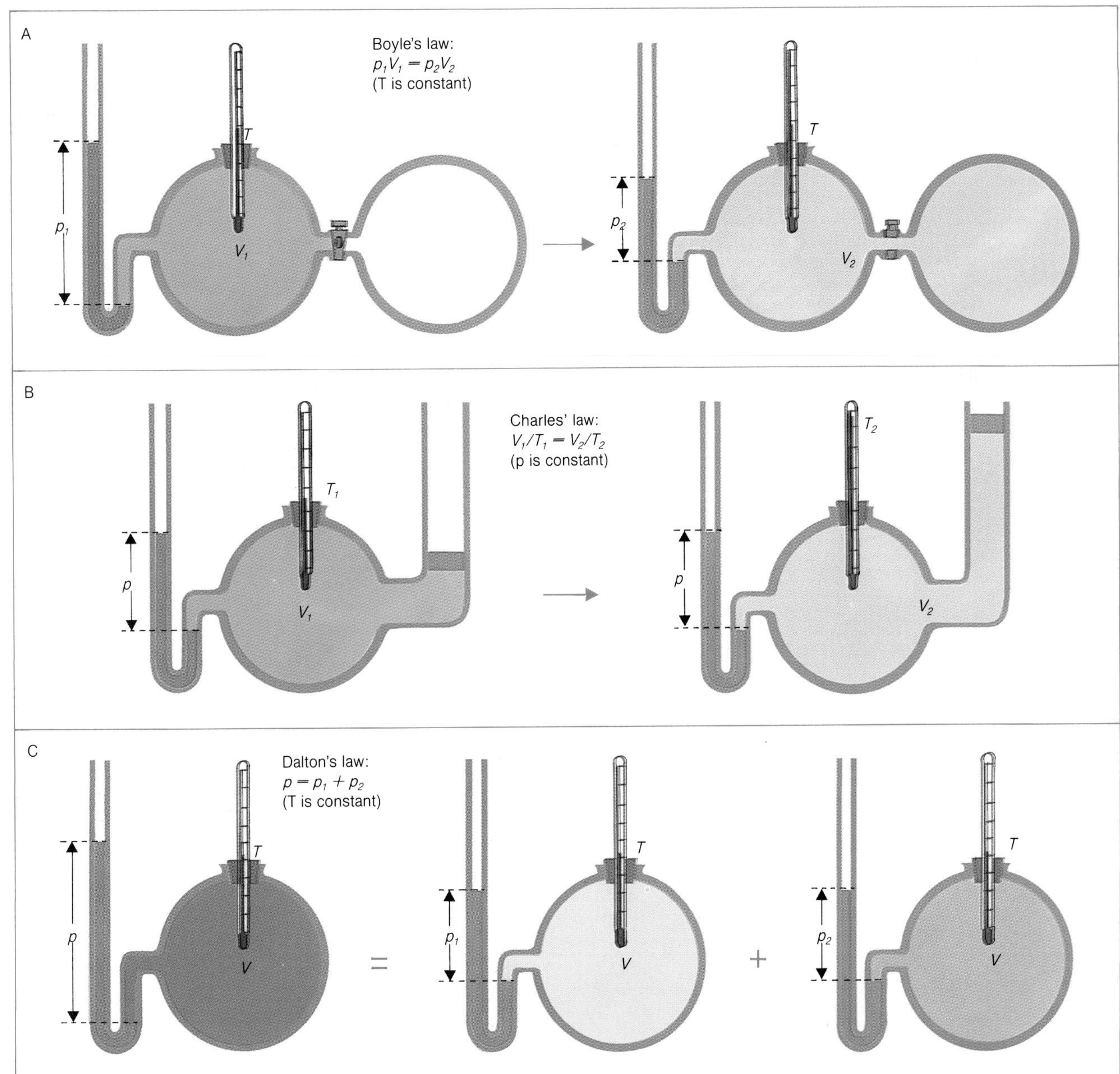

A laboratory experiment graphically demonstrates how heat causes a gas to expand. The balloon on the left is submerged in a beaker of ice-cold water. When the water is boiled *(right)*, the gas is heated and it expands.

each one exerts a partial pressure that is the same as the pressure it would exert if it was by itself. This fact was first formulated by John Dalton in 1802. Dalton's law of partial pressures states that the pressure of a mixture of gases (which do not react chemically) is the sum of the partial pressures of its constituents.

When two gases mix, they diffuse through each other. This diffusion is not due to the motion of the gas as a whole (as in the movements of air we call wind). Instead it is caused by the motion of the individual molecules of gas, which are moving at high speed in all directions. The average velocity of the molecules at any given temperature depends on their mass and hence, because density is mass per unit volume, on the density of the gas. The study of the diffusion of gases was undertaken by Thomas Graham in the 1820's. Graham's law states that the rate of diffusion of a gas is inversely proportional to the square root of its density.

Old and new forms of transport make use of expanding gases. In the engine of the steam tractor *(left)*, steam pressure works the piston mounted on top of the boiler. The engine of the tanker truck *(above)* works by using the pressure of explosively burned diesel fuel. The tanker carries liquid air, a liquefied gas with many industrial applications.

Measuring pressure

The pressure of the air around us is known as atmospheric pressure. Variation in this pressure can be measured using a mercury barometer. At its simplest, this instrument consists of a container of mercury and a long tube that is sealed at one end.

The tube is filled with mercury and then placed open end down in the container of mercury. Provided that the tube is long enough, the mercury in the tube drops to about 30 inches (about 75 centimeters) above the level of mercury in the container, leaving an airless space—a vacuum—in the sealed end. Atmospheric pressure, which acts on the surface of the mercury in the container, supports the mercury in the tube.

In physics, atmospheric pressure is thus usually measured in millimeters of mercury (mm Hg); 1mm Hg is approximately equal to 133.3 newtons per square meter. Standard (normal) pressure is regarded as being 760 millimeters Hg. In customary units, this is equal to 29.92 inches of mercury or 14.7 pounds per square inch.

In meteorology, pressure measurement is usually based on the bar (1 bar = $10^5 Nm^{-2}$), the practical unit being the millibar.

An aneroid barometer works on the same principle as the mercury barometer. The difference is that, instead of having mercury between the vacuum and the outside air, it has the thin walls of a hollow drum. The vacuum is contained within the drum, and changes in air pressure cause the drum to change shape. The movement of the walls of the drum is transmitted to a pointer, which registers the pressure on a scale.

The vacuum in an aneroid barometer is in fact only a partial vacuum. A true vacuum is the complete absence of any molecules. In practice this is impossible to obtain, as the best pumps cannot remove all the air from a container.

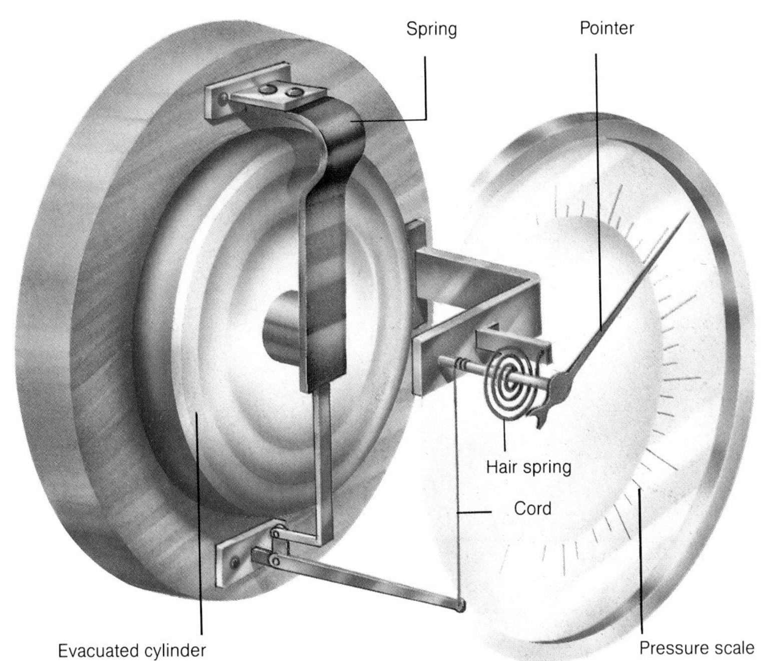

An aneroid barometer *(above)* has a circular evacuated container that changes shape in response to variations in atmospheric pressure. A small movement of the cylinder is magnified by levers and made to rotate a pointer against a calibrated scale.

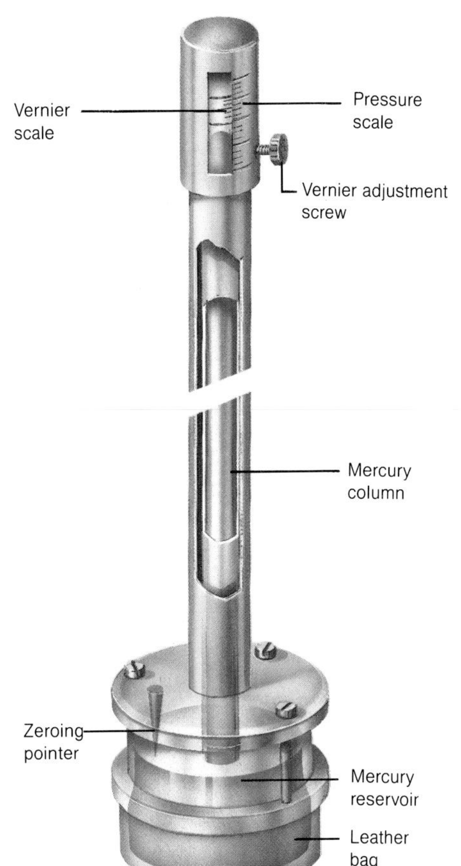

A Fortin barometer *(left)* measures atmospheric pressure by its effect on a vertical column of mercury. At the bottom of the instrument mercury is contained in a leather bag. The adjusting screw is turned to zero on the instrument by making the mercury surface touch a fixed pointer. The height of the column and hence the atmospheric pressure (in inches or millimeters of mercury) can then be read on the vernier scale at the top of the barometer.

Alloys, solutions, and mixtures

When two substances are mixed together, they may react to form a chemical compound. If they do not react, they form a mixture. The constituents of a chemical compound cannot be separated by physical or mechanical methods. The constituents of a mixture, on the other hand, can be separated. For example, a mixture of salt and sand can be separated by dissolving the salt in water. And a solution of salt in water can be separated by evaporating or distilling off the water.

Alloys

An alloy consists of a metallic element and one or more other, usually metallic, elements. For example, most gold used in jewelry consists of an alloy of gold and smaller amounts of copper or nickel. Most alloys are solidified solutions, but in some alloys chemical bonds form between the constituents.

Alloys are used in the making of everyday metal objects because most pure metals are too weak. For example, aluminum, which is popular because of its lightness and resistance to corrosion, is rather weak in the pure state—it is easily bent or stretched. But an alloy of aluminum containing about 4 per cent magnesium and about 1 per cent copper or chromium is considerably stronger.

Alloys are usually made by melting two or more different metals together. While cooling, most alloys do not solidify at a particular temperature, as do pure metals. Instead they solidify over a wide temperature range. For example, a 50/50 alloy of copper and nickel solidifies between 2,394° F. and 2,278° F. (1,311° C and 1,248° C), and between these temperatures it has a pasty texture. A few alloys, such as soft solders (alloys of tin and lead), do behave like pure metals and have definite melting points.

Alloys can also be formed by mixing nonmetals with metals. Silicon, for example, is used in some aluminum alloys. The most important alloys are steels, which are basically alloys of iron and carbon.

Alloys are made by melting together a metal and one or more other substances (usually also metals) to produce a new material with different physical properties. Alloying aluminum with copper, for example, produces an alloy with a greater hardness, although with less than 60 per cent copper the melting point is lowered. All alloys of zinc and copper, on the other hand, have melting points higher than that of pure zinc.

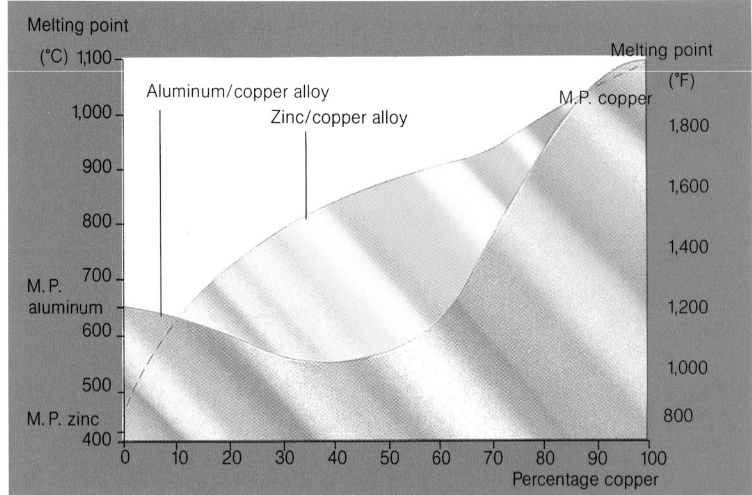

Modern industry makes use of a wide variety of alloys. Here the different types of steel in the stores of an engineering works are identifiable by color coding, and each type is held in a variety of sizes of bar (square or rectangular section) and rod (circular section).

Solutions

When one substance dissolves in another, the two substances are known respectively as the solute and the solvent. The solute dissolves because the attractive forces between its molecules and the solvent molecules are greater than those that hold its own molecules together. When a substance such as sugar is added to water, molecules of sugar begin to leave the crystals and mix with the water molecules. Ionic substances such as salts break up into positive and negative ions; common salt, for example, breaks up into sodium (Na^+) and chloride (Cl^-) ions.

The process of dissolving is not all one way. At any one time there are molecules entering and leaving solution. To begin with, of course, there are more molecules or ions entering solution than there are rejoining the crystals. But if enough solute is present, a stage is eventually reached at which the solvent can hold no more solute molecules. At this stage the solute and solvent are in dynamic equilibrium (that is, the number of molecules entering solution is the same as that coming out of solution), and the solution is described as being saturated. To obtain solid crystals from a saturated solution it is necessary only to evaporate the solvent, which forces the excess solid out of solution.

An increase in temperature generally increases both the rate at which a solute dissolves and the amount of solute that the solvent will hold. Cooling produces the opposite effect, and if a saturated solution is cooled, solid crystals form. It is possible, however, to cool some saturated solutions without producing crystallization. This results in the formation of a supersaturated solution.

In some respects, dissolving a substance is like melting it. The molecules or ions in solution are mixed with liquid molecules and therefore themselves behave like liquid molecules. When a substance melts, it absorbs heat (the latent heat of fusion). And when some substances, such as ammonium chloride, dissolve in water, they too absorb heat, although this is usually less than the latent heat of fusion.

But other substances give out heat when

they dissolve—for example, sodium hydroxide and sulfuric acid both give out a lot of heat when dissolved in water. This results from the fact that a certain amount of chemical bonding takes place between the molecules of these substances and the molecules of the water, releasing energy as heat.

Liquids and gases can also be dissolved in liquids. Alcohol, for example, dissolves in water, as do oxygen and carbon dioxide gases. Unlike solids and liquids, gases are less soluble at higher temperatures. So a gas can be removed from solution by increasing the temperature. Reducing the pressure also releases a gas from solution, as happens when a carbonated drink bottle is opened.

Properties of solutions

Dissolving a substance in a liquid raises the boiling point and lowers the freezing point of the liquid. This is why antifreeze is added to the water in a car radiator. And, following the same principle, salt is sometimes used to melt ice on roads.

Pure water begins to freeze at 32° F. (0° C). As it does so, the ice formed is in dynamic equilibrium with the water around it—that is, as many water molecules are entering the solid phase as are leaving it. But if molecules of a dissolved substance (solute) are also present they reduce the concentration of water (solvent) molecules and so the number of water molecules entering the solid phase at 32° F. (0° C) is reduced. The dynamic equilibrium is achieved only by a lowering of temperature, which reduces the rate at which water molecules leave the solid phase.

A solute raises the boiling point of a solvent because it reduces the vapor pressure above the liquid. Solute molecules at the liquid surface tend to prevent solvent molecules leaving, while still allowing solvent molecules to return from the vapor phase at the normal rate. Thus, when a solution of salt in water reaches 212° F. (100° C), it does not boil, because its vapor pressure is still lower than atmospheric pressure. In order to increase the vapor pressure and cause boiling to take place, the temperature has to be raised.

Another effect associated with solutions is known as osmosis. When two solutions of different concentrations are separated by a semipermeable membrane, there is a tendency for the solvent to pass from the weaker solution to the stronger solution in order to try to equalize the concentrations. The flow of solvent can be prevented by exerting an external pressure on the stronger solution. The weaker solution must therefore be exerting a pressure, which is known as osmotic pressure. It is similar to the pressure exerted by a gas. And in fact a substance in solution exerts an osmotic pressure that is equal to the pressure that the solvent would exert if it were a gas at the same temperature and occupying the same volume.

Trees obtain water by osmosis through their roots, the surface layer of which is a semipermeable membrane through which water passes into the more concentrated contents of the cells within. The water then passes up the tree and is lost through tiny pores in the leaves. This process ensures that the roots do not become waterlogged.

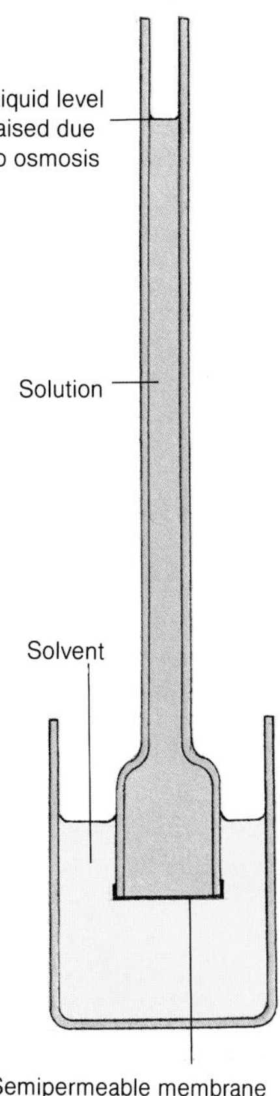

Osmosis can be demonstrated by the simple apparatus *(above right)* consisting of an inverted tube closed by a semipermeable membrane. The tube contains a solution (green) and the beaker a solvent (yellow). Osmosis causes the solvent to pass through the membrane and raise the level of liquid in the tube.

The Camargue in southern France *(below)* is renowned for its salt marshes and lagoons, some of which are commercially exploited for salt. As the water evaporates, the solution becomes saturated and the solute (salt) separates from the solvent, dropping to the bottom as solid salt that can easily be scooped up.

Mist is an unusual state of matter, despite being a familiar phenomenon. It is a colloid in which the dispersion medium is a gas (air) and the disperse phase is a liquid (water droplets). In physical terms, fog and cloud are similar types of colloids.

Unusual states of matter

Some substances exist in states that do not comply with the normal definitions of a gas, a liquid, or a solid. For example, jelly is neither a true solid nor a liquid, and smoke is neither a pure gas nor a solid. Matter in stars and in the tails of comets exists as a plasma, a mixture of charged particles that is outside the normal definition of a gas. In general, a plasma can exist only at extremely high temperatures. At extremely low temperatures, approaching absolute zero, some materials take on remarkable properties. Although they are strictly not different states of matter, their exceptional behavior is also described in this article.

Colloids

In 1861, the physical chemist Thomas Graham discovered that some substances in solution, such as salt, sugar, and copper sulfate, diffuse through parchment, whereas others, such as glue, gelatin, and gum arabic, do not. He therefore divided substances into two groups. Those that would diffuse through parchment he called crystalloids; those that would not diffuse he called colloids. And he believed that the difference between a crystalloid and a colloid depended largely upon particle size. We now know that Graham was broadly correct. But we also know that most crystalloids can be brought into the colloidal state. A colloid is a solution in which the component particles are large molecules or clumps of small molecules, ranging in size from about 4 millionths to 4 hundred-millionths of an inch (10^{-4} to 10^{-6} millimeters) across.

The solvent in a colloid is termed the dispersion medium, and the particles are collectively known as the disperse phase. Colloids are classified according to the physical state of the dispersion medium and the disperse phase. Many substances form colloids naturally and are described as being lyophilic ("solvent-loving"). If the dispersion medium of a lyophilic colloid is evaporated, the colloid can be re-formed simply by adding more dispersion medium. Lyophobic ("solvent-hating") colloids, on the other hand, do not form naturally, and the disperse phase has to be introduced into the dispersion medium.

Colloids have different properties from those of true solutions or suspensions, because of the size of the particles. In a true solution the particles are small and behave like the particles of a liquid, moving randomly about. In a suspension the particles are too large to behave like liquid particles; they are bombarded by molecules of the dispersion medium equally on all sides, with the result that they do not move about randomly. And, because they are relatively heavy, they tend to settle under the influence of gravity. A colloid, however, is intermediate between a solution and a suspension. Small enough to be affected by molecular collisions, the particles move about randomly—a phenomenon called Brownian motion. But, although large enough for gravity to affect them, they show little or no settling, because of the viscosity of the dispersion medium and the presence of minute convection currents.

Because the particles in a colloid are larger than those in a true solution, they will not pass

Crystalloids and colloids can be distinguished by a simple experiment. If an aqueous solution is placed on one side of a parchment membrane, with water on the other side, a crystalloid (A) diffuses into the water because its molecules are small enough to pass through the pores in the membrane (B). A colloid (C) does not diffuse, because its molecules are too large (D).

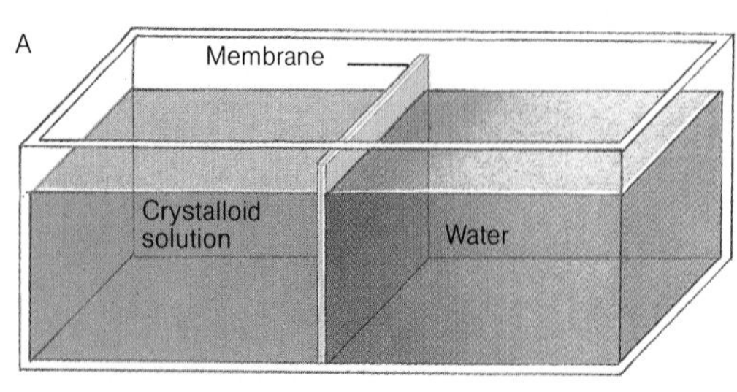

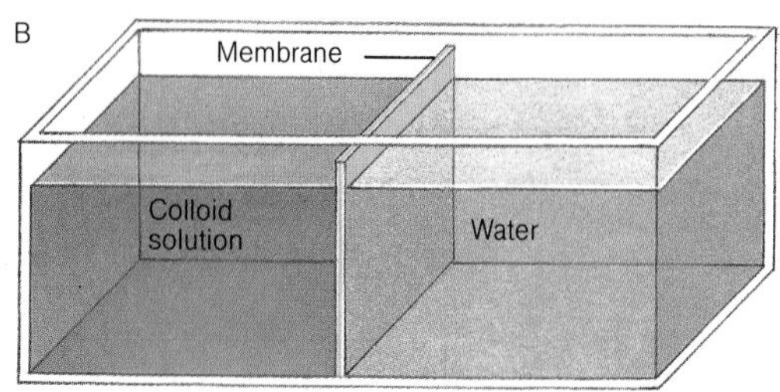

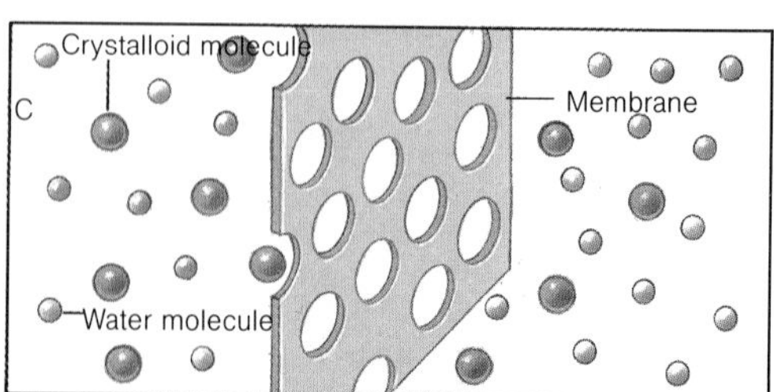

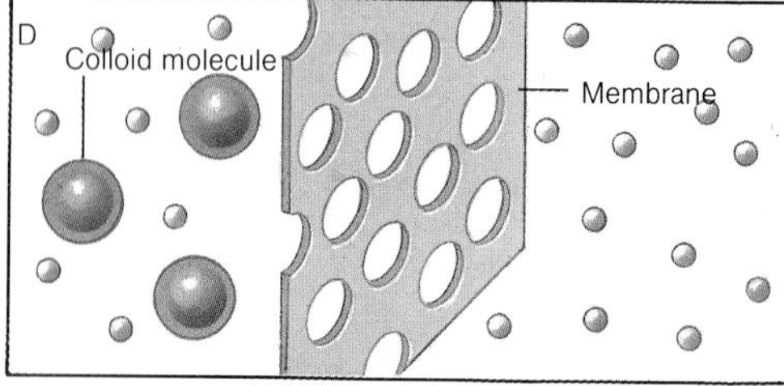

Liquid foam, a form of colloid, is used to extinguish certain types of fires—oil fires, for example—because it is very effective at smothering flames. Physically, foams consist of a liquid dispersion medium and a gaseous disperse phase. In fire-fighting foams, the gas is often carbon dioxide, in which combustion is impossible.

through a semipermeable membrane. This fact is used in the process known as dialysis, when a solution of a crystalline substance is separated from a colloid. Dialysis is used in kidney machines to remove soluble waste substances from the colloidal solution of proteins in the blood.

Like true solutions, colloids show what are known as colligative properties, such as osmosis, lowering of vapor pressure, raising of boiling point, and lowering of freezing point, which all depend on concentration. But because colloidal particles are larger than those of a true solution, there are fewer particles per unit volume. As a result, the colligative properties of a colloid are hardly noticeable.

In a colloid in which the dispersion medium is a solid and the disperse phase is a gas or a liquid, the solid dispersion medium has a very large surface area. All solids can, to some extent, adsorb other substances because of the presence of "unused" Van der Waals forces or chemical bonds at the surface. But because they have such a large surface area, some colloids are particularly good adsorbers. Charcoal, for example, can adsorb many gases and is, therefore, used in gas masks and extractor hoods for cookers. Colloidal aluminum hydroxide is used as a mordant to "hold" dyes to the fibers of fabrics by adsorption.

Sometimes adsorption can occur at a liquid-gas interface. Some mineral ores are separated from the rocks in which they are found by a process called froth flotation, in which a crushed ore is removed by causing it to become part of a foam.

Colloidal particles nearly always carry an electrical charge, which is the same for every particle in a colloid. The presence of these like charges helps to keep the particles apart and prevent the colloid from coagulating. The fact that a colloid is electrically charged can sometimes be useful. Smoke particles, for example, can be removed by using electrically charged plates in flues and chimneys, thus reducing atmospheric pollution.

Liquid crystals

The liquid state is intermediate between the solid state and the gaseous state. Some substances, however, are intermediate between solids and liquids. In a liquid crystal the atoms or molecules are arranged in a pattern, like those of a solid crystal. But the pattern is not completely fixed; it can be altered by heat or an electric field. Some liquid crystals change color at certain temperatures and can, therefore, be used in liquid crystal thermometers. In others an electrical voltage causes a change in pattern that alters the plane of polarized light. Such types are used to make liquid crystal displays for watches and calculators.

Liquid crystal displays are used in some pocket calculators and watches, and—as here—in thermometers. The color of this type of liquid crystal depends on temperature, indicated in degrees Celsius.

Plasmas

A plasma is sometimes described as being the fourth state of matter—that is, one phase farther on from a gas. In fact a plasma is created by heating a gas to such a high temperature that its atoms or molecules lose electrons and become ions. The gas is almost fully ionized and becomes a very good electrical conductor. The gases that are involved in the thermonuclear reactions of the sun and other stars are in the form of a plasma. Scientists are now trying to recreate such a plasma on earth by heating the gases deuterium and tritium (isotopes of hydrogen) to tens of millions of degrees in special installations. In this way, they hope to produce a controlled thermonuclear fusion reaction as a source of power.

Supercooled liquids

When a liquid cools to its freezing point, atoms or molecules usually begin to join up to form a solid. But they can do this only if there

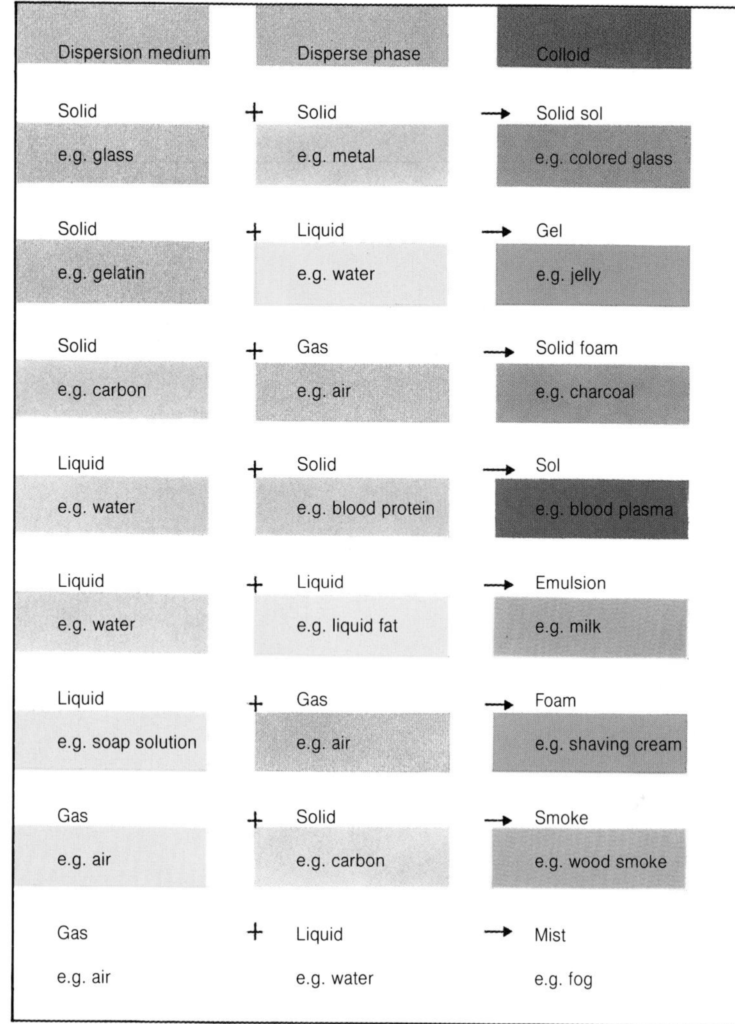

Dispersion medium		Disperse phase		Colloid
Solid e.g. glass	+	Solid e.g. metal	→	Solid sol e.g. colored glass
Solid e.g. gelatin	+	Liquid e.g. water	→	Gel e.g. jelly
Solid e.g. carbon	+	Gas e.g. air	→	Solid foam e.g. charcoal
Liquid e.g. water	+	Solid e.g. blood protein	→	Sol e.g. blood plasma
Liquid e.g. water	+	Liquid e.g. liquid fat	→	Emulsion e.g. milk
Liquid e.g. soap solution	+	Gas e.g. air	→	Foam e.g. shaving cream
Gas e.g. air	+	Solid e.g. carbon	→	Smoke e.g. wood smoke
Gas e.g. air	+	Liquid e.g. water	→	Mist e.g. fog

There are eight types of colloids *(listed above)*, which depend on various combinations of dispersion medium—solid, liquid, or gas—and disperse phase.

Three categories, particularly gels, foams, and emulsions, have various practical applications in the manufacture of plastics and paints.

Snowflakes may form as a result of supercooling in clouds. Water droplets at very low temperature (down to −40° F or −40° C) evaporate and crystallize onto small particles of ice and grow into feathery snowflakes.

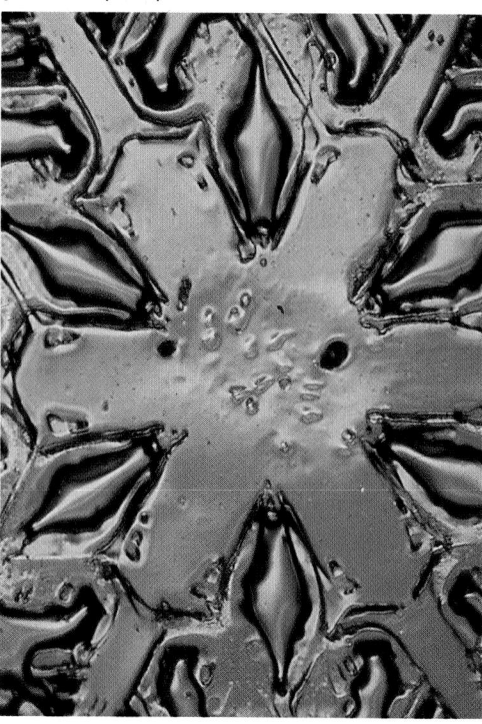

is a small piece of solid or an impurity present for them to start building on. A liquid in an extremely pure state can sometimes be cooled to below its normal freezing temperature without any solid forming. If a small particle of the solid is then added, the supercooled liquid freezes, and its temperature rises to the freezing point. Shaking a supercooled liquid may also have the same effect.

In temperate regions of the world supercooling occurs in clouds, which may be at temperatures as low as −40° F (−40° C). There supercooled droplets of water form, as well as minute ice crystals. The supercooled water droplets evaporate, and water vapor condenses on the ice crystals. These grow in size and fall as snow or, if they melt on the way down, rain.

Superconductivity

While measuring the electrical resistance of mercury at temperatures a few degrees above absolute zero (0 K, −459.67° F or −273.15° C), the Dutch physicist Heike Kamerlingh Onnes discovered that its resistance suddenly vanished at 4.2 K (4.2 kelvins or degrees Celsius above absolute zero). It was later found that this peculiarly sharp transition to perfect conductivity occurs with several other metals and alloys; the phenomenon is called superconductivity.

Superconductivity sets in at a fixed temperature (the transition temperature), which is a characteristic of the substance concerned. For zinc the transition occurs at 0.88 kelvins, whereas for lead it is 7.2 kelvins. The transition to the new state takes place over less than one-thousandth of a degree. With these characteristics, superconductivity has proved to be of immense value in the production of very powerful electromagnets.

Explaining the phenomenon took nearly half a century and won a Nobel Prize for three American physicists. In metals, electrical conductivity depends basically on particle (electron) scattering processes: the fewer the number of scatterings, the greater the conductivity. In a superconducting metal, interactions between the free electrons within the lattice and nearby positive ions cause deformation of the lattice structure. Its effect is to produce an indirect attractive force between the mutually repulsive electrons, which pair up. Once some pairs have been formed, others are produced. The resulting "Cooper pairs," as they are called, are difficult to split apart by scattering processes. The pairs also have quantum properties that make them difficult to scatter. Conductivity thrives on there being few collisions, and so superconductivity results, with the transition temperature marking the onset of quantity production of Cooper pairs.

Many materials besides metals and alloys have been found to be superconductors given the proper conditions. In particular, a group of ceramic materials have been found to have the highest transition temperatures yet recorded. These materials are poor conductors at room temperature, but many become superconducting at temperatures approaching 150° below zero Celsius, which is 100° C warmer than most metals.

Superfluidity

Superfluidity is a state of zero or near-zero viscosity. As with superconductivity, superfluidity sets in at a sharply defined temperature. For the helium isotope helium-4 it is about 2 kelvins—only 2 degrees below the temperature at which it liquefies. Below 2 kelvins, the liquid behaves as if it were made up of two components, a normal fluid and a superfluid component, whose proportions vary with temperature. The superfluid has no viscosity and infinite thermal conductivity. One demonstration of the unusual properties of superfluid helium-4 is provided by putting the liquid in an unglazed pot. Because of its extremely low viscosity, the liquid simply passes through the pores in the pot until it is empty.

There is as yet no complete explanation for the behavior of superfluids, although the theory of superconductivity in metals has provided some insight. Other theories have depended on the observation that although superfluids are liquids, they are very weak ones and have much in common with gases. For example, their densities are low (helium-4 has a density only one-seventh that of water) and their viscosities are similar to that of air at room temperature, even in the largely normal-fluid regime.

Noting this, some scientists have suggested that the helium-4 superfluid can be likened to a gas in which there are very weak interatomic forces. This "gas" is composed of helium-4 atoms which, having integral spin, are classified in the group of particles called bosons. One property of collections of bosons is that below a certain temperature they suddenly take up a state of motionlessness, or zero kinetic energy. Calculations ignoring interatomic forces predict a condensation temperature of 3.2 kelvins, similar to the transition temperature. If it proves to be correct, this theory goes some way to explaining the peculiar properties of superfluid helium-4.

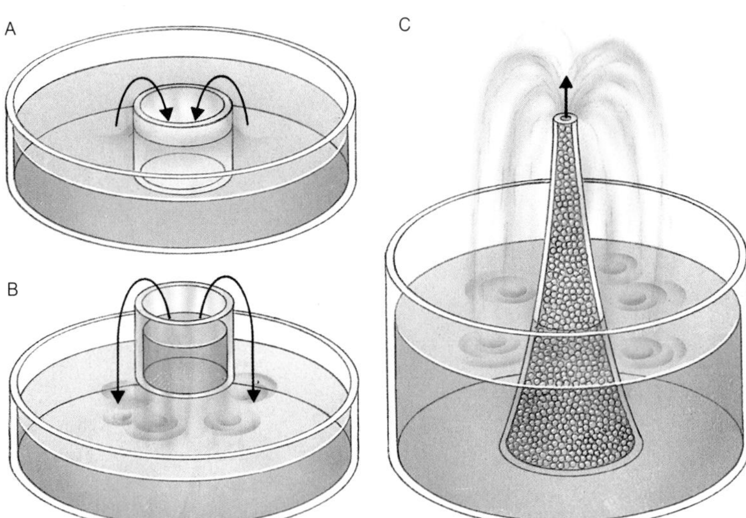

The remarkable properties of a superfluid are exhibited in these three experiments. In (A) an empty beaker in a bowl of liquid helium gradually fills as the liquid "crawls" over the sides. With the beaker above the liquid (B), the helium gradually "crawls" out again. In (C) a funnel-shaped container of powder is placed in a bath of liquid helium. Light causes superfluid helium in the upper part to revert to the normal form; more superfluid rushes up the tube and creates a fountain.

Gases in the sun reach such an enormously high temperature that they exist as a plasma—often termed the fourth state of matter.

BREAKING THE COLD BARRIER: *superconductivity*

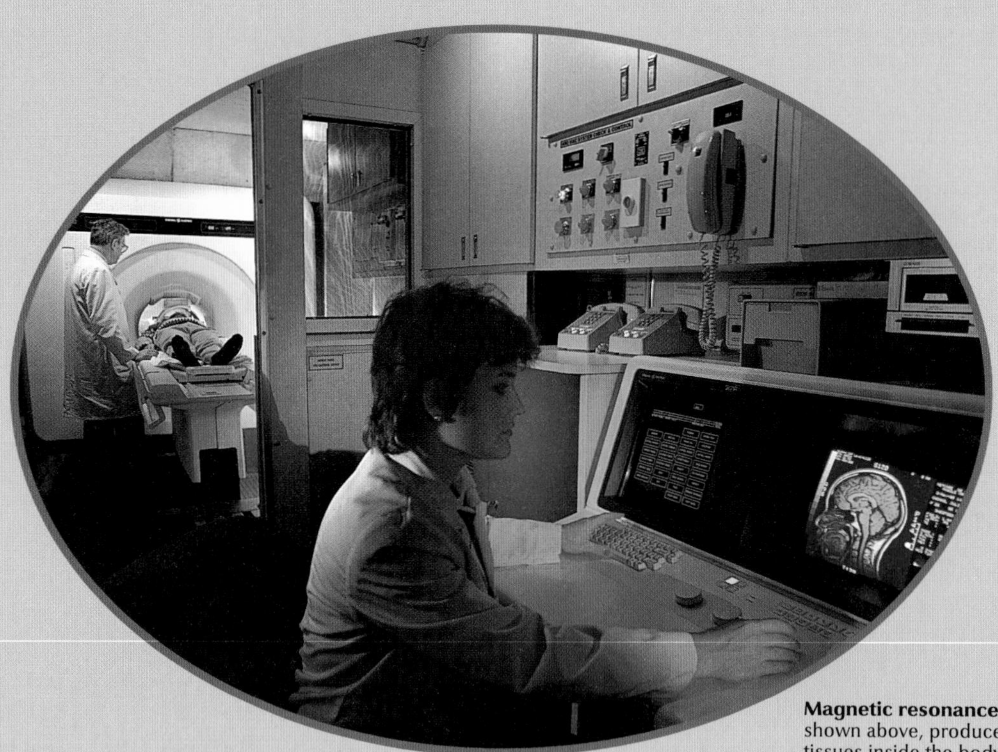

Magnetic resonance imaging (MRI), shown above, produces images of tissues inside the body. Doctors use these images to diagnose diseases and injuries.

A train floats along, frictionless, on a magnetic cushion above the tracks. Electrical wires deliver power to homes and industries over long distances without any loss of energy. Supercomputers perform calculations with 50 times the speed of the fastest computers available today. These are a few of the advances scientists believe are coming closer to reality as advances in high-temperature superconductivity continue.

The phenomenon of superconductivity, in which certain metals, alloys, organic compounds, and ceramics conduct electricity without resistance, has been recognized since the early 1900's. That was when Dutch physicist Heike Kamerlingh Onnes found that mercury chilled to a temperature near absolute zero conducted an electrical current without resistance. Absolute zero is a theoretical temperature at which the atoms and molecules of a substance have the least possible energy. Absolute zero, or 0 Kelvin, is equal to –459.67 °F, or –273.15 °C.

Research on superconductivity remained largely in the background of physics throughout the first half of the 1900's, partly because the difficulties and expense of cooling materials to near absolute zero made practical applications seem remote. However, in 1986, physicists developed certain ceramic materials that became superconductors at much higher temperatures, about 32 K, or -402.07 °F (-241.15 °C). Researchers eagerly predicted that room-temperature superconductors were just around the corner, making levitating trains and other futuristic inventions a reality.

However, scientific progress rarely comes swiftly or easily, and until the early 1990's materials that become superconductors above 133 K (more than 150 °C below room temperature) still eluded researchers. Part of the problem is that scientists don't really understand how superconductivity works. The conventional theory, called the BCS theory, developed in 1957, seems to explain low-temperature superconductivity but does not explain how superconductivity occurs at higher temperatures.

The BCS theory involves the behavior of electrons. Resistance occurs in an ordinary conductor because imperfections in it, along with the vibration of atoms due to thermal, or heat, energy, impede the flow of electrons through the material. According to the BCS theory, at temperatures near absolute zero two conditions occur that turn a metal into a superconductor. First, since the atoms in the material have almost no movement, there is no vibration to "scatter" electrons around and impede their flow. Second, at such low temperatures, electrons in the material form pairs, which do not scatter like the unpaired electrons. Instead, the pairs flow through the material without resistance.

Without a theory to explain how superconductivity occurs at higher temperatures, physicists have nothing to guide them in their search for other superconducting materials. Instead they use a hit-and-miss approach based on their experience with various materials.

Researchers are also faced with more practical difficulties in turning available superconducting materials into usable devices. Because today's higher-temperature superconductors are made of ceramics, they are brittle and very difficult to shape into the kind of flexible wires usually needed to conduct electricity.

Even with these difficulties, a number of devices in wide use today employ superconducting materials. Magnetic resonance imaging (MRI) machines, for example, which create images of tissues inside the human body, and particle accelerators, used by physicists to break apart atoms, use superconducting magnets that create a powerful magnetic field with relatively little electricity.

Other examples of the application of superconductivity are devices called superconducting quantum interference devices, or SQUIDS. SQUIDS are used to make sensitive measurements of magnetic fields. They have been used to detect the weak magnetic signals that emanate from the heart and brain.

In 1979, an experimental magnetic levitating train using superconducting magnets achieved a speed of 310 miles (517 kilometers) per hour on a test track in Japan. In Germany, "maglev" test vehicles have been under evaluation at a demonstration facility since the early 1980's. These trains routinely achieve speeds of up to 270 miles (450 kilometers) per hour.

However, most scientists admit that large-scale uses for superconductors will have to wait until better superconducting materials are developed. Progress is being made; in 1993, a group of Swiss researchers demonstrated a material that became superconducting at 134 K. Soon after, researchers at Geophysical Lab at the Carnegie Institution in Washington, D.C. raised the temperature to 164 K (about −164 °F, or −109 °C). These advances encourage scientists to keep searching for a room-temperature superconductor.

The rainbow colors appearing in the computer-enhanced image below show a magnetic field surrounding a super-conducting crystal. The characteristics of this magnetic field, produced in an experiment by IBM researchers, support the theory that some materials are superconducting at relatively high temperatures because they cause electrons to pair up through the formation of magnetic bonds.

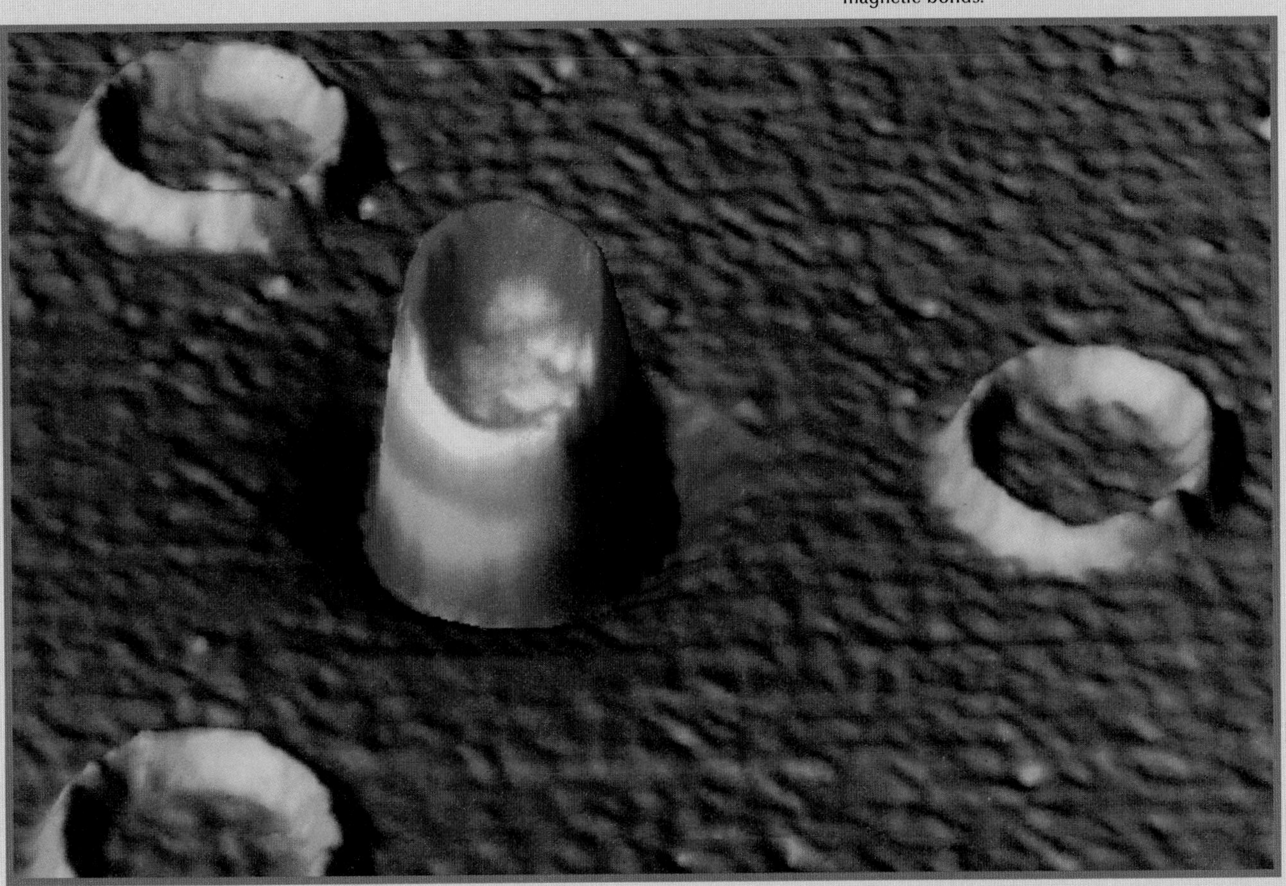

Statics

Terms, units, and abbreviations
The following units are used:

- **length** in meters (m)
- **mass** in kilograms (kg)
- **weight** in newtons (N)
- **density** in kilograms per cubic meter (kg m^{-3})

The equivalent customary units and conversion factors are as follows:

- 1 meter = 1.09 yards or 3.28 feet
- 1 kilogram = 2.20 pounds
- 1 newton = 7.23 poundals
- 1,000 kilograms per cubic meter = 62.43 pounds per cubic foot

Time is measured in seconds (s) in both systems of units.

Statics is the branch of physics that deals with the analysis of bodies that are held stationary under the influence of a system of forces. Using statics it is possible to predict what will occur when the forces acting on an object are changed, the size of the forces needed to keep an object stationary, and many other phenomena of interest to physicists and engineers.

Forces

The study of statics depends crucially on an understanding of the concept of force, which can be defined as an agent that is capable of altering the state of rest or motion of an object. A force acting on an otherwise free object will accelerate it.

A common force is that of gravity, which on earth attracts everything downward to its surface. Other forces include friction, which in some cases can act against another force to prevent movement, and electric forces, which bind atoms together.

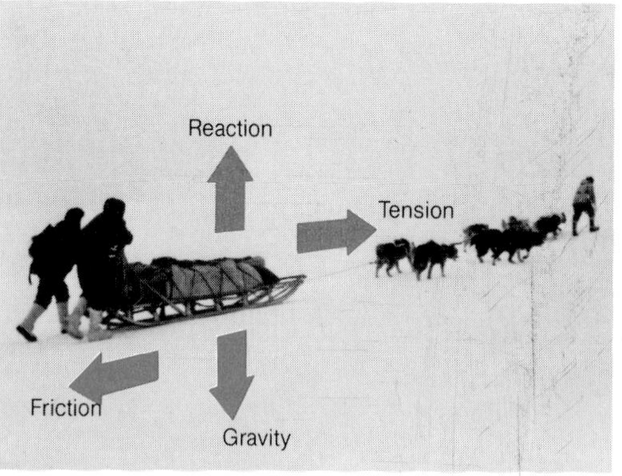

The forces involved when an object such as a sledge is pulled include gravity, which pulls it downward; reaction, which pushes upward on the sled; friction, which is exerted by the ground on the sledge in the opposite direction to the pull; and tension, which is in the trace with which the sledge is pulled.

Mass and weight

Mass is defined as the resistance of an object to any change of its state of motion or rest by the action of a force. Mass is, therefore, a measure of the inertia of an object. The greater an object's mass, the smaller is the acceleration a given force produces. The mass of an object is proportional to the amount of matter making up the object.

The weight of an object is not, however, an unchanging intrinsic property, but is the downward force exerted by an object resulting from it being acted on by the force of gravity. So, although an object's mass is the same everywhere, its weight depends on the strength of the gravitational field acting on it. The moon's gravitational field, for example, is only one-sixth as powerful as the earth's, and as a result the weight of an object on the moon is only one-sixth of its weight on earth.

Density and specific gravity

One of the major areas of study in statics is the behavior of objects in fluids (that is, in liquids and gases). Much of this work revolves around the concept of density, which is the mass of an object per unit volume. If the mass is measured in kilograms and the unit of volume is the cubic meter, then density is measured in kilograms per cubic meter (kg m^{-3}).

Another useful concept is that of specific gravity, or relative density. This property is given by the density of an object divided by the density of water. Because the density of

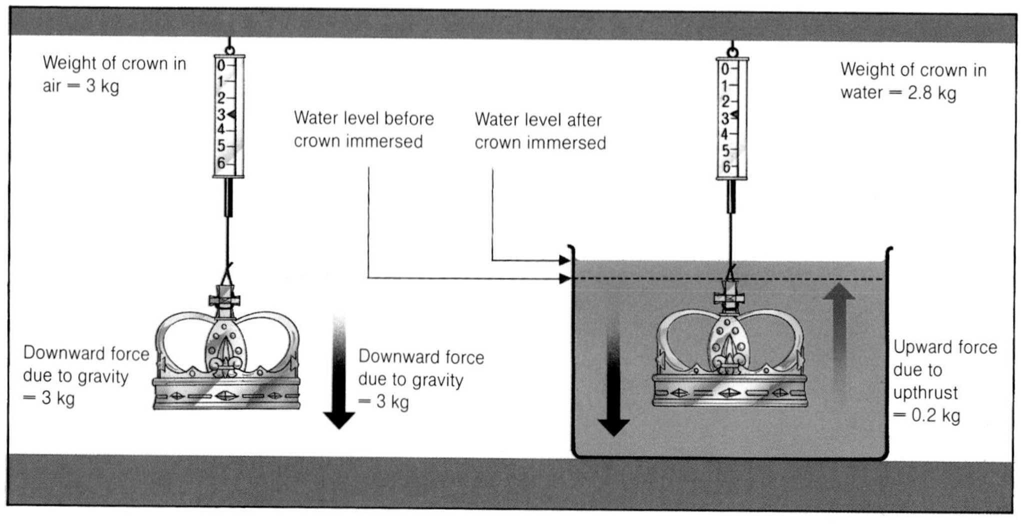

Archimedes' principle enables the specific gravity (numerically equal to density) of an object to be determined. In this example *(right)*, the crown weighs 3 kg in air but only 2.8 kg in water. The upthrust, which causes the weight difference, equals the weight of water displaced—0.2 kg. Specific gravity equals the weight of an object divided by its apparent loss of weight in water, hence the crown's specific gravity is $\frac{3}{0.2}$ = 15. But the specific gravity of pure gold is 19.3, so the crown must be made of a material with a lower specific gravity.

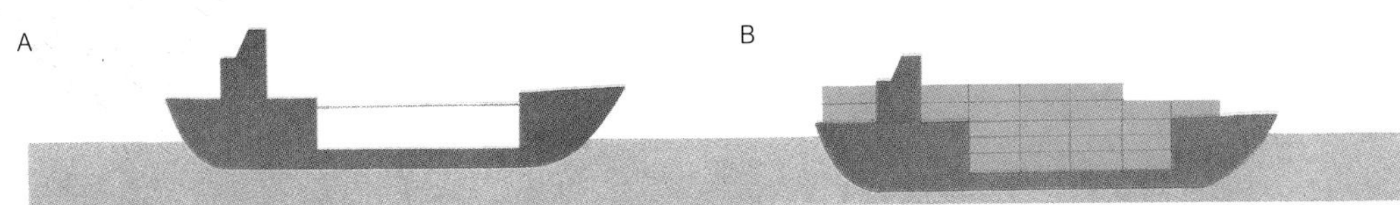

A floating object, such as a ship *(left),* floats at such a depth that the upthrust due to the liquid's displacement equals the weight of the object. An unladen ship (A, *top left*) floats high in the water because relatively little of the ship's volume needs to be below the waterline to produce the comparatively small upthrust necessary to keep the ship afloat. A heavily-laden ship (B, *top right*) floats much lower in the water because a substantially greater displacement is needed to produce the larger upthrust required to counteract the weight of the ship and its cargo.

water is 1,000 kg m^{-3}, copper for example (which has a density of 8,960 kg m^{-3}) has a specific gravity of 8.92.

An important law in the statics of fluids is Archimedes' principle, which states that the upward force acting on an object immersed in a fluid is equal to the weight of fluid that is displaced. The value of the principle lies in providing a way of finding the specific gravity, and therefore the density, of an object without having to measure both its mass and volume. Legend tells of Archimedes applying the principle after being asked by Hiero II, the King of Syracuse, to prove that a gold crown was indeed made of solid gold. The crown was intricately worked and finding its volume would have been extremely difficult. So Archimedes found its density directly by putting the crown and then an equal weight of pure gold in a bowl of water and measuring the amount of water displaced by each. He found that the crown displaced more water than did the gold weight, which meant that it had a greater volume and therefore consisted of a material of lower density than pure gold.

The density of liquids is measured by hydrometers. These instruments take advantage of the principle of flotation, which states that when an object floats, its weight is equal to the weight of fluid displaced. In other words, the upward "flotation" force exactly balances the downward gravitational force. When a hydrometer is put in a high-density liquid, the upthrust is sufficient to prevent a large part of the instrument becoming immersed, so that it floats high up in the liquid. In a liquid of lower density, the upthrust is less and so the hydrometer floats at a lower height. Densities are read off a scale on the stem of the hydrometer.

Density is one of the characteristic properties of a substance. This table *(below left)* lists the densities of various solids (at 68° F. or 20° C, except for ice, whose density is given at 32° F. or 0° C), liquids (at 68° F. or 20° C) and gases (at 32° F. or 0° C and normal atmospheric pressure). A vivid example of density differences among liquids is shown below. Six liquids have been made to float on each other, the densest at the bottom and progressively "lighter" ones above.

Densities of various solids

Solid	**Density** (kg m^{-3})	**Liquid**	**Density** (kg m^{-3})	**Gas**	**Density** (kg m^{-3})
Balsa wood	110-140	Gasoline	660-669	Hydrogen	0.0899
Lithium	534	Diethyl ether	714	Helium	0.179
Rubber, natural	910-1,190	Acetone	780	Methane	0.717
Ice	917	Alcohol (ethanol)	791	Ammonia	0.771
Polyethylene	920-955	Kerosene	800	Neon	0.899
Ebony (a hardwood)	1,110-1,330	Turpentine	870	Acetylene	1.173
Glass	2,600-4,200	Benzene	879	Carbon monoxide	1.250
Aluminum	2,700	Olive oil	920	Nitrogen	1.251
Steel, mild	7,860	Water pure	998	Air	1.293
Iron, pure	7,870	sea	1,025	Oxygen	1.429
Brass (70:30 Cu:Zn)	8,500	Milk	1,030-1,035	Fluorine	1.696
Silver	10,500	Phenol	1,073	Argon	1.784
Lead	11,350	Glycerin (glycerol)	1,262	Carbon dioxide	1.977
Gold	19,300	Carbon tetrachloride	1,632	Sulfur dioxide	2.927
Platinum	21,450	Bromine	3,100	Chlorine	3.214
Osmium	22,480	Mercury	13,546	Radon	9.730

Equilibrium

Statics is concerned with bodies that are in equilibrium, which is the state of an object when it is not accelerated; a body that is at rest or moving at constant velocity is therefore in equilibrium.

For an object to be in equilibrium it is necessary for all the forces acting on it to cancel each other out exactly.

Center of gravity

The concept of center of gravity is used in determining the stability of an object's equilibrium. The center of gravity is the point where the entire weight of an object can be considered to be concentrated. A disk, for example, has its center of gravity at the center, whereas that of a rigid triangular sheet lies at the point of intersection of the lines that join the vertices of the triangle to the midpoints of the opposite sides. The center of gravity of an irregularly-shaped flat object can be found by suspending it from any two points on it and marking the point of intersection of a plumb line suspended from each point in turn.

For an object to be in stable equilibrium when resting on a surface, the vertical line passing through the object's center of gravity must also pass through the base of the object. If it does not, the object is in unstable equilibrium and the slightest displacement will make it topple. Generally, an object is stable if its center of gravity (or center of mass) is near its base. The center of mass is the same as the center of gravity in a uniform gravitational field (which, for practical purposes, our planet can be considered to have).

The principle of moments

The reason an object falls over when it is disturbed from its unstable equilibrium is that the force of gravity acts as a torque, turning the object on an axis. The strength of this turning effect, the moment of the force, can be measured at any point on the object and varies according to the distance of the point from the object's axis. For any given point, the moment of the force equals the magnitude of the force multiplied by the perpendicular distance from the point to the line along which the force acts. The principle of moments states that when a body is in rotational equilibrium, the algebraic sum of all the torques acting on the body about all its axes is zero.

A low center of gravity makes this London bus very stable. The tilt test shows that a ballasted double-decker can be tilted to at least 28° without toppling over.

An object is unstable when a slight displacement from its unstable equilibrium position causes it to fall over. The diagram below shows stable (returns to original position) and unstable positions (denoted by S and U respectively) for various objects.

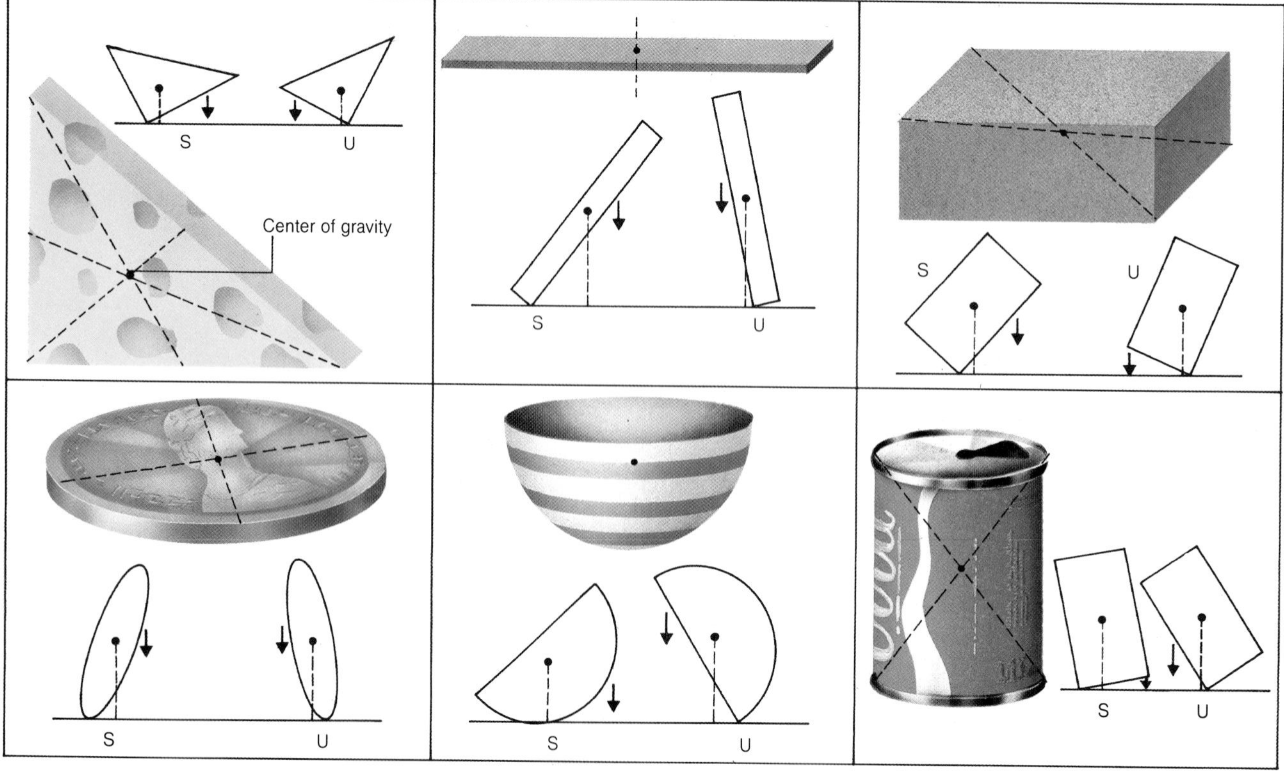

Tightrope walkers seem to balance extremely precariously but in fact they are in less danger than it appears—although in physical terms they are still in unstable equilibrium. The long, drooping poles lower the centers of gravity of the tightrope walkers and can also exert a turning force so that if a tightrope walker begins to overbalance to one side, he or she can move the pole to the other side and use the resultant turning force to regain equilibrium. The inertia of the pole helps the tightrope walker maintain equilibrium.

The principle of moments can be seen in the action of the balances that are used for weighing. If an object is put in one pan, the beam of the balance swings down at that end. This happens because the weight of the object produces a downward force on the pan and, therefore, a torque on the beam. The torque due to the weight of the other empty pan is less (because it is lighter) and so the equilibrium is disturbed as the beam swings downwards on the heavier side. Removing the object, or placing one of equal weight on the empty pan, equalizes the torques and the beam returns to the horizontal.

If the forces acting on an object are equal in strength but acting in opposite directions, a couple is produced. This is a system of forces that, instead of producing motion in a straight line, produces only rotational motion, as in the handle of a faucet operated by a turning movement.

The man and child carrying a heavy bag provide an example of several forces acting on one point: gravity is pulling the bag downward but this is offset by the lifting forces exerted by the man and child.

Practical applications

The principle of moments is used for working out the strength of materials needed to construct bridges and buildings.

For example, frameworks consisting of structural members are often found in the roofs of houses and in some forms of bridges; using the principle of moments and knowing the forces at work on the framework or truss, it is possible to discover which members are under compression (called struts) and which are in tension (called ties). Materials of suitable strength and thickness can then be chosen for the construction.

On simple bridges that have beams that carry most of the load, it is also possible to calculate what forces are required to ensure that the bridge does not collapse, as well as the amount of deflection produced by a given load.

Such calculations always include a safety factor, to make sure that the real strength of a particular structure exceeds the maximum that is expected of it.

Terms, units, and abbreviations

These metric units are used:

force, effort or load in newtons (N)
work in joules (J)
power in watts (W)

The equivalent customary units and conversion factors are as follows:

1 newton = 7.23 poundals
1 joule = 23.73 foot poundals
1,000 watts or 1 kW = 1.34 horsepower

Multiplying forces

A machine is defined in physics as a device for applying or transmitting mechanical power. Its function may be to overcome resistance to motion or to change of shape or size at one point of an object by applying a force, often at some other point. The multiplication of forces is possible in mechanical devices, enabling work to be performed that human strength alone could not do. Two terms commonly used in describing machines are work and power; both of which have been given precise, technical meanings. In a simple machine, the work done is equal to the product of the load (a force) and the distance it moves in the direction of the force, irrespective of the time taken; it is usually measured in joules (abbreviation J—1 joule = 1 newton meter). Power is the rate of doing work, and is expressed in watts (abbreviation W).

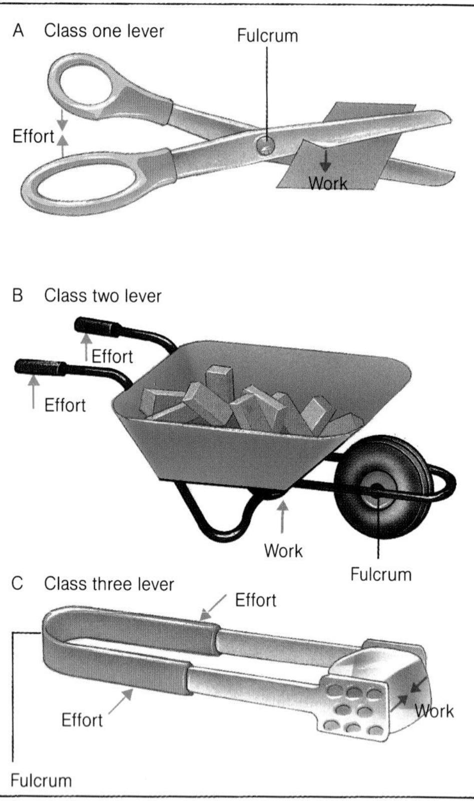

Levers are among the most commonly used means of multiplying a force, but doing so usually involves a greater movement at the point of effort than at the point of work. A class one lever has the effort and load points on opposite sides of the pivot, or fulcrum. A pair of scissors (A) consists of two class one levers with a common fulcrum. A class two lever, as in a wheelbarrow (B), has the points of effort and work on the same side of the fulcrum but with the work point closer to it. A class three lever, as in the tongs (C), also has the effort and work points on the same side of the fulcrum, but with the effort closer. As a result, the work point moves farther than the effort point.

Levers

Archimedes, the Greek mathematician, discovered the theory behind the simplest of all machines, the lever. Although unsophisticated, levers are nevertheless extremely versatile. They are classified in three basic types. In a class one lever, the point at which the work is done is separated from the point at which the effort is applied by the pivot point, or fulcrum; an example is a pair of scissors. In this tool the effort is exerted at the looped ends, bringing them together or moving them apart. The motion is transmitted through the central pivot to the cutting edges, thereby opening and closing the blades. Material that is difficult to cut is held nearer to the pivot, where the cutting force is greatest.

A class two lever has the points of work and effort on the same side of the fulcrum, with the effort point farther away. A wheelbarrow is an example of this type of lever. Lifting the handles involves applying a force against gravity, and the barrow pivots about the fulcrum at the axle.

A class three lever also has both the work and effort points on the same side of the fulcrum, but with the work point farther away. Tweezers are class three levers.

A power shovel, operated by hydraulics and with sets of levers to augment the force generated by the hydraulic system, can lift a tremendous weight of soil. Such machines make use of several of the mechanical ways of multiplying forces.

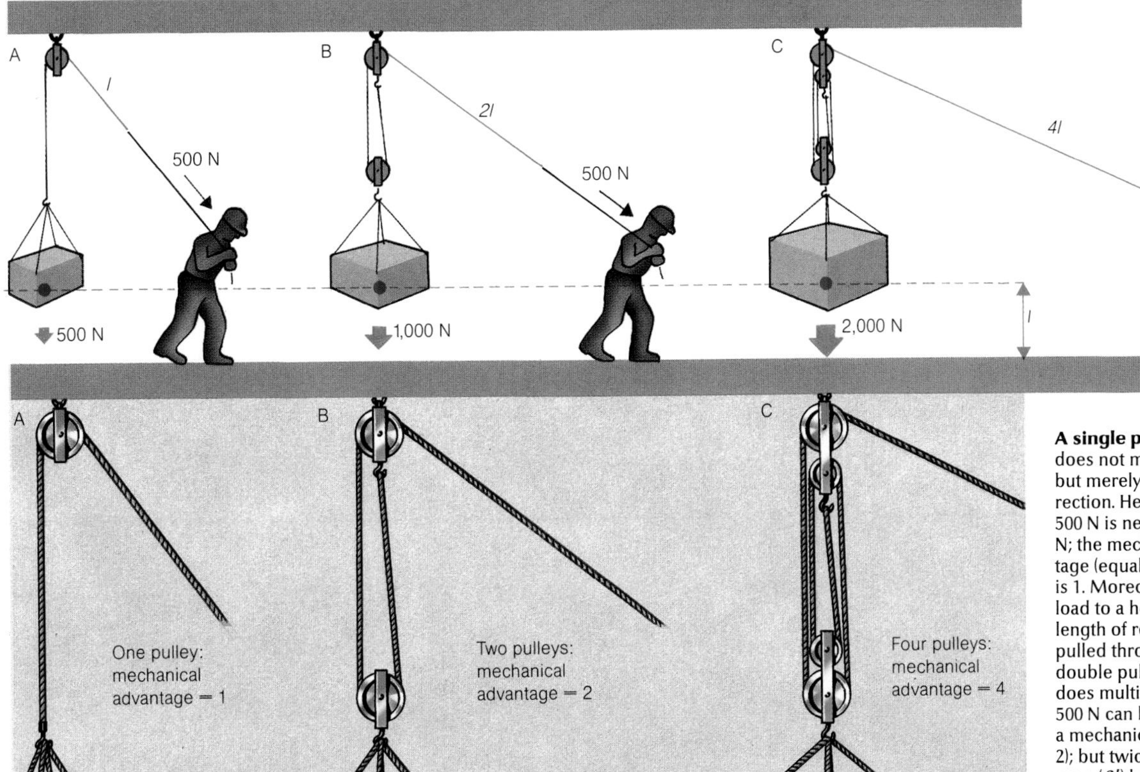

A single pulley system (A) does not multiply the effort but merely changes its direction. Hence a force of 500 N is needed to lift 500 N; the mechanical advantage (equal to load/effort) is 1. Moreover, to lift the load to a height *l*, the same length of rope must be pulled through the pulley. A double pulley system (B) does multiply effort, so that 500 N can lift 1,000 N (giving a mechanical advantage of 2); but twice the length of rope (*2l*) has to be pulled through the pulleys to lift the load to a height *l*. The more pulleys there are, the greater is the mechanical advantage. In C, the mechanical advantage is 4, but to lift the load to a height *l* necessitates pulling a length 4*l* of rope through the pulleys.

The efficiency and the advantage gained from the use of a particular lever, or any machine, can be quantified by calculating simple ratios. The efficiency of a machine is the work obtained from it divided by the work put into it. The mechanical advantage is the load divided by the effort needed to move it; the larger the load moved by a given amount of effort, the greater is the mechanical advantage. The velocity ratio is the distance moved by the effort divided by the distance moved by the load in the same time. The ratio of the velocity ratio to the mechanical advantage of a machine equals its efficiency (usually expressed as a percentage).

Friction in the fulcrum and elsewhere reduces the mechanical advantage (and efficiency) from its theoretical maximum, but the velocity ratio is not affected because it is related only to the distance moved.

Pulleys

Hiero II, the King of Syracuse, is reported to have challenged Archimedes to demonstrate the power of the simple machines (pulleys) that he was studying. In response, Archimedes arranged a system of pulleys so that single-handedly he dragged a fully-laden ship out of the water and onto dry land.

Pulleys can be arranged so as to provide some mechanical advantage in the block that houses them, aided by the lifting tackle. The velocity ratio of a set of pulleys equals the number of pulleys in the set, which usually consists of a lower and an upper system of pulleys. Hence the greater the number of pulleys, the greater the load that can be lifted by a given amount of effort, ignoring friction and the weights of the pulleys themselves.

To find the mechanical advantage provided by a particular pulley system, a known weight (representing the load) is attached to the bottom set of pulleys. Weights are successively added to the point at which the effort is applied until the load-weight rises steadily; the weights needed to cause this rise then represent the effort necessary to lift the load. The mechanical advantage is calculated by dividing the load by the effort. The velocity ratio is found by measuring the distance the load rises and the distance the weights (that represent the effort) fall, and dividing the rise by the fall. Dividing the mechanical advantage by the velocity ratio then gives the efficiency.

A pole-vaulter uses a combination of momentum and leverage to launch himself into the air. The springiness of the pole and the energy stored in it when bent also contribute to the upward force.

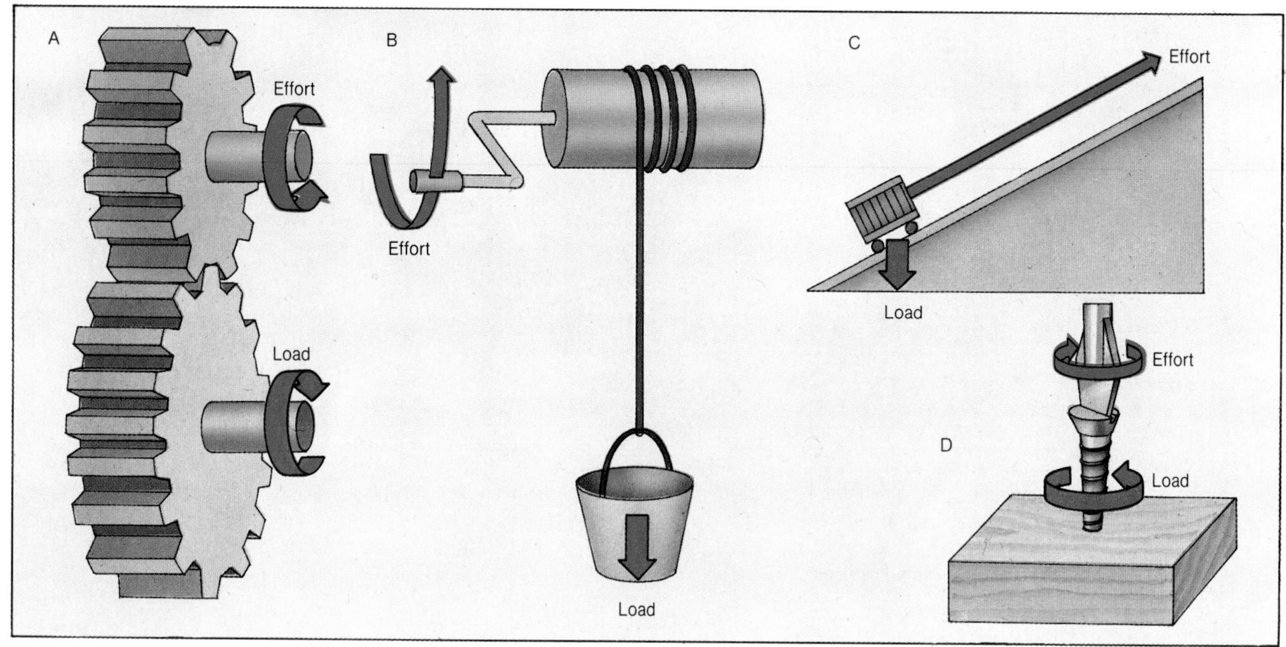

Four simple machines multiply forces in different ways. The blue arrows show the direction of movement of the effort, and the red arrows show the direction in which the load acts. The approximate mechanical advantages in the illustrated machines are: gear wheels (A) 1.33, windlass (B) 1.8, inclined plane (C) 2.1, and screwdriver (D) 50.

Other simple machines

The wheel-and-axle arrangement, and gearing systems, also supply mechanical advantage. In a wheel and axle, one turn of the (larger) wheel results in several turns of the connected axle. The velocity ratio (and therefore the mechanical advantage, if the system is totally efficient) is equal to the radius of the wheel divided by the radius of the axle.

With a pair of gear wheels, the velocity ratio is also equal to the ratio of the radii of the wheels, or the ratio of the numbers of teeth on each gear wheel. A small gear driving a large one, or a train of intermeshing gears, produces a large mechanical advantage.

Even a wedge, or inclined plane, can be considered to be a simple machine. By striking a wedge under an object, or by pushing a load up the slope of an inclined plane, an effort is applied to raise a load through a vertical distance. The velocity ratio is the distance moved by the effort (along the slope) divided by the vertical distance moved by the load, and is equal to $\frac{1}{\sin}\Theta$, where Θ is the angle of the inclined plane to the horizontal. In the absence of friction, the mechanical advantage equals the velocity ratio, and therefore the smaller the angle of incline, the greater the mechanical advantage.

Many seemingly complex machines make use of simple mechanisms. This aircraft loader consists basically of several class one levers, pivoted in pairs so that they work in a similar way to scissors. The loading platform is raised and lowered by means of hydraulics.

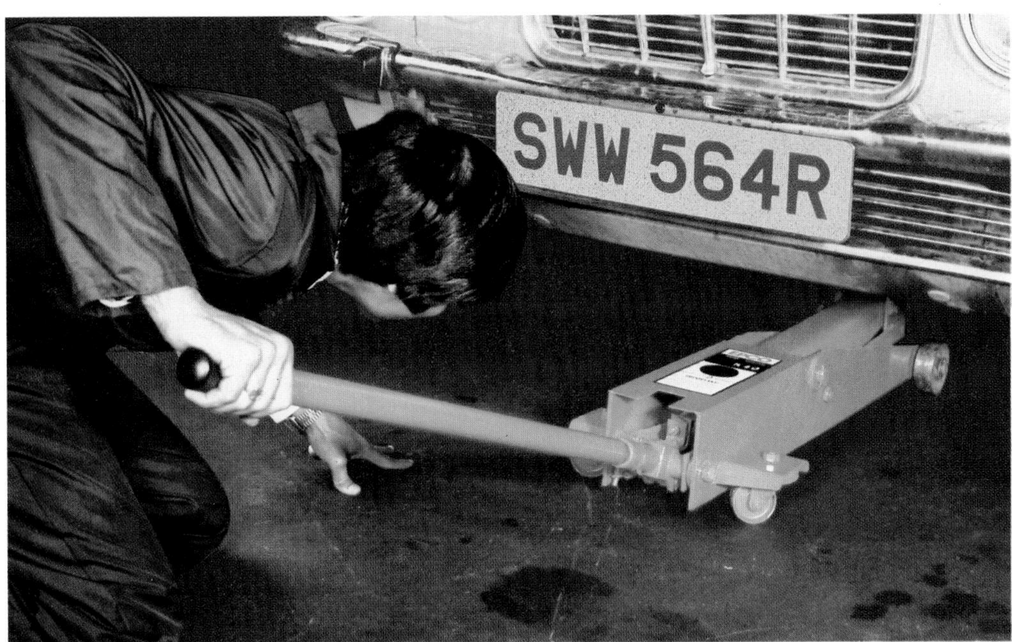

A floor jack uses hydraulics to multiply the effort applied so that it is great enough to lift a car. When the jack is in position, its handle is moved repeatedly up and down, which forces a fluid (usually oil) from a small reservoir cylinder through a nonreturn valve and into a larger cylinder. At one end of this second cylinder is a large piston that, because it has a large area, multiplies the force transmitted to it through the fluid. The piston is connected in turn to the lifting arm, which gradually raises the car as the pressure on the piston increases.

The spiral screw is another form of machine that, although simple, can amplify an applied force. The rotation of the screw produces a movement along the axis of rotation, which can be transmitted to a load. The mechanical advantage depends on the pitch—the distance between successive threads—and on the radius of the turning handle; the smaller the pitch and the larger the radius of the turning handle, the greater is the mechanical advantage.

Hydraulics

Hydraulics is the study of the flow of fluids and how they can be used in machines capable of giving a mechanical advantage. This is possible because of the property of liquids described by Pascal's principle, which states that the force per unit area (pressure) applied to an enclosed liquid is transmitted without change to every part of it, whatever the shape of its container.

The transmission of force by an incompressible fluid is a property that is central in the application of hydraulics to engineering problems, the simplest application being the hydraulic press. The basic form of this press consists of a narrow cylinder that is joined to a wider cylinder, with the hydraulic fluid between them. Two pistons—one larger than the other—are fitted into the two cylinders, the larger one to carry the load, and the smaller piston to take the effort applied to it.

The application of a force to the smaller piston generates a pressure in the fluid that is transmitted to the larger, load-carrying piston in the bigger cylinder. Because the latter cylinder has a larger area, a greater force is generated and so a mechanical advantage is obtained. But the movement of the load-carrying piston is correspondingly smaller.

In theory, the mechanical advantage is given by the area of the large piston divided by that of the smaller piston, but in practice frictional forces slightly reduce the lifting power. The hydraulic press, although simple, can be modified to perform a wide variety of tasks, from lifting heavy objects with little effort, to driving presses and molding machines.

The most common application of Pascal's principle is in the hydraulic braking system of motor vehicles. When the brake pedal is depressed, brake fluid is forced out of a reservoir and along pipes to the four wheels. Near each of the wheels there is a cylinder, and the cylinders contain pistons connected to the brake shoes (for drum brakes) or brake pads (for disk brakes). The pistons are forced outward, so that the shoes or pads make contact with the revolving drums or disks, and the friction that is produced slows the vehicle.

A worm-and-wheel is a gear system that redirects the turning effort through a right angle. This arrangement can also give a mechanical advantage.

Dynamics

Terms, units, and abbreviations
These metric units are used:
- **distance** (length) in meters (m) or kilometers (km)
- **mass** in kilograms (kg)
- **velocity** in meters per second (m s^{-1}) or kilometers per hour (km hr^{-1})
- **acceleration** in meters per second per second (m s^{-2})
- **force** in newtons (N)

The equivalent customary units and conversion factors are as follows:
- 1 meter = 3.28 feet
- 1 kilometer = 0.62 miles
- 1 kilogram = 2.20 pounds
- 1 newton = 7.23 poundals

Time is measured in seconds (s) or hours (hr).

Dynamics is the study of the ways in which objects behave when they are acted on by forces. Such forces are all around us—for example, the gravitational force acting on a falling object, the air resistance that offsets the full effect of gravity on it, and the frictional force that makes it difficult to drag an object along the ground.

Just as there are many different types of forces, there are also different types of motion produced by those forces. In linear motion, an object moves in a straight line—a falling body is an example. Circular motion is produced when an object is acted on by a force that originates from a central point. An everyday example is the circular motion of a small weight attached to a string and swung around.

If an object is held in equilibrium by two forces, and the extra force resulting from slightly moving the object from its equilibrium position is directly proportional to the distance moved, then the object oscillates regularly in what is termed simple harmonic motion. A small weight attached to a vertical spring initially takes up a position at which the elastic force of the spring matches the downward force resulting from the weight of the object. Moving the object upward or downward by a small amount (that is, slightly disturbing the equilibrium of the object) produces an extra force proportional to the size of the displacement. The weight bobs up and down with simple harmonic motion until the combined effects of air resistance and friction in the spring cause it to stop.

Kinematics

Kinematics is the study of motion without taking into account what causes the motion. The most basic concept in kinematics is that of the displacement of an object; this is the length and direction of the line along which the object moves from some fixed point, the origin. Because the displacement is both the length and direction of this line, it tells us more about the object than merely its distance from the fixed point. Such a quantity, in which both magnitude and direction are specified, is called a vector.

Velocity is defined as the rate of change of the displacement with time. For most everyday purposes velocity is measured in miles or kilometers per hour, but in physics it is usually measured in meters per second (m s^{-1}). Because it depends on another vector quantity—displacement—velocity is itself a vector. The commonly used word *speed* is reserved by physicists to denote only the magnitude of an object's velocity, and not its direction. Any quantity that measures only the magnitude of some phenomenon and not its direction as well is known as a scalar quantity. Multiplying the velocity of an object by its mass gives us another very useful vector quantity, momentum, a concept much used in dynamics.

When two objects move relative to each other, the concept of relative velocity is useful; for example, a car travelling at 85 km hr^{-1} is moving at that velocity relative to the earth's surface. Yet the earth itself is moving at about 107,200 km hr^{-1} relative to the sun.

The way in which velocities are combined in the case of two objects moving in parallel directions can be easily verified from a car moving along a busy road. Another car overtaking your own is clearly moving rapidly relative to the earth, but appears to be moving much more slowly relative to one's own car. However, a car moving at the same speed as one's own, but in the opposite direction, appears to move very quickly relative to one's own car.

The general rule is that, for objects traveling parallel to each other and in the same direction, the relative velocity of the faster one to the slower one is found by subtracting the velocities. If two objects are moving in parallel but opposite directions, the relative velocity is found by adding the two velocities. But, although this law works at low speeds, it does not apply at speeds near that of light.

When the velocity of an object changes, it is said to accelerate. The rate of change of velocity with time is measured in meters per second per second (m s^{-2}) or an equivalent unit such as kilometers per hour per hour (km hr^{-2}). For example, if an object's velocity increases from 10 m s^{-1} to 15 m s^{-1} in one second, its average acceleration is 5 m s^{-2}; alternatively, if its velocity decreases from 260 m s^{-1} to 10 m s^{-1} in 20 seconds, its average retardation (negative acceleration) is 12.5 m s^{-2}.

Acceleration itself can vary with time; air resistance, in which the acceleration produced depends on the velocity of the object, is an example of variable acceleration.

Newton's cradle, the name given to the device that consists of several balls suspended so that they just touch, can be used to demonstrate energy transfer. Swinging one of the end balls so that it hits the next in line causes the ball at the other end to swing outward; the first ball's energy is transferred through the other balls to the one at the opposite end.

Dynamics

The three basic laws governing the motion of an object are called Newton's laws of motion, after the English mathematician and physicist Isaac Newton (1642-1727), who formulated them in the seventeenth century, along with the law of universal gravitation.

Newton's first law of motion states that an object will remain at rest or will continue to move uniformly in a straight line at constant speed unless acted on by a force. So, in the special case of no forces acting on an object, its velocity will not change.

Newton's second law describes in a quantitative way how the motion of an object will change if there are forces exerted on the object. This law states that the change in momentum of an object is directly proportional to these forces. Thus, if no forces act on an object (or if any forces acting on an object are balanced out by identical forces acting in the opposite direction) then the object will remain still (if it is not already moving), or it will con-

An orbiting spacecraft provides a good example of Newton's first law of motion; although its engines are not firing, the spacecraft maintains a constant speed because there is no air friction to slow it down.

Gravity causes an object to accelerate toward the ground when it is dropped. If there is little air resistance, then the acceleration is equal to 32.16 feet per second per second (9.8 meters per second per second). This is vividly shown in the multiple-image photograph on the right, which was taken by dropping a small ball and illuminating it once every 40 milliseconds (40×10^{-3}s). The distance through which an object falls can be calculated using the equation $d = \frac{1}{2}gt^2$ (d is the distance, g is the acceleration due to gravity, and t is the time). The ball in the photograph, for example, fell about 0.3 inches (8 millimeters) during the first 40 milliseconds that elapsed between the first and second positions in the photograph.

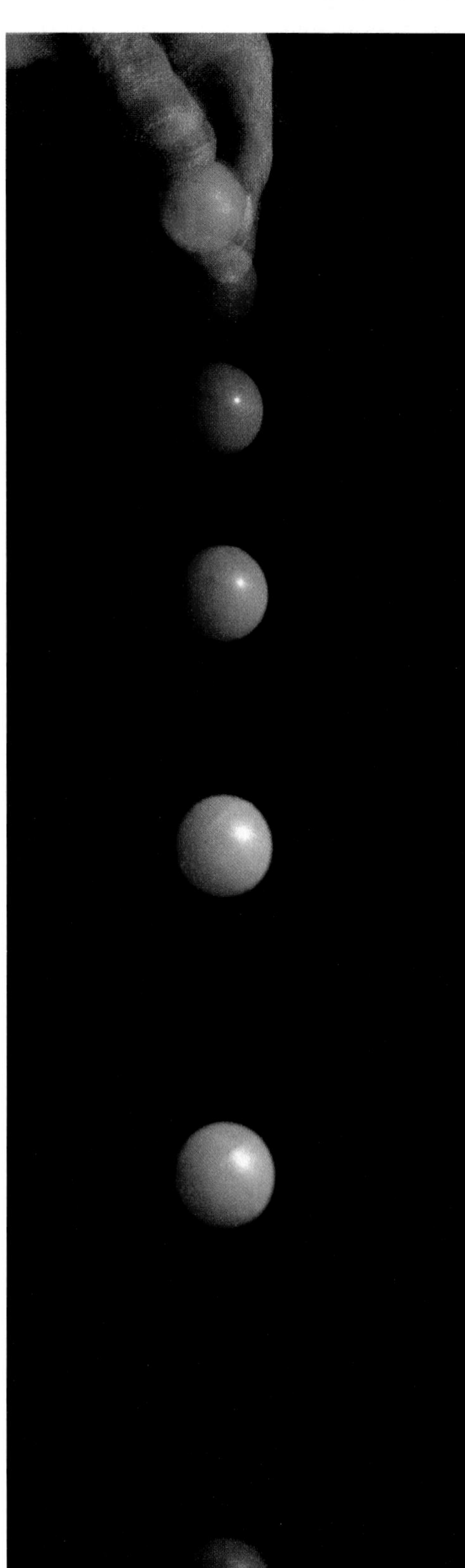

tinue to move in a straight line at constant speed (assuming it was already moving). This is identical to Newton's first law. However, if there are forces exerted on an object which are not balanced out by identically opposite forces, then the momentum of the object will change, and the amount it changes can be exactly predicted by this second law.

The momentum of an object is given by the product of the mass and the velocity of the object. Assuming that the mass of the object does not change, then the momentum of the object can only change if the velocity changes. A change in velocity over time is defined to be an acceleration. Therefore, in this case, Newton's second law can be written in the familiar form $F = ma$, where F is the force exerted on the object, m is the mass of the object, and a is the acceleration of the object.

Newton's second law gives a means to predict the changes occurring in any system if the forces acting on the objects in the system are known. Thus, it is often considered to be the most fundamental law in all of classical physics. In practice, however, many systems of objects are so complicated that it is difficult or impossible to apply Newton's second law directly. In these cases, other methods must be used to predict the changes that will occur. In addition, Newton's second law must be modified when objects are moving with velocities close to the speed of light. In these cases, an analysis must be made using the theory of relativity.

Newton's third law of motion concerns the interaction of forces produced by objects. It states that if one object exerts a force on another, that second object exerts an equal and opposite force on the first; this is the principle of action and reaction. A common example of this principle in operation occurs when someone tries to step onto land from a boat. In getting out of the boat, the person exerts a force on it, so that as he goes forward onto land, the boat is pushed in the other direction.

Rocket motors provide a more sophisticated example of Newton's third law in operation. In such a motor, fuel is burned and ejected at high velocity from the combustion chamber. The mass of fuel lost produces a backward force or thrust (the action), which, by Newton's third law, produces in turn an equal but opposite forward force (the reaction). When the rate of burning of fuel is high enough, the forward force is great enough to overcome the force of gravity, and the rocket lifts off.

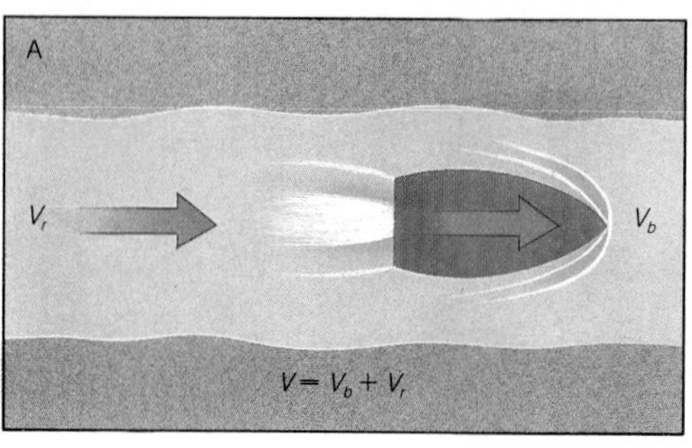

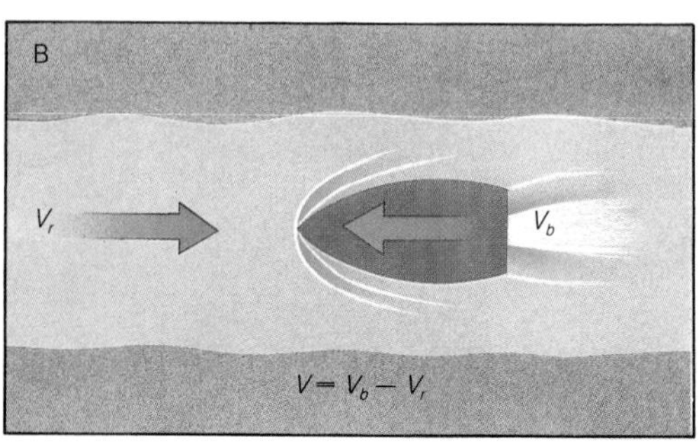

A boat moving downstream (A) with a velocity V_b relative to the river (itself flowing with a velocity V_r) is traveling at a velocity $V_b + V_r$ relative to the land. When moving upstream (B), the boat's velocity relative to land is $V_b - V_r$.

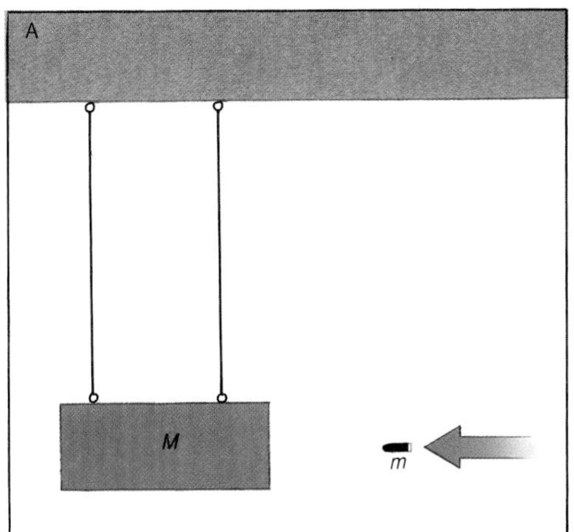

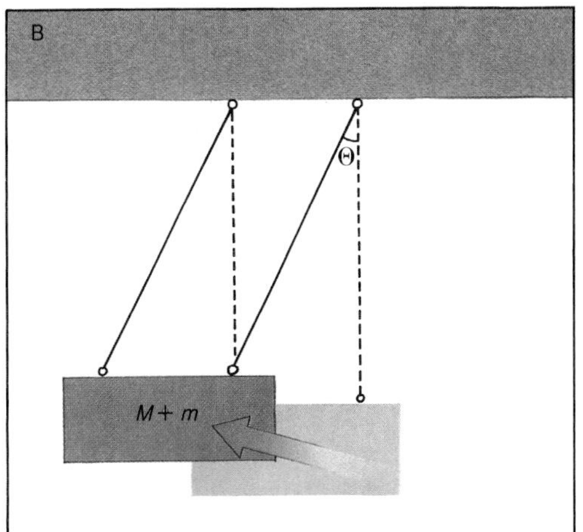

The ballistic balance is a simple device for measuring the velocity of a bullet. A bullet of known mass (*m*) is fired into a block of wood whose mass (*M*) is also known (A). By measuring the angular displacement (θ) of the block (B) it is possible to find the momentum, and therefore velocity, of the bullet.

A hovering helicopter acts in a similar way, with the rotors producing a downward thrust that, again by the third law, results in an upward force that cancels out the force of gravity. By increasing the speed of the rotors, a greater downward thrust can be produced, enabling the helicopter to accelerate upward.

Collisions

When two objects collide, their subsequent motions are changed as a result of the collision—as is immediately apparent when watching a game of pool, for example.

There are several laws governing collisions, the principal one being the law of conservation of linear momentum, which states that the total momentum of an isolated system is constant. In the above example, the isolated system is the table and the balls; and the law then implies that the total momentum of the balls just before they collide equals the total momentum just after the collision.

Thus, if we know the masses of two colliding objects, the velocity of one of them before and after the collision, and the velocity of the other before the collision, we can in principle calculate the final velocity of this second object after it has collided.

The type of collision is characterized by what is called the coefficient of restitution. This quantity is approximately constant for a collision between two given objects and can be determined experimentally. If the relative velocities of the two objects are the same before and after impact, the coefficient is equal to 1, and the collision is elastic. In practice, however, such perfectly elastic collisions occur only on an atomic scale; most collisions are therefore inelastic, with a coefficient of restitution of less than 1.

Dropping a ball so that it collides with a hard surface causes the ball to bounce, the height of each successive bounce being determined by the coefficient of restitution for the particular ball and surface material used. The height of successive bounces diminishes when the restitution coefficient is less than one (that is, when the collision is less than perfectly elastic), as shown in the multiple-image photograph on the left.

Terms, units, and abbreviations
These metric units are used:
- **distance** in meters (m) or kilometers (km)
- **mass** in kilograms (kg)
- **velocity** in meters per second ($m\ s^{-1}$)
- **acceleration** in meters per second per second ($m\ s^{-2}$)
- **force** in newtons (N)
- **weight** in newtons (N)

The equivalent customary units and conversion factors are as follows:
- 1 meter = 3.28 feet
- 1 kilometer = 0.62 miles
- 1 kilogram = 2.20 pounds
- 1 newton = 7.23 poundals

Time is measured in seconds (s) in both systems.

Gravity

Gravity is a force—the mutual attraction between objects that have mass. Its influence in everyday life is so pervasive that it is taken for granted.

The seventeenth-century mathematician and physicist Isaac Newton was the first to formulate the law of gravitation, expressing it in mathematical terms that could be used in calculations. Called Newton's law of universal gravitation, it states that the gravitational force between two objects is directly proportional to the product of the masses of the two objects, and inversely proportional to the square of the distance between them.

To make calculations possible, it is necessary to include a constant, G, which is known as the universal constant of gravitation. Experimental measurements of the force of attraction between two objects of known mass and separation have shown that the gravitational constant G has a value of about 6.67×10^{-11} N m^2kg^{-2}.

Physicists characterize the strength of a gravitational field in terms of the acceleration produced on a falling object. The result (the acceleration due to gravity, or the gravitational field strength) is denoted by g, which for the earth has a value of about 9.8 m s^{-2} (equivalent to 9.8 N kg^{-1}). In comparison, the acceleration due to gravity on the moon is only about 1.6 m s^{-2} (approximately one-sixth of that on Earth), and on Jupiter is about 24.9 m s^{-2} (approximately two-and-a-half times that on Earth).

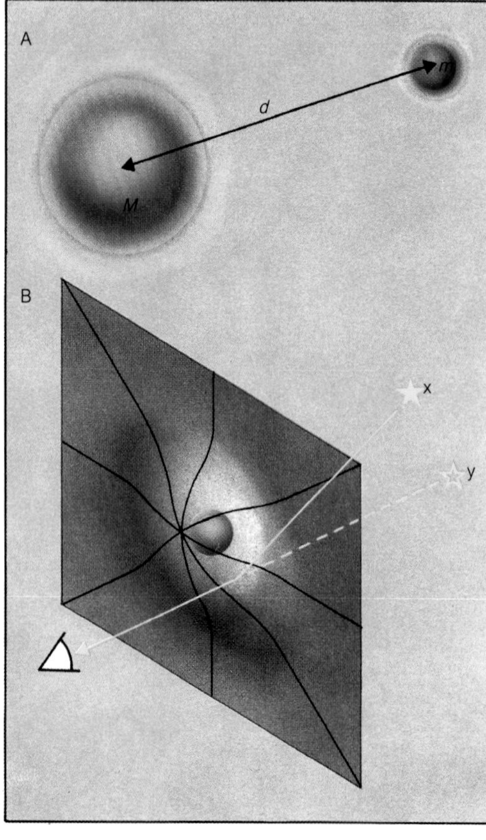

In classical (Newtonian) physics (A), the gravitational attraction between two objects of masses M and m is proportional to the product of their masses and inversely proportional to the square of the distance (d) between their centers of mass. In relativistic (Einsteinian) physics, however, massive objects, such as stars, are thought to deform space, forming a "gravity well" (B). Evidence supporting the Einsteinian view comes from the observation of stars that appear closer to the sun than they really are because light rays from them have been bent by the sun's gravitational field (a star actually at position x appears to be at y).

Relativity and gravity

Newton's mathematical description of gravity is adequate for weak gravitational fields. But it fails when applied to large masses confined to small regions in which the force of gravity is very strong, for which Einstein's theory of gravitation (the General Theory of Relativity) has to be used. In most situations, however, Newton's law suffices for calculating gravitational forces.

According to Einstein's General Theory, the mutual attraction between objects is the result of the curving of space and time around them. The curvature has observable effects on rays of light that pass close to massive objects, such as the sun or other stars, and studies of these effects indicate that the idea of space-time curvature is valid.

One prediction of the General Theory is the existence of gravitational waves, a phenomenon produced whenever an object is accelerated. The waves are in the form of ripples of curved space time that travel at the speed of light (approximately 186,000 miles per second or 3×10^8 m s^{-1}). But gravitational waves are so weak that they have not yet been detected directly; at present there is only indirect evidence for their existence.

Gravity and orbits

Gravity is a non-contact force, meaning that two objects will exert an attractive force on each other due to gravity even if they are not touching. Thus, the sun pulls the earth toward itself (and the earth pulls the sun toward itself) even though the two bodies are nearly 100 million miles apart. This mutual attraction through gravity exists between all objects in the universe. In most cases, however, these bodies in space do not collide due to this force because the objects are already moving away from each other. Thus, the earth does not hit the sun, because while it is pulled toward the sun, it is also moving to the side. The resulting orbit of the earth is an elliptical path around the sun. Kepler's third law mathematically describes the orbit of all objects in space as they move around more massive objects. The time for one complete orbit is proportional to the average distance between the two objects raised to the $\frac{3}{2}$ power.

To maintain an orbit of a certain radius about an object of a particular mass, a satellite must maintain a specific velocity, called the orbital velocity. If the velocity of a satellite is great enough, it has enough energy to escape from the effective gravitational field of the parent body. The minimum velocity necessary for this to happen is called the escape velocity; it is approximately equal to one-and-one-half times the orbital velocity at the particular orbital radius. For an object launched from the earth's surface, the escape velocity is about 11 km s^{-1}.

When in orbit, any occupants of a spacecraft become weightless. This effect occurs because the spacecraft and its occupants are in free fall; that is, they are both being subjected to a "downward" force equal to the acceleration due to gravity (g). And because the

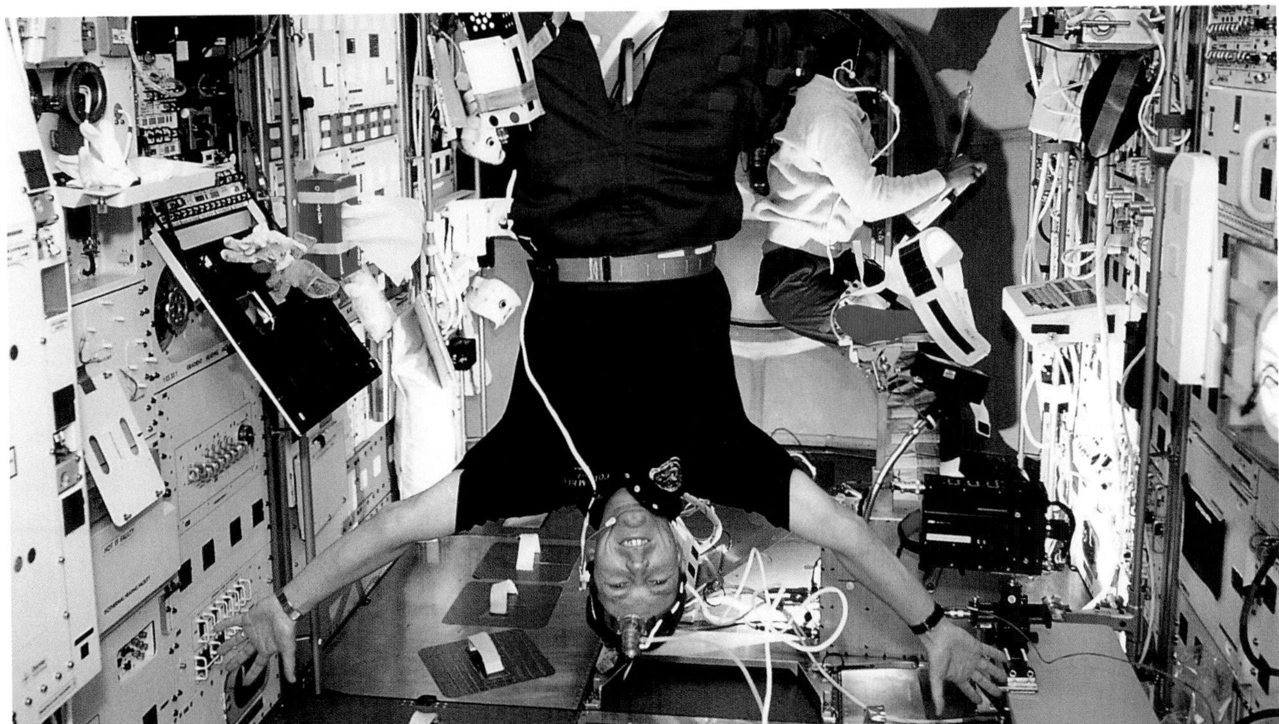

occupants are "falling" toward the earth at the same rate as their spacecraft, they are motionless relative to the spacecraft's interior, so they feel weightless.

Gravity and tides

A commonplace manifestation of the gravitational attraction between massive bodies is the phenomenon of the tides, which results from the unequal attractions of the moon and sun on the water in the oceans of the earth.

Although quite complicated in detail, tides basically arise because the oceans of the earth that happen to be closest to the moon (or sun) are more strongly attracted to the moon (or sun) than are the oceans on the opposite side of the earth. Thus, certain bodies of water are pulled up more at any given time than are other bodies of water. To make a complete analysis of tides, however, effects due to the spinning of the earth also have to be considered.

Two tides every 24 hours are produced by the sun, and 2 every 25 hours by the moon. When the two combine, a spring tide (which is higher than a normal tide) results; when the sun offsets some of the moon's effect, a neap tide (a lower than normal high tide) is produced.

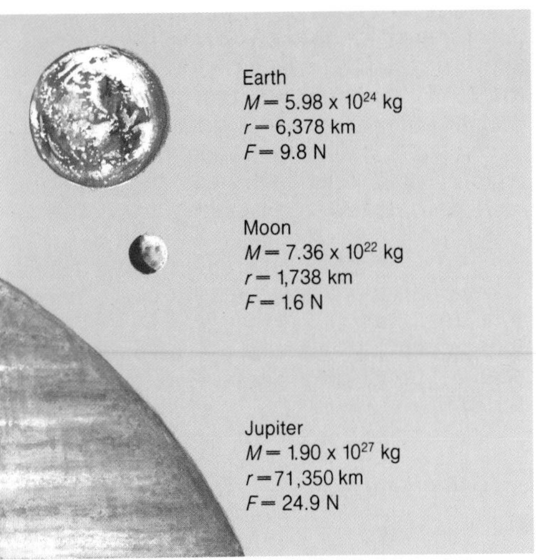

Weightlessness training for astronauts *(above)* is carried out in aircraft that climb to a high altitude, then accelerate earthward; when the aircraft's acceleration equals that due to gravity, its occupants are in free fall. On the moon, gravity is low—only about one-sixth that on Earth—and so the astronaut *(above)* can carry a large amount of bulky equipment. For any planet, the force of gravity F acting on a mass m at the surface is given by $F = GMm/r^2$, where G is the universal gravitational constant (6.67×10^{-11} N m^2 kg^{-2}), M is the planet's mass and r is its radius. The diagram *(left)* shows the force of gravity acting on a mass of 1 kg for the earth, moon, and Jupiter, obtained by substituting the appropriate values in the above equation.

Circular motion

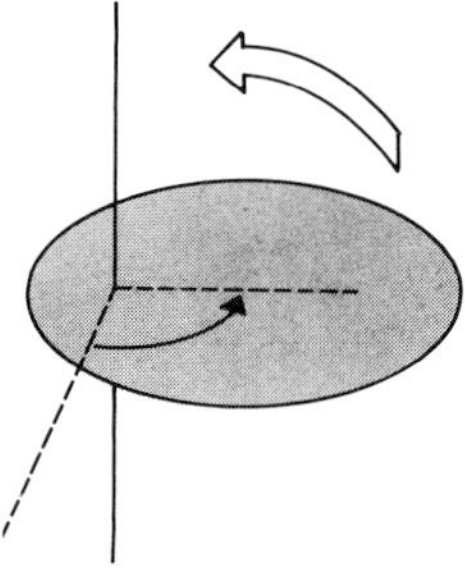

The angular velocity of a rotating object is the angle (in radians or degrees) through which it moves per unit time (in seconds). There are 2π radians in a circle (360°), so if an object rotates once in 60 minutes, its angular velocity is 1.75×10^{-3} rad s^{-1} (or 0.1 deg s^{-1}).

Circular motion is an extremely common phenomenon among natural and manmade objects. Planets spin on their axes, and many travel in near-circular orbits around the sun; satellites orbit planets; wheels spin on axles. Circular motion often is treated as a special branch of dynamics, sometimes called rotational dynamics.

Fundamental concepts

Many of the terms used to describe linear motion (such as velocity, acceleration, kinetic energy, and momentum) have counterparts in the description of objects that move in a circle. Much of the study of circular motion concerns the behavior of ideal bodies that are rotating. A rigid body is one of these ideal concepts, being defined as an object with a definite shape that does not deform under the action of forces involved in rotational motion (in practice, all real objects deform to some extent).

When a rigid body rotates about an axis, its angular velocity can be defined—just as it is possible to define linear velocity for an object moving in a straight line. Angular velocity is the angle through which the object rotates divided by the time it takes to do so. The angle is measured in radians or degrees (2π radians are equal to 360°), and angular velocity is therefore expressed in radians (or degrees) per unit time (rad s^{-1} or deg s^{-1}).

A rigid body acted on by a torque (a turning force—the rotational analogue of normal force in linear motion) undergoes angular acceleration. This is defined as the rate of change of angular velocity and is measured in radians (or degrees) per second per second (rad s^{-2} or deg s^{-2}).

The rotational analog of inertia is called the moment of inertia but, unlike inertia (which is the same for all objects of the same mass), the moment of inertia depends also on the shape of the object and on the position of its rotational axis. For example, a flat square plate rotating about its center has a lower moment of inertia than the same plate rotating about one corner. Applying Newton's second law of motion to circular motion gives: torque equals moment of inertia multiplied by angular acceleration.

The identification of the moment of inertia with the mass (or inertia) term in linear motion relationships can also be used when deriving equations for angular momentum and rotational kinetic energy. Thus the angular momentum of a rotating object equals its moment of inertia multiplied by its angular velocity, and rotational kinetic energy equals half the product of the moment of inertia and the square of the angular velocity.

There are two other important concepts with analogs in linear motion. The first concerns the work done by a torque acting on a rotating object: the work done (given by the product of the torque and the angle through which the object rotates) equals the change in rotational kinetic energy. The second concept is the law of conservation of angular momentum, which states that if no resultant external torque acts, then the total angular momentum of a system remains constant.

Centripetal and centrifugal forces

Circular motion is considered to be unnatural motion, meaning that an object cannot move in a circle unless there is a force causing this to happen. The force which causes circular motion is called centripetal force. It is a force exerted on the object toward the center of the circle. Without this force the object could only move in a straight line.

Centripetal force also causes centripetal acceleration toward the center of the circle. This acceleration is proportional to the square of the velocity of the object and inversely proportional to the radius of the circle.

Ice skaters spinning around demonstrate the principle of conservation of angular momentum. In the photograph below, the skaters are moving in a circle with a certain angular velocity. As the woman is pulled inward by the man, their moment of inertia decreases and, because angular momentum is conserved, they spin faster (that is, their angular velocity increases).

A car cornering exemplifies many concepts used to describe circular motion.

Simple harmonic motion

Terms, units, and abbreviations
length, displacement, or amplitude in feet or meters (m)
time in seconds (s)
acceleration in feet per second per second or meters per second per second ($m\ s^{-2}$)

Simple harmonic motion (SHM) is a special type of regular oscillation. A common and familiar phenomenon, it can occur when an object that is in equilibrium is disturbed from its equilibrium position. The small to-and-fro movement of a pendulum bob is one example of SHM; another is the up-and-down movement of a weight attached to a vertical coil spring and displaced slightly from its equilibrium position. (These are actually examples of damped SHM, because friction gradually diminishes the extent of the motion.)

SHM is important not only because it is common, but also because any motion that is not pure SHM can be broken down into several components that do execute SHM. Thus it is useful for analyzing a wide range of different motions.

A metronome is basically an inverted compound pendulum and therefore executes SHM. The period of swing can be altered by varying the position of a small weight on the arm of the metronome.

Definitions of terms

SHM is defined in terms of the movement of an object: an object moves with SHM when it oscillates along a line whose midpoint is the point of equilibrium, the acceleration of the object towards that point being proportional to its distance from it.

This definition can be visually represented in a simple way. If a graph is plotted of the displacement against time, the resultant curve is a sine wave. The displacement is a vector with positive values when the object is one side of the equilibrium position and negative values when it is the other side.

The amplitude of the motion is the magnitude of the greatest displacement from the equilibrium position. On the sine-curve graph, amplitude can therefore be measured at the peaks and troughs of the curve.

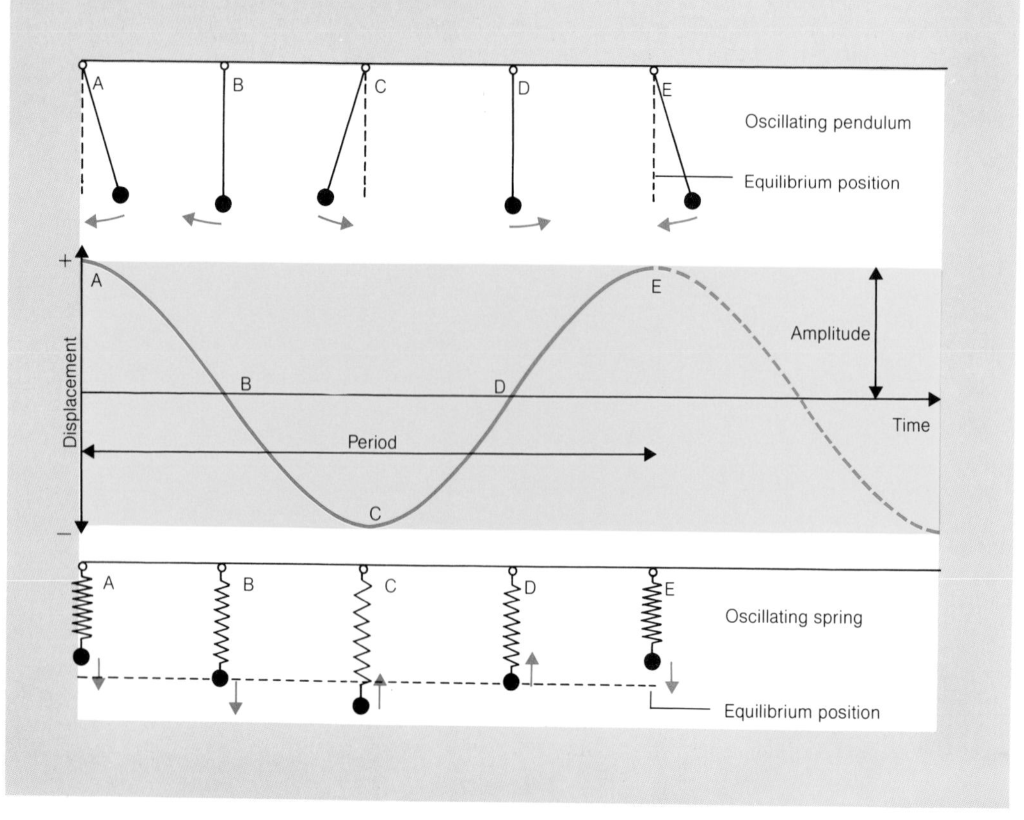

Simple harmonic motion is characterized by a sine curve representing a variation of displacement against time. This diagram shows the sine curve generated by a simple pendulum and a weight oscillating on a coiled spring. Positions A to E show different displacements of the pendulum bob and oscillating weight and the corresponding part of the sine curve. Position E is equivalent to A and marks the end of one complete cycle and the beginning of the next. The amplitude is the maximum displacement from the equilibrium position.

Because an object executing SHM is oscillating regularly, it has a period and frequency. The period is the time taken for the object to complete one cycle—such as from one point of maximum displacement through the equilibrium to the other point of maximum displacement and back to the first point. The frequency is the number of such cycles completed every second; thus frequency is the reciprocal of the period.

The pendulum

A simple pendulum consists of a small object on a lightweight suspension, such as a length of string. When undisturbed, the string and object (called the pendulum bob) hang vertically, with the tension in the string balancing the downward force produced by the weight of the bob. This is the equilibrium position. When the equilibrium is disturbed by moving the bob slightly to one side and then releasing it, the bob executes SHM. This is because, for small angular displacements (less than about 10°), the acceleration of the bob back to the equilibrium position is proportional to the angle of displacement and, therefore, to the distance the bob is from the equilibrium position. (Having been displaced and then released, the bob does not simply return to the equilibrium position and stay there because its momentum carries it past the equilibrium position to another point of maximum displacement on the other side of the swing.)

If instead of a lightweight string attached to a comparatively heavy bob, a relatively large, solid object is pivoted about a point within it so that the object can swing, SHM is still produced (with small angles of swing). This arrangement, commonly used in clocks, is called a compound pendulum.

Because it relies on the force due to gravity, a simple pendulum can be used to determine the value of g. The period (T) of a pendulum of length l is given by the equation $T = 2\pi(l/g)^{\frac{1}{2}}$ (which can be re-expressed as $g = 4\pi^2 l/T^2$). So by using a pendulum of known length and measuring its period of swing, it is possible to calculate g. (In practice, the time taken for a large number of swings is measured and the result averaged to achieve greater accuracy.)

SHM and energy

In an ideal situation, an oscillating system (such as a pendulum or a weight on a spring) is undamped—that is, it does not have to work against resistive forces such as friction. In this perfect case the total energy of the system, which remains constant, has two components: potential energy, which results from the disturbed object not being at its equilibrium position, and kinetic energy, which results from the object's movement.

For an object attached to a spring and executing SHM, the potential energy depends on the amount by which the spring is extended or compressed from its equilibrium length (assuming that the system is frictionless). Hence the potential energy is greatest at the positions of maximum compression and elongation. At these two points, however, the object is stationary for an instant and, because it has zero velocity, also has zero kinetic energy. At the instant the object is at the equilibrium position, while moving from one point of maximum displacement to the other, it has zero potential energy but its kinetic energy is a maximum because it is moving with its greatest velocity.

For any given system oscillating with SHM, the sum of the potential and kinetic energies is always the same.

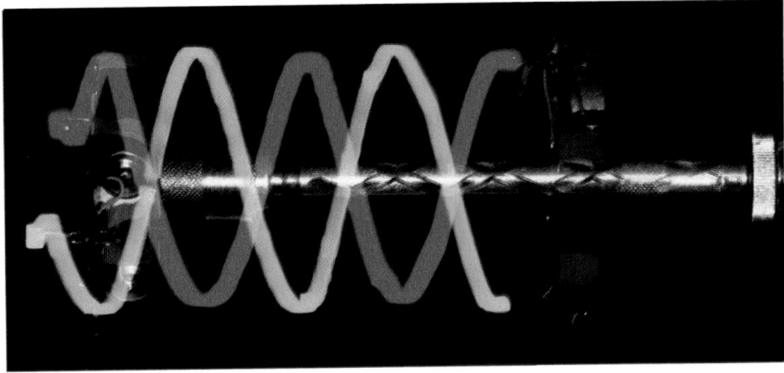

The two sine curves above were produced by mounting two colored light-emitting diodes on the chuck of a spiral-drive screwdriver and then retracting the chuck. A plot of displacement against time (which, in effect, the photograph is) of any object executing SHM generates a sine curve.

A spring diving board oscillates with SHM after the diver has started his dive. Before this, when the diver is bouncing to gain maximum height, the board undergoes forced oscillation.

Fluid flow

Terms, units, and abbreviations
These metric units are used:
distance in meters (m)
velocity in meters per second (m s^{-1})
force in newtons (N)
The equivalent customary units and conversion factors are as follows:
1 meter = 3.28 feet
1 newton = 7.23 poundals
Time is measured in seconds (s) in both systems.

Hydrodynamics is the study of fluids in motion (a fluid is any substance that flows—a gas, such as air, or a liquid, such as water). It also includes the study of the interactions between fluids and objects immersed in them, when the object or the fluid—or both—are moving. Hydrodynamics is therefore of considerable importance in everyday life—for example, moving cars and flying aircraft are both basically objects moving through a fluid (air). The term aerodynamics is generally used when the fluid is air.

Viscosity

One of the most important concepts in hydrodynamics is that of viscosity. In everyday terms it is the thickness or stickiness of a fluid—gases and mobile liquids, such as alcohol, have comparatively low viscosities, whereas sticky fluids such as syrup and tar have high viscosities. Viscosity results from friction that exists within fluids; a substance's coefficient of viscosity is an indication of the strength of this internal friction and a measure of the substance's resistance to flow. (The coefficient of viscosity, usually called simply viscosity, is symbolized by η and is measured in N s m^{-2}.) Fluids with a low viscosity flow easily, whereas those with a high viscosity exhibit much greater resistance to flow.

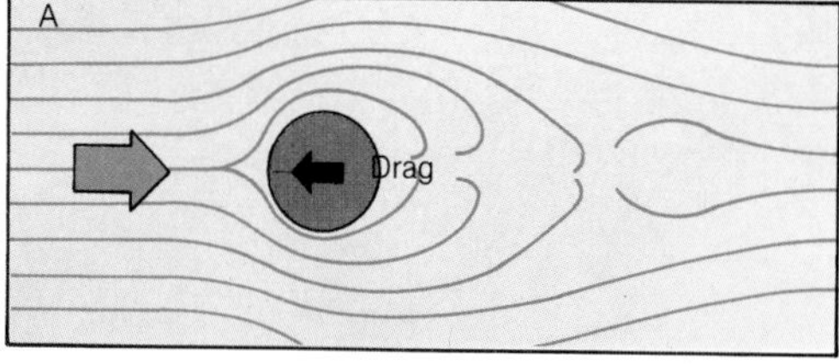

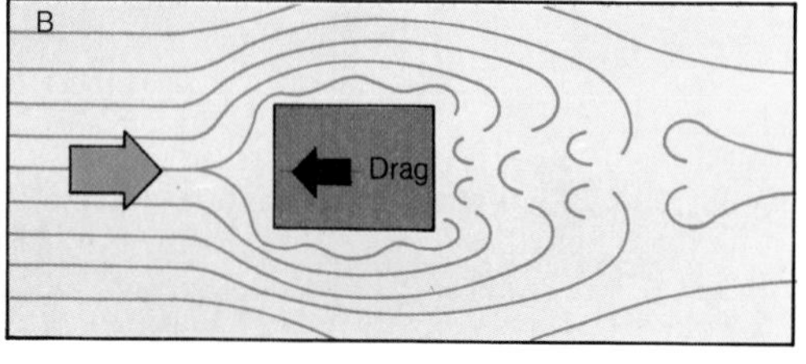

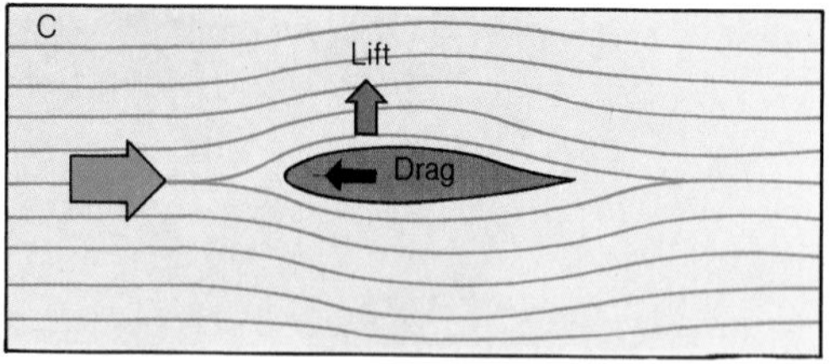

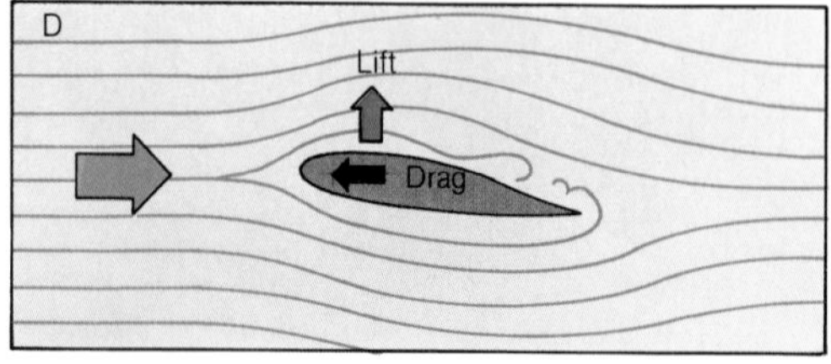

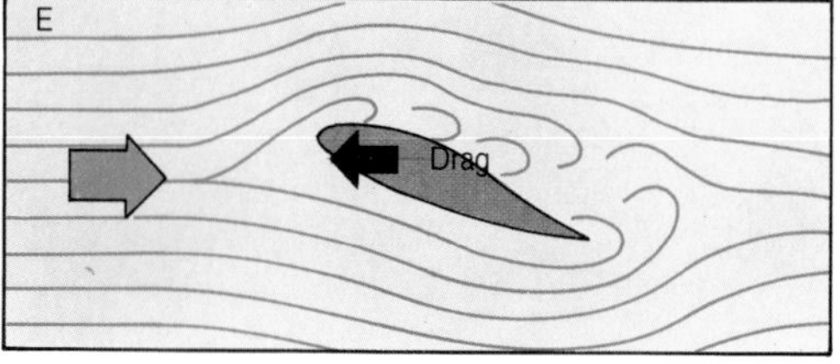

Air is a fluid, and whenever an object moves through air—or moving air flows round an object (as shown in the diagrams, right)—the airflow is modified. For a spherical object (A) the flow is initially streamline (laminar), but the air becomes turbulent to the rear of the sphere. The resistance to flow is called drag, shown by the black arrows. The flow round a rectangular object (B) is nonlaminar, with much turbulence and high drag. The flow round an aerofoil such as an aircraft's wing (C), is streamline and such that the air pressure is higher below the wing than above it. The result is an upward force called lift (red arrow). If the wing is tilted (D), at a steeper angle of attack, turbulence sets in and drag increases; but the pressure difference between the upper and lower surface of the wing also increases (up to a critical angle of attack), and the increased drag is more than offset by increased lift. At a very steep angle (E) there is no lift at all and an aircraft would stall.

Types of flow

When a fluid flows through a pipe, its rate of flow is not the same at all points across the pipe. The outermost layer of fluid is almost stationary because of the drag exerted on it by the wall of the pipe. This outer layer retards the next inner layer—and so on, to the center of the pipe. If the velocity of the fluid is not too great in relation to its viscosity, the flow is described as laminar (or streamline), with layers of fluid sliding past each other at different velocities and with the maximum velocity at the center of the tube. If, however, the maximum velocity of the fluid exceeds a critical value, the fluid's motion becomes irregular; vortices may develop within the fluid and the resistance to flow increases. This type of flow is described as turbulent (or nonlaminar).

Objects within a fluid experience viscous drag as a result of the fluid's viscosity. The exact strength of this drag is given by Stokes' law, which shows that for a spherical object, the viscous drag depends on the viscosity of the fluid, the size of the sphere, and its velocity relative to the fluid.

If it were not for viscous drag due to the atmosphere, raindrops would hit the ground at a velocity of more than 150 m s^{-1}. In practice, the droplets accelerate to a maximum velocity (called the terminal velocity, considerably slower than 150 m s^{-1}) before they hit the ground. This phenomenon occurs because the viscous drag on the droplets increases as their velocity increases, and eventually cancels out the downward acceleration due to gravity.

Below a certain velocity, the flow of fluid around an object is smooth and streamlined, with the viscous forces providing a resistance proportional to the relative velocity of the fluid and object. But when this relative velocity exceeds a critical value, the streamlines start to break up and turbulence occurs. At this stage the resistance produced by the fluid no longer depends on the fluid's viscosity but on the square of the relative velocity, and so is of greatest significance at high velocities.

Turbulence can be seen in water coming out of a faucet or a pipe. At slow speeds the water flow is streamlined. As the speed increases, vortices develop within the column of water and break through the boundary layer. The velocity at which this effect occurs depends on the radius of the pipe, the density of the liquid, and its viscosity.

Bernoulli's principle

Fundamental to much of hydrodynamics is the relationship between a fluid's pressure, kinetic energy, and potential energy, which was discovered in the eighteenth century by the Swiss mathematician Daniel Bernoulli. His principle makes it possible to calculate pressure differences resulting from changes in the velocity and height of a fluid flowing through

Water gushing from open sluices in a dam demonstrates turbulent flow on a grand scale. The stability of flow from a pipe can be quantified by Reynold's number R_e, given by the expression $R_e = 2rv\rho/\eta$, where r is the radius of the pipe, and v is the fluid's velocity, ρ its density and η its viscosity. When R_e exceeds 2,200, the flow is turbulent.

a tube. The fluid's velocity is affected by changes in the diameter of the pipe; because the same mass of fluid must enter and leave the whole pipe in a given period, it must speed up when passing through a constriction. Because the fluid's velocity changes, its kinetic energy changes. If, in addition, the fluid changes height (because of a slope in the pipe), there is also a change in the potential energy of the fluid. Bernoulli showed that the sum of the pressure of the fluid and the two types of energy (potential and kinetic) is always constant. Strictly, Bernoulli's principle applies only to a fluid flowing steadily without turbulence; it must also be incompressible—that is, its density must remain constant under pressure. But although real fluids do compress slightly under pressure, the principle is sufficiently accurate to be useful in practice.

The general principle of an increase in velocity causing a decrease in pressure is used in the Venturi flowmeter. Fluid passes through a pipe with a constriction in it, which causes an increase in speed of the fluid and (by Bernoulli's principle) a decrease in pressure. If the pressure difference can be measured and the diameter of the constriction is known, then the increase in speed of the fluid can be determined, as can its rate of flow.

Sky divers fall until they reach a constant terminal velocity, at which their downward acceleration due to gravity is balanced by the viscous drag resulting from the air flowing past them. When their parachutes open, the drag increases and they decelerate to a safe landing speed.

Energy, work, and power

Terms, units, and abbreviations
These metric units are used:
- **distance** in meters (m)
- **mass** in kilograms (kg) or grams (g)
- **energy** in joules (J)
- **power** in watts (W)

The equivalent customary units and conversion factors are as follows:
- 1 meter = 3.28 feet
- 1 kilogram = 2.20 pounds
- 1 gram = 0.035 ounces
- 1,000 joules = 0.95 British thermal units
- 1,000 watts = 1.34 horsepower

Time is measured in seconds (s) in both systems.

Energy can exist in many different forms, some of which include sound, light, electrical, nuclear, and thermal energy. For every form of energy known to exist, a mathematical expression has been derived which can be used to calculate the amount of energy existing in this form.

Energy

The concept of energy was developed in science as an attempt to quantitatively measure the amount of change which can occur in a system. It is believed that the total energy of the universe cannot change. Therefore, by measuring the amount of energy initially existing in a system, one can predict how much the system can change over time as energy changes from one form to another.

Kinetic energy is possessed by a moving object by virtue of its motion. It equals the work done to accelerate the object to a particular velocity; it also equals the work done to bring a moving object to rest. Thus in a car, the engine works to increase the car's speed and therefore its kinetic energy; the brakes absorb this energy, thus reducing the car's speed.

The two principal forms of kinetic energy are known as translational and rotational. The first is possessed by an object moving from one position to another. The second is possessed by rotating objects, which revolve about an axis and therefore periodically return to the same position.

An object has potential energy by virtue of its position; it is not apparent until released (mainly as kinetic energy). Two common types are gravitational and elastic potential energy. An object gains gravitational potential energy as work is done to raise it against the force of gravity. When the object falls, it loses potential energy (because its height decreases) but gains kinetic energy because of its motion. The amount of gravitational potential energy depends on the relative positions of the object and an arbitrarily chosen base level (usually taken as ground level), where its potential energy is considered to be zero.

Elastic potential energy is gained as work is done to stretch or compress an elastic object such as a spring. This type is converted into kinetic energy when the spring is released and it regains its former shape.

An object possesses heat, or thermal, energy by virtue of its temperature. It is, in fact, merely a form of kinetic energy, because the temperature of a substance depends on the motion of its component atoms or molecules; the higher its temperature, the faster the molecules move. (Heat radiation is not, however, classified as thermal energy but as radiant energy.)

Radiant energy consists of electromagnetic radiation and includes radio waves, visible light, ultraviolet and infrared radiation, and X rays. A form of energy that can exist in the absence of matter, it consists of a wave motion in electric and magnetic fields. Radiant energy is emitted when electrons within atoms fall from a higher to a lower energy level and release the "excess" energy as radiation.

Sound energy consists of moving waves of pressure in a medium such as air, water, or metal. They consist of vibrations in the molecules of the medium, and sound can therefore be considered as a special form of kinetic energy.

Matter that has gained or lost some electric charge has electrical energy. It can be regarded as a form of electrostatic potential energy, because work has to be done to move electric charges toward an object or away from it. The movement of charges constitutes an electric current, which flows between two objects at different potentials when they are joined by a conductor, because the charges

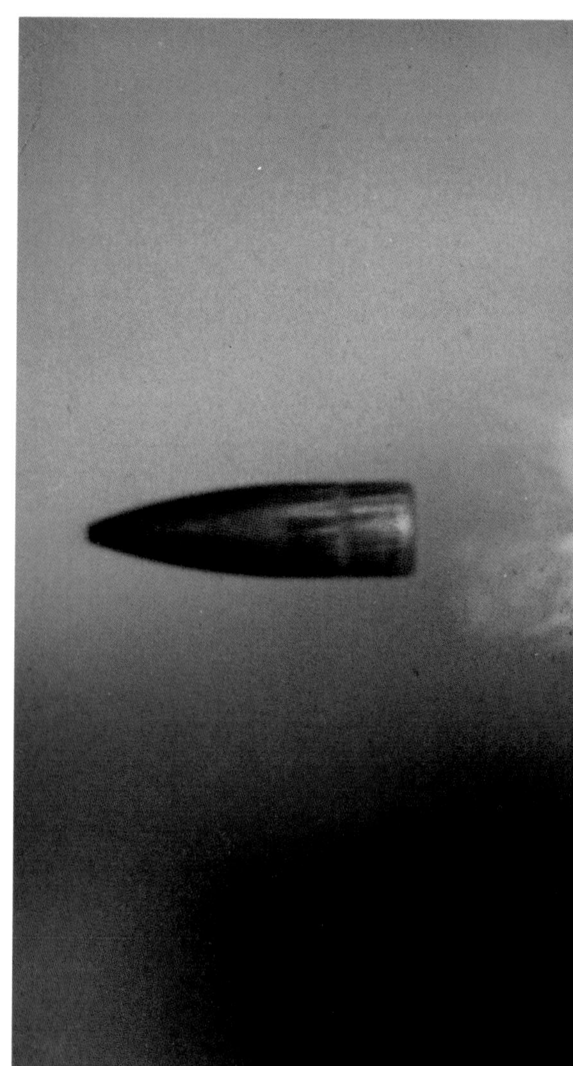

move from one object to the other object until an equal potential is restored to each.

Chemical energy is possessed by substances that undergo a chemical reaction, such as combustion. It is stored in the chemical bonds between the atoms that make up the molecules of a substance. During a reaction, the atoms of the reactants rearrange themselves to form the different molecules of the products. If the products have less chemical energy than the reactants, energy is released during the reaction—usually in the form of heat, light (or both, as in combustion), or electrical energy (as in a battery). If, on the other hand, the products have more chemical energy than do the reactants, then energy is absorbed during the reaction; energy (often in the form of heat) may have to be supplied to make the reaction take place.

Nuclear energy is produced when the nuclei of atoms change, either by splitting apart or joining together. The splitting process is known as nuclear fission, the joining together as nuclear fusion. Such changes can be accompanied by the release of enormous amounts of energy in the form of heat, light, and radioactivity (the emission of atomic particles or gamma radiation, or both). The resulting motion of the nuclei and particles also causes an increase in the thermal energy of the substance that is undergoing the nuclear reaction, and in the thermal and kinetic energy of its surroundings—as occurs in an explosion, for example.

Energy conversions

The total amount of energy possessed by an object (or a system of interacting objects) must always remain the same. This phenomenon is the principle of conservation of energy, which states that energy can neither be created nor destroyed but only converted into other forms.

A pendulum, for example, continually swings to and fro, with the bob effectively moving up and down in height and changing its speed from a maximum at the lowest point to zero at the highest. In doing so, its kinetic energy is greatest at the lowest point of the swing and zero at the highest. But its potential energy is at a maximum at the highest point and zero at the lowest. Kinetic energy changes to potential energy as the bob rises, and the reverse conversion occurs as it falls.

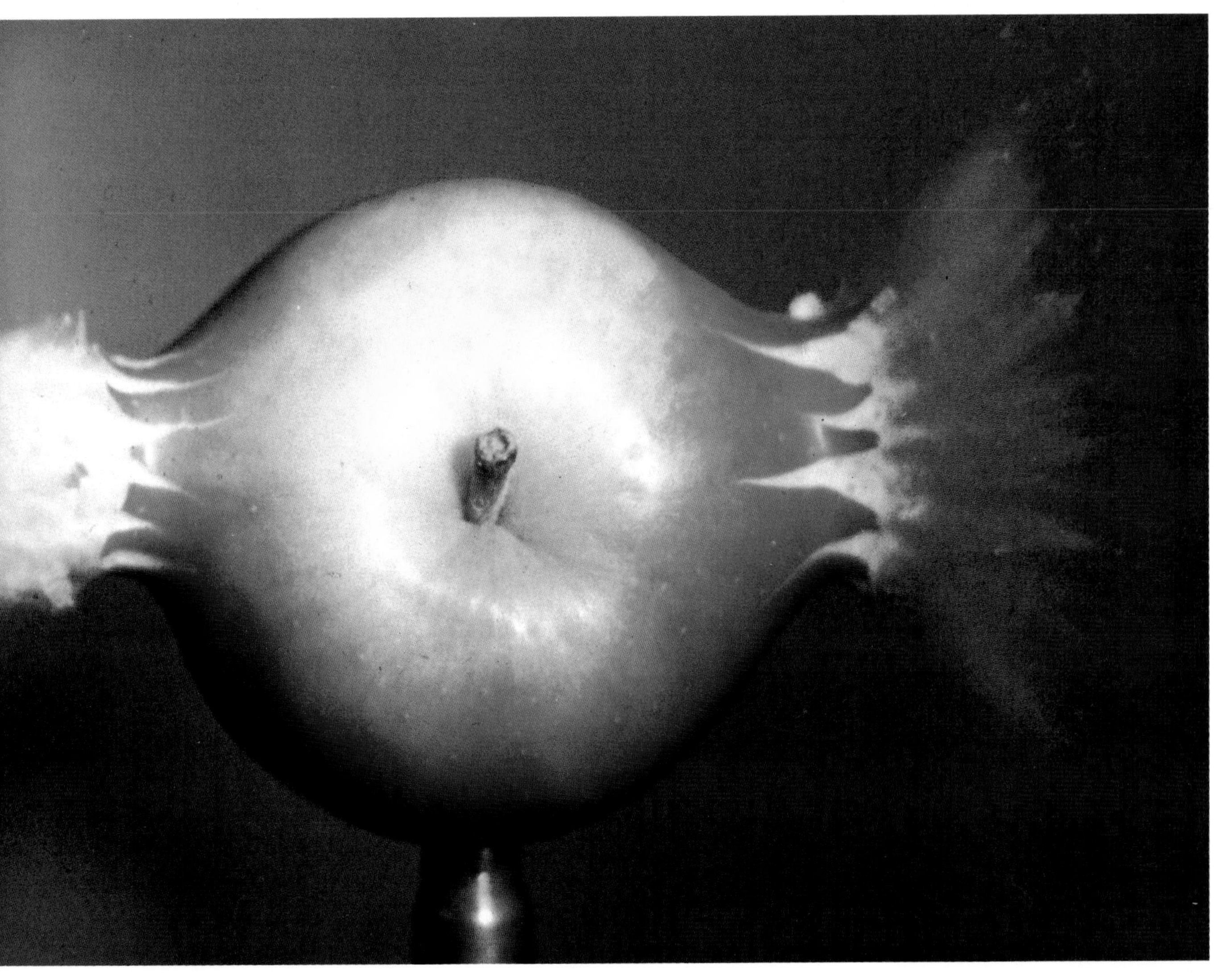

A moving bullet has a large amount of kinetic energy despite its relatively small mass because its velocity is high. The kinetic energy of a moving object equals half its mass multiplied by the square of its velocity; hence a 15 g bullet moving at 500 m s^{-1} has a kinetic energy of 1,875 joules.

When nuclei change in a nuclear reaction that produces energy, the total mass of the products is less than that of the original nuclei. This occurs because the "lost" mass has been converted into energy; a small decrease in mass produces a vast amount of energy. If mass and energy are considered together, however, the total amount of mass and energy remains the same. Consequently the principle of mass conservation has been modified into what is called the principle of conservation of mass-energy. The Theory of Relativity shows that mass and energy can be considered to be totally interconvertible, and the amount of energy produced when matter is destroyed is given by the well-known equation $E = mc^2$ (E is the energy released, m is the mass changed, and c is the velocity of light). Thus, mass is considered to be a form of energy.

An electric light bulb filament becomes hot and glows when an electric current passes through it, a common example of interconversion between different forms of energy (in this case of electrical energy into heat and light).

Work and power

Work is done when a force makes an object move (and is defined as the product of the force and the distance through which the object moves). If the object moves in the same direction as the force, then the work that is done equals the magnitude of the whole force times the distance that the object has moved.

But if the force acts in a different direction to the movement of the object, then the work done equals the component of the force in the direction of movement (which is less than the total force) multiplied by the distance through which the object moves.

If the force (or its component) acts in the same direction as the motion, then positive work is done. If the force acts in the opposite direction to the motion (such as a force applied to slow down an already moving object), then the work done is negative. Zero work is done when no motion (or change in existing motion) results, as happens, for example, when someone holds up an object without moving it.

Power is the rate of doing work (and is measured in watts, one watt being equivalent to a rate of working of one joule per second). Thus, although it takes the same amount of work to lift 10 kg through 10 m in 30 seconds as it does to do it in 60 seconds, it involves twice as much power to perform the task in the shorter time. Put another way, the amount of work that a machine can do depends both on its power and on the length of time for which it operates.

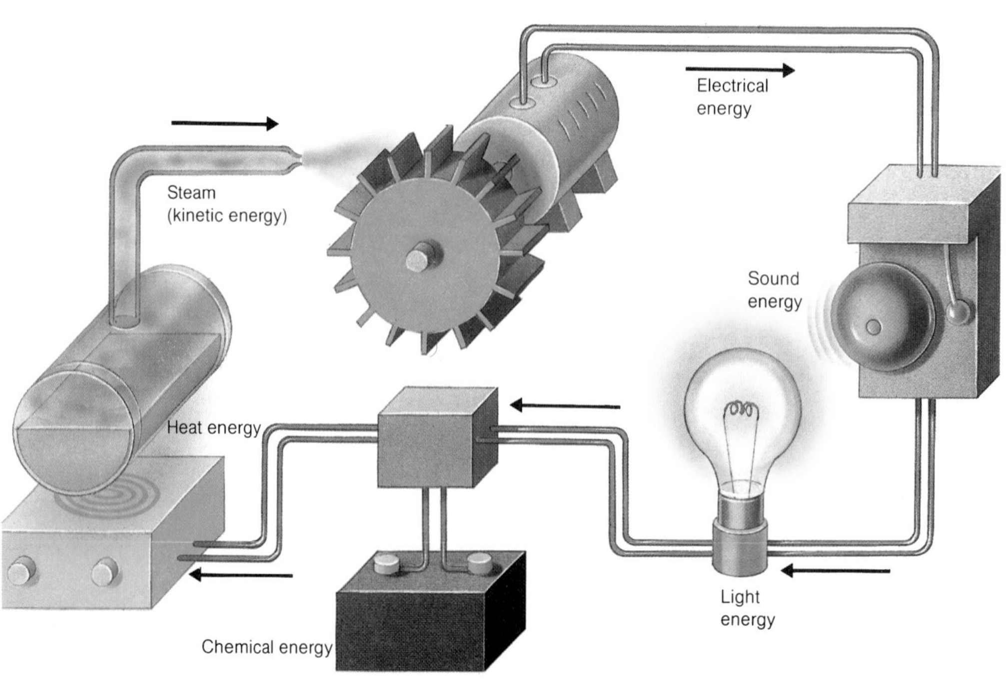

The various forms of energy are interconvertible, and this diagram shows some of the interconversions possible. From the left, heat energy is used to boil water and generate steam, which drives the rotor of a generator (the steam and rotor are moving so both have kinetic energy) and thereby produces electricity. This electrical energy can then be converted into sound (by an electric bell) or light (by an electric light bulb); it can also be used to charge an accumulator, thus being converted into chemical energy. The electrical energy can be converted back to heat by passing it through a heating element.

The transmission of energy by waves

Energy is often transmitted by wave motions, and for this reason the study of waves is of crucial importance in physics—from the wave mechanics of the atom to the study of gravitational waves produced by black holes. In general, a traveling wave is the movement of a disturbance from a source, and energy is transported as the disturbance moves outwards.

If the disturbance produced is parallel to the direction of energy travel, the wave is said to be longitudinal; sound waves are of this type. If the disturbance is perpendicular to the direction of energy travel—as in electromagnetic radiation and waves on the surface of water—then the wave is transverse.

Four properties of a wave can be distinguished and described mathematically: wavelength, frequency, velocity, and amplitude.

Wavelength is the distance between successive crests (or troughs) of the wave.

Frequency is the number of vibrations the wave completes in each second (or the number of complete waves that move past a stationary point in each second).

The velocity of a wave equals its wavelength multiplied by the frequency. And the amplitude is the greatest departure from equilibrium it undergoes. For a wave in water, the amplitude is the height of the crest above the normal water level—or the depth of the trough below it.

The amount of energy a wave transmits is determined by its amplitude, its frequency, and its velocity; the greater each of these is, the more energy is transmitted. It is partly in the form of potential energy (which is determined by the amplitude) and partly in the form of kinetic energy (which results from the velocity).

A tennis service involves many energy interchanges. When the player throws up the ball, it is moving and hence has kinetic energy. At the top of its trajectory the ball is momentarily stationary and so has zero kinetic energy (but maximum potential energy). Many tennis players time their service to hit the ball at this point; when this happens, the kinetic energy of the fast-moving racket is transferred to the ball.

Thermal energy

Terms, units, and abbreviations

These metric units are used:

- **temperature** in kelvins (K) or degrees Celsius (° C)
- **quantity of heat** in joules (J)
- **frequency** in hertz (Hz)
- **heat capacity** in joules per kelvin ($J\ K^{-1}$)
- **specific heat capacity** in joules per kilogram per kelvin ($J\ kg^{-1}\ K^{-1}$)

The customary units equivalent to the basic units used and their conversion factors are as follows:

- a degree F. = kelvin $\times \frac{5}{9}$ or degree C $\times \frac{5}{9}$
- temperature in ° F. = (temperature in K $\times \frac{9}{5}$) $-$ 459.67 kelvins, or (temperature in ° C $\times \frac{9}{5}$) + 32 degrees Celsius
- 1,000 joules = 0.95 British thermal units
- 1 kilogram = 2.20 pounds
- 1 hertz = 1 cycle per second

Thermal energy is a form of energy that is possessed by all material things due to the random motion of the constituent atoms and molecules. When a hot object cools, it loses thermal energy. But it does not lose all its thermal energy; it simply has less thermal energy than it had before. Only at the temperature of absolute zero would an object have no thermal energy at all. This temperature cannot be attained in practice, however, and so everything has a certain amount of thermal energy, no matter how hot or cold it is.

Temperature is a measure of how much thermal energy the average atom or molecule in a substance possesses. Temperatures also indicate how heat will flow when one object is placed in contact with another: heat always passes from an object at a higher temperature to one at a lower temperature. Thus, heat is simply the flow of thermal energy from one object or substance to another. When two objects are at the same temperature, heat exchange between them stops.

When something is heated, the energy is absorbed by its molecules. If the object is a solid, the molecules vibrate more energetically; if it is a gas or a liquid, they travel faster. In gases, the faster-traveling molecules strike the container's walls with more force, causing the gas to increase either in pressure (if its volume is fixed) or in volume (if the pressure remains constant).

Thermal energy can also be transferred between objects that are not making contact. This is because thermal energy can radiate through space (at a frequency of approximately 10^{14} Hz, or a hundred million million vibrations per second). All atoms vibrate at about this frequency, and resonance occurs in any substance that absorbs the heat radiation, making its atoms vibrate faster and thereby raising its temperature.

The formation of heat

Thermal energy is produced by the conversion of other forms of energy. For example, heat is formed from kinetic energy by friction. As two surfaces move over each other or as a solid moves relative to a liquid or gas, the motion causes the molecules in the surface regions to move faster, thereby increasing the temperature. Electrical energy is converted into heat in an electric stove. In this case, the motion of electrons through the wire of the heating element causes its metal atoms to vibrate faster, giving heat. Chemical energy is changed into heat during combustion, as when a fuel burns. The atoms in the molecules taking part in the reaction rearrange themselves into new molecules, and the interatomic forces released act to increase the speed of the molecules that are formed. Sound also produces heat when it is absorbed, although in such small amounts that the temperature rise is generally unnoticeable. The waves of pressure that make up sound waves agitate the molecules and cause them to move faster.

Just as it is possible to convert various forms of energy into thermal energy, so thermal energy itself can be converted into other types of energy. The heat-formation conversions proceed in reverse, absorbing heat and producing energy. A thermocouple converts heat into electricity by freeing electrons, and heat is turned into kinetic energy in any heat engine, such as a gasoline or steam engine. A fire converts some of its heat into light, and a lightning flash turns it into sound and light. Many chemical reactions take up heat to form products: a common example is cooking, dur-

ing which heat is converted into chemical energy that is stored in the food.

Temperature scales

There are several different temperature scales, but the one used in physics is the absolute thermodynamic (or Kelvin) scale, which is based on the amount of thermal energy that bodies possess. It has two basic fixed points, which are values given to precise temperatures at which particular effects occur; divisions between the fixed points define the intervals of the scale (called degrees). The lower fixed point is absolute zero, the temperature at which an atom has zero thermal energy. The unit of temperature is the kelvin (K), so that absolute zero is 0 K. The upper fixed point is the triple point of water, a unique temperature at which ice, liquid water, and water vapor can exist in equilibrium. This is given the value 273.16 K.

Related to the thermodynamic scale is the Celsius (or Centigrade) scale, which is more convenient to use because it avoids large figures at normal temperatures. On this scale, the triple point of water is 0.01° C. The size of a kelvin degree is the same as a degree on the Celsius (or Centigrade) scale, so absolute zero in this scale is −273.15° C.

Because absolute zero cannot be attained in practice, other fixed points have to be determined to define a practical temperature scale. In addition, thermometers have different ranges of temperature over which they can be used effectively and a wide range of fixed points is necessary to calibrate them. The International Practical Temperature Scale has eleven principal fixed points, extending from −259.34° C, the triple point of hydrogen (the temperature at which solid, liquid, and gaseous hydrogen exist together in equilibrium), to 1,064.43° C, the melting point of pure gold.

The Fahrenheit scale is another scale of temperature that is still in use. On this scale, the freezing point of water is 32° F. and the boiling point of water is 212° F. The unit for temperature, the degree Fahrenheit (° F), is equal to $\frac{5}{9}$ of a degree Celsius or kelvin. Absolute zero is −459.67° F.

Heat is one of the most useful forms of energy—it can also be one of the most destructive. Every year throughout the world fire destroys millions of dollars worth of goods, property, grassland, and forests. Over a large fire, huge volumes of hot gases rise into the atmosphere; more air rushes in to fill the void and an explosive fire storm can result.

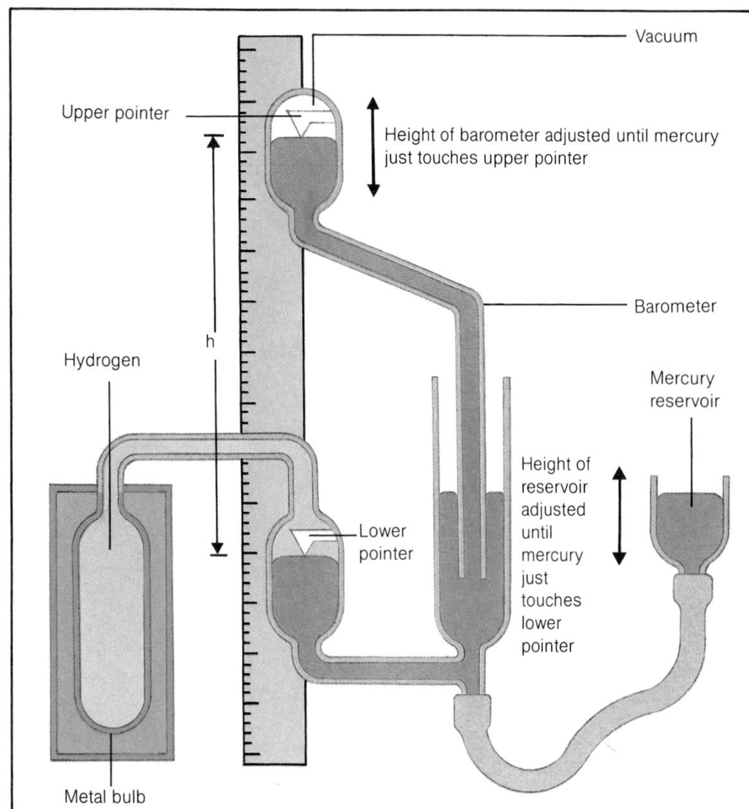

In a constant-volume gas thermometer, the volume of hydrogen in a metal bulb is made constant by raising or lowering a reservoir. The height of mercury in the barometer is then adjusted until it just touches the upper pointer, and the difference in levels (h) indicates the gas pressure and, therefore, its temperature.

Thermometers

Several different types of thermometers are used in physics, depending on the range of temperature involved or accuracy required. All depend on a thermometric property of a substance—that is one that changes continuously with temperature (such as the length of a column of liquid or the pressure of a constant volume of gas).

Liquid-in-glass thermometers are probably the most familiar types—the mercury-in-glass thermometer is widely used in medicine to measure body temperature, and room thermometers are usually of the (colored) alcohol-in-glass type. Mercury-in-glass thermometers can be made to work in the range −38.37° C (the freezing point of mercury) to 356.58° C (its boiling point) and have the advantages of being portable and of producing a direct reading. They are not, however, very accurate for scientific purposes. The (colored) alcohol-in-glass thermometer is also portable but is even less accurate; nevertheless, it is valuable where high accuracy is less important than convenience. It has the advantage of registering temperatures from −114° C (the freezing point of ethanol—the alcohol used) to 78° C (its boiling point), thus covering the complete range of temperature normally encountered in the environment.

The constant-volume gas thermometer is very accurate and has an extremely wide range—from −270° C to 1,477° C, but it is cumbersome and is therefore employed chiefly as a standard against which other thermometers can be calibrated. The constant-volume gas thermometer consists of a bulb of gas—helium, hydrogen, or nitrogen, according to the desired temperature range—and a manometer to measure pressure. The bulb of gas is placed in the environment whose temperature is to be measured, and the column of mercury (the manometer) connected to the bulb is then adjusted so that the gas in the bulb has a set volume. The height of the mercury column indicates the pressure of the gas. From this the temperature can be calculated.

The platinum resistance thermometer depends on the variation of electrical resistance with temperature of a coil of platinum wire. It is the most accurate thermometer over the range −259° C to 631° C, and it can be used to measure temperatures as high as 1,127° C. But it is slow to react to changing temperatures because of its large heat capacity and low conductivity, and is therefore best used for measuring steady temperatures.

A thermocouple contains two wires made

The temperature of red-hot steel can be measured using a radiation pyrometer (the cylindrical instrument with cables, *far right*). Heat radiation is focused onto a thermocouple, generating an electric current that is recorded on an ammeter calibrated to read temperatures directly.

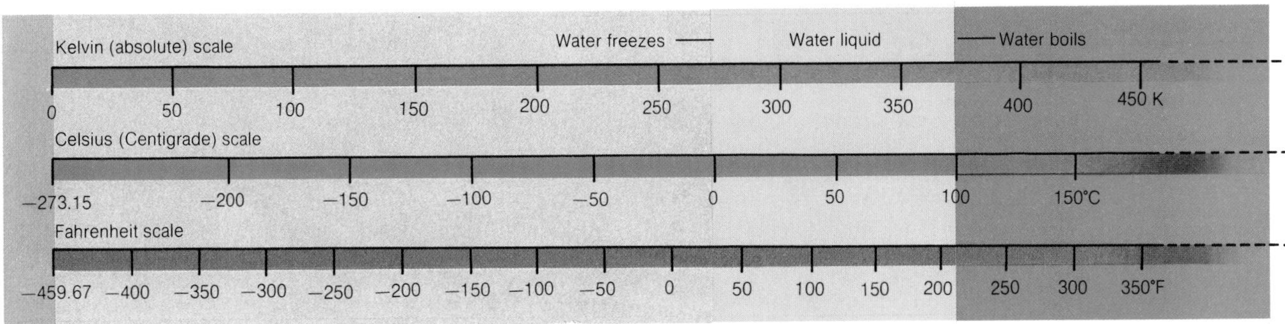

Three temperature scales compare the freezing and boiling points of water.

of different metals joined together, which produce a voltage that varies with the temperature of the junction. Different pairs of metals are used for different temperature ranges, and the overall range is very wide: from −248° C to 2,800° C. The thermocouple is the most accurate thermometer in the range 631° C to 1,064° C and, because it is very small, can respond quickly to changing temperatures.

The radiation pyrometer is used to measure very high temperatures. It depends on the heat or visible radiation emitted by hot objects, measuring either the heat of the radiation with a thermocouple or the brightness of visible radiation by comparison with a glowing tungsten filament connected to an electric circuit. A pyrometer is the only thermometer that can measure temperatures above 1,477° C.

Heat capacity

Different substances contain different amounts of thermal energy depending on their nature, temperature, and mass. The heat capacity of an object is the amount of heat required to raise its temperature by one degree. The specific heat capacity is the amount of heat required to raise the temperature of unit mass of a substance by unit temperature; it is measured in joules per kilogram per kelvin ($J\ kg^{-1}\ K^{-1}$). With gases, the molar heat capacity (the heat capacity of 1 mole of a gas at constant pressure or constant volume) is generally more useful than the specific heat capacity.

Specific heat capacities vary widely from one substance to another and from one range of temperatures to another. At room temperature, the specific heat capacity of water is 4,200 $J\ kg^{-1}\ K^{-1}$, whereas that of copper is 390. The temperature of a kilogram of water therefore increases 1 K when it receives 4,200 J of heat, whereas a kilogram of copper reacts to the same amount of heat with a temperature rise of more than 10 K. The specific heat capacity of a substance can be found by measuring either the rise or fall in temperature of a sample of known mass under controlled conditions in a calorimeter. The total amount of thermal energy that any body gains or loses is the product of its mass, its specific heat capacity, and its change of temperature.

Cooling

As an object cools, it loses thermal energy by transferring heat to its immediate surroundings. This normally continues until the object has the same temperature as its surroundings. It is possible, however, to remove so much thermal energy from an object that it becomes colder than its surroundings. This is done in refrigeration.

With so-called "freezing mixtures," such as ice and salt, the salt dissolves and lowers the temperature of the solution (adding a substance to a solvent also lowers the freezing point of the mixture; salt water, for example, freezes below 0° C). The evaporation of a volatile liquid also produces cooling because the vapor molecules have greater energy than the molecules that remain in the liquid. The overall amount of energy in the liquid is diminished, and its temperature falls.

If a stream of gas is made to flow through a small aperture into a larger container so that its pressure falls, then its temperature usually decreases too. This phenomenon, called the Joule-Kelvin (or Joule-Thomson) effect, occurs because moving the molecules farther apart requires energy to overcome the attractive intermolecular forces. The energy comes from the gas molecules themselves and the temperature of the gas therefore falls. If the larger container is not insulated (or if the gas merely expands into normal air, as when using an aerosol spray), then the gas quickly absorbs heat from its surroundings and warms up again. If, however, the expansion container is very well insulated, then the expanded gas remains at the lower temperature. This method is widely used to liquefy gases and to produce very low temperatures.

An aerosol spray demonstrates the Joule-Kelvin effect, in which the rapid expansion of a gas through a small jet results in a sharp fall in temperature. The liquid droplets quickly evaporate, producing a further lowering of temperature. Both effects can be felt when an aerosol is sprayed onto the skin.

Terms, units, and abbreviations
These metric units are used:
temperature in kelvins (K) or degrees Celsius (° C)
length in meters (m) or millimeters (mm)
heat energy in joules (J)
mass in kilograms (kg)
The customary units equivalent to the basic units used and their conversion factors are as follows:
degree F = kelvin × $\frac{5}{9}$ or degree Celsius × $\frac{5}{9}$
temperature in ° F. = (temperature in K× $\frac{9}{5}$)−459.67 kelvins,
or
(temperature in ° C × $\frac{9}{5}$) + 32 degrees Celsius
1 meter = 3.28 feet
1 millimeter = 0.039 inches
1,000 joules = 0.95 British thermal units
1 kilogram = 2.20 pounds

Effects of heat

Because the properties of the molecules of an object depend on the thermal energy, it is not surprising that the gain or loss of heat has several marked effects. Most notable is the possibility that a change of state may occur—a solid may melt or a liquid may boil and vaporize. Less dramatic, but no less important, are the expansion or contraction that takes place, and the way in which heat moves through a substance. At very low temperatures the unusual effects of superconductivity and superfluidity occur. All these effects are brought about by changes in the motion of the molecules or in the motion of electrons moving among and within the atoms of which the substance is formed.

Expansion and contraction

As the molecules in a solid or liquid become more agitated with a rise in temperature, the extent of their vibrations increases. The forces governing the separation of one molecule from another act so that the centers of vibration usually move farther apart. This increase in the separation of the molecules occurs throughout the entire object as it gets hotter, so that it expands in all directions. Any two points within the solid or liquid get farther apart, so that any hole or space enclosed by a solid body usually also gets bigger. On cooling, the reverse effect occurs and the solid or liquid contracts. When it reaches the same temperature that it had before it expanded (or contracted), it is exactly the same size.

Linear expansion is expansion in any one dimension, and the amount of that expansion is related to the rise in temperature. Every substance has a particular linear expansivity, which is the proportional increase in length per degree Celsius (° C) or kelvin (K). The measure of a material's linear expansivity is called its coefficient of linear expansion. For most metals, the linear expansivity varies between $1 \times 10^{-5}\,K^{-1}$ and $3 \times 10^{-5}\,K^{-1}$. The area of a solid increases by approximately twice this amount per degree, and the volume by about three times.

Linear and superficial (area) expansion have no significance for liquids, because they freely change their shape to that of the container, which may also expand (or contract) with a change in temperature. The cubical (volume) expansivity of most liquids is approximately $1 \times 10^{-4}\,K^{-1}$.

When a gas is heated, its pressure or volume—or both—increase, depending on how much its container expands. At 0° C (273.15 K) the expansivity of a gas either at constant volume or constant pressure is approximately 3.7×10^{-3}, which is equal to $\frac{1}{273}$. The pressure of a gas decreases or its volume contracts by this proportion with every degree drop in temperature, so that a gas should in theory have zero pressure and zero volume at absolute zero (0 K or −273.15° C). In practice, however, gases liquefy and then solidify before this point can be reached.

Almost all substances expand when they are heated and contract when they cool. There are some, however, that do not. Invar nickel-steel alloys have very low or even zero expansivities (depending on the exact composition of the alloy); and water, which normally has a low cubical expansivity, actually expands as its temperature drops from 4° C to freezing point. This happens because the molecules rearrange themselves so that they are farther apart.

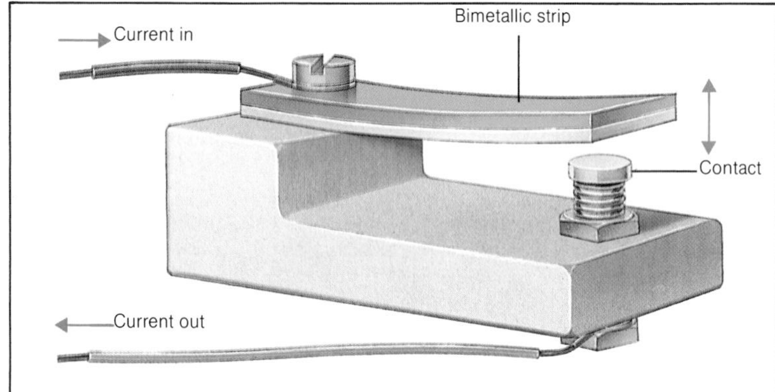

A bimetallic strip, the basis of some thermostats, consists of two dissimilar metals, one of which expands more than the other on heating. As the strip bends, it breaks electrical contact; on cooling it straightens and re-makes the contact.

Icebergs float because ice is less dense than water. Water contracts as its temperature falls from 100° C to 4° C (at which temperature its density is greatest), then expands slightly on cooling to freezing point (0° C). When water actually freezes to form ice, it expands considerably; it is the pressure of this expansion that sometimes bursts water pipes in winter or even splits rock faces apart. Below 0° C ice contracts slightly with further cooling.

Expansion and contraction are readily apparent in large structures—metal bridges expand on a hot summer's day, for example, and heat generated by friction with the air causes supersonic airliners to expand several inches or centimeters in length at top speed. Such structures and machines have to be designed so that the expansion can be taken up without affecting safety.

Expansion effects may also have positive uses. Bimetallic strips are made of two metals of different expansivities, such as brass and iron, welded together along their entire length. As the temperature changes, the metals expand or contract by different amounts, causing the strip to bend (the amount of bending depends on the size of the temperature change). The effect is utilized in some thermostats.

Conduction

Heat energy flows through solids by the process of conduction. If an object such as a metal bar is heated at one end and no heat is lost from the sides, there is a temperature gradient along the bar from the hot end to the cooler end. The exact fall in temperature (or thermal gradient) depends on the thermal conductivity of the substance involved, which is a constant property of each specific substance and is independent of the temperature drop and length. Most metals conduct heat well and therefore have high thermal conductivities. Their conductivities are generally about 1,000 times those of other solids (such as wood and cloth) and liquids, and about 10,000 times greater than the conductivities of gases. In practical terms this means that if there is a temperature difference between two points 1 m apart in a metal bar, the points between which the same temperature difference occurs are only about 1 mm apart in a liquid and 0.1 mm apart in a gas. Generally, the higher the thermal conductivity, the lower is the thermal gradient required for a fixed quantity of heat to flow from the high temperature point to the low temperature point.

Metals conduct heat well because of their atomic structure; they have electrons that are free to move among the atoms that make up the metal as a whole. When a metal is heated, the free electrons move faster and transmit the energy by colliding with the metal atoms more frequently and with greater force. Because the electrons are free to move, they can transmit energy quickly throughout the metal. (It is because of their free electrons—which carry electric charges—that metals conduct electricity well. In fact, at any given temperature, most metals conduct heat and electricity equally well.) In solids that conduct heat badly, all the electrons are bound to individual atoms or molecules, and heat is transmitted by the increased vibrations of their constituent atoms or molecules. These are much less free to move and so heat energy spreads relatively slowly through the solid.

Convection

Heat spreads through liquids and gases principally by convection—that is, by gross movements of currents of hot liquid or gas. (Fluids also conduct heat by collisions between their atoms or molecules, but this process does not result in a great transfer of energy to liquids; and gases are poor conductors.) Convection is mainly responsible for spreading heat from a radiator throughout a room. Atmospheric winds are basically convection currents caused by differential heating of the atmosphere by the sun and by heat from the earth's surface.

As a liquid or gas is warmed, the hot region of the fluid expands and becomes less dense than the surrounding fluid. As a result it rises and cooler liquid or gas flows in to take its

Heat losses from a house are revealed in this thermograph. The parts of the image colored pink, red, or yellow are hot; blues and greens are cool. Much of the heat escapes from the roof, which could be prevented by using an insulating material of low thermal conductivity.

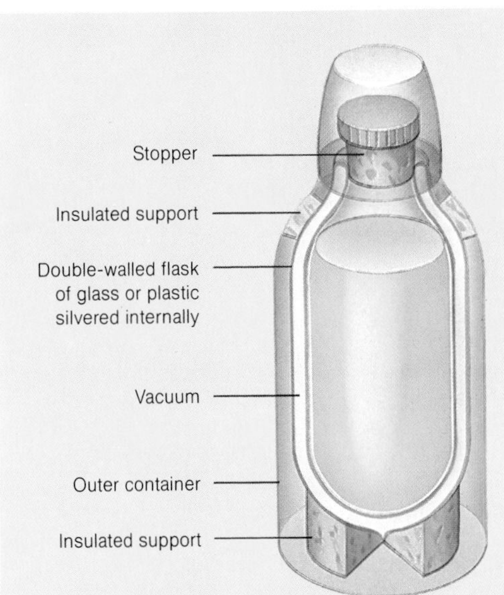

A vacuum bottle keeps hot liquids hot (or cold liquids cold) by minimizing heat transfer between the contents and the surroundings. The silvering on the inner flask reduces heat transfer by radiation, and the vacuum prevents transfer by conduction or convection.

A scuba diver can remain submerged in cold water without dangerous loss of body heat because the wet suit, and the thin layer of water it maintains next to the skin, insulates the body from the surroundings.

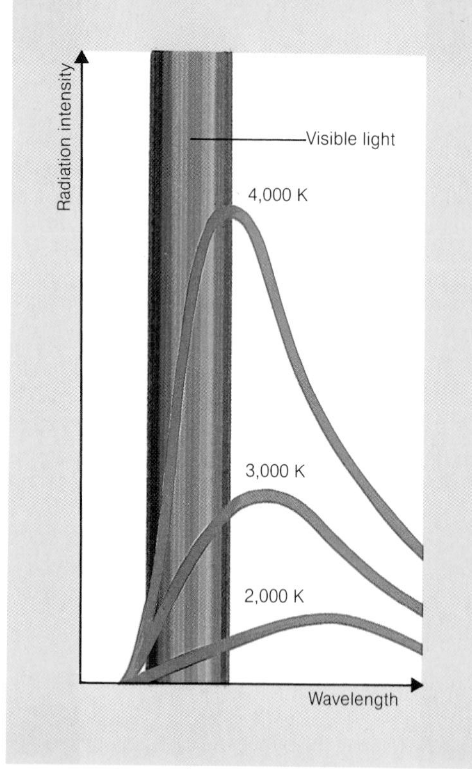

The intensity of radiation emitted by an object varies over a range of wavelengths. Also, the hotter the object, the greater is the total amount of radiation emitted. At temperatures of about 4,000 K and above, the peak intensity is in the visible spectrum.

place. This cool fluid is then warmed and rises, and yet more cool liquid or gas flows in. As the process continues, a convection current of warm fluid circulates and eventually the entire body of liquid or gas becomes warmer.

Radiation

In addition to conduction and convection, heat can also be transferred by radiation. All objects are constantly emitting heat radiation, which consists of radiant energy of frequencies that generally lie in the infrared region of the electromagnetic spectrum. Infrared radiation is emitted by an object as a result of changes in the energy states of its constituent atoms or molecules when they become cooler; thus heat energy is converted into radiant energy. The amount of energy radiated depends on the temperature of the object: the hotter the object, the greater is the rate at which radiant energy is emitted.

All objects also constantly absorb heat radiation. An exchange of heat radiation occurs continuously between an object and its immediate surroundings, both of which simultaneously gain and lose radiant energy. The gains and losses balance when both reach the same temperature. A cold object in warmer surroundings absorbs heat until its temperature increases to the point at which it emits as much energy as it receives. It then remains at the same steady temperature as its surroundings.

The amount of thermal energy that a body emits and absorbs depends not only on its temperature but also on the nature of its surface. Heat radiation behaves in a similar way to light (because both are forms of electromagnetic radiation), and passes through transparent substances without producing a rise in temperature. It is also reflected by white and polished surfaces and absorbed by dark surfaces. Objects that do not themselves emit light do radiate heat; white and polished objects radiate relatively little heat, whereas dark objects emit a large amount.

Objects that radiate the maximum amount of heat and absorb all the heat energy they receive are known as "black bodies." A hole in the side of a black container is the most perfect black body that can be constructed. Black body is not a descriptive term, however, because at high temperatures all bodies begin to emit light as well as heat radiation—as occurs, for example, in the element of an electric heater. A very hot object therefore emits a range of wavelengths. The most intense of these wavelengths depends on the object's absolute temperature and, therefore, on its energy. This is why a very hot object emits light of different colors as it gets even hotter, initially radiating red (when it is relatively cool) and passing through the spectral range until it gives out white light when it is extremely hot.

Latent heat

Pure substances have precise melting and boiling points—fixed temperatures at which they change from one state to another. Most alloys and mixtures, on the other hand, undergo a decrease or increase in temperature as they change state. The possession of a specific melting or boiling point is, therefore, a criterion of the purity of a substance.

All substances require heat to make them change from solid to liquid and from liquid to vapor; they give out heat as they change back from vapor to liquid and from liquid to solid. This heat is called the latent heat of a substance.

The specific latent heat of fusion is the amount of heat required to melt a unit mass of the substance at its melting point, which is

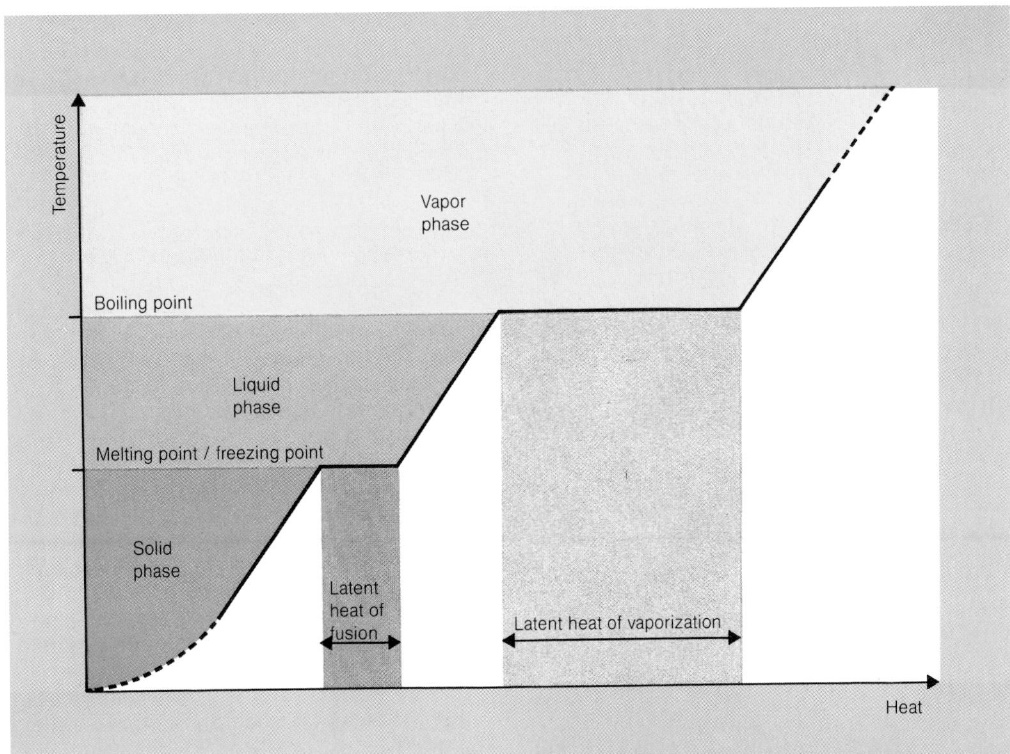

When a pure solid is heated, its temperature rises until it is at its melting point, when an additional quantity of heat (the latent heat of fusion) causes it to melt. Further heating raises the temperature of the liquid to its boiling point, and then another additional heat input (the latent heat of vaporization) turns it into a gas or vapor. The process is reversed on cooling; a gas gives out latent heat of vaporization as it condenses to a liquid, and a liquid gives out latent heat of fusion as it freezes to a solid. For example, the specific latent heat of vaporization of water is 23×10^5 joules per kilogram, and its latent heat of fusion is 33×10^4J kg^{-1}.

the same as the amount of heat it releases as it freezes. The specific latent heat of vaporization is the amount of heat required to make a unit mass of a liquid turn to vapor at its boiling point, which is equal to the amount of heat released as it condenses back to a liquid. Specific latent heats are measured in joules per kilogram (J kg^{-1}). The exact values of the various latent heats depend on the pressure at which they are measured, and they are normally stated for standard atmospheric pressure (which is 760 mm of mercury).

As a substance changes state and takes up or gives out latent heat energy, the average kinetic energy of the molecules does not change and so the temperature remains constant. When a substance melts, the heat energy it absorbs is used to break the attractive forces between its constituent atoms or molecules so that they are relatively free to move. Nevertheless, there remains some interatomic or intermolecular attraction. When a substance is converted to vapor, the heat absorbed is used to free its atoms or molecules completely.

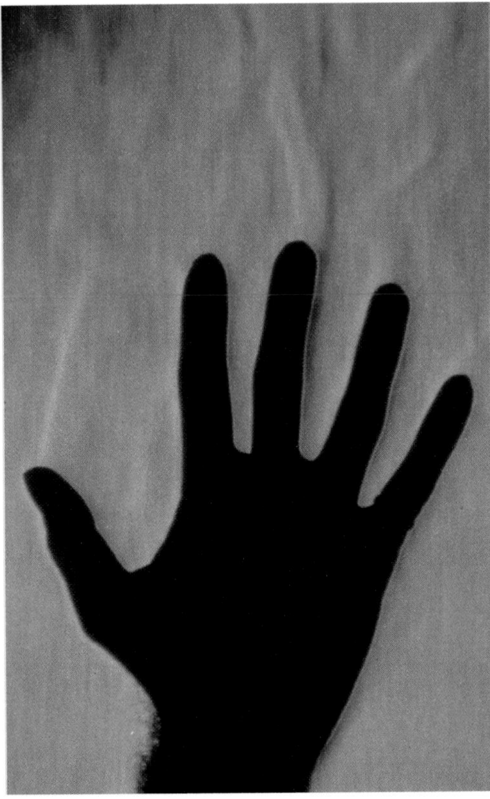

Convection currents transfer heat in a fluid. In this Schlieren photograph, the fluid is a gas (air), and the source of heat is a human hand. Air immediately around the hand is heated; it therefore becomes less dense and rises, to be replaced by cooler air from the surroundings. Convection currents are set up and the air circulates.

Thermodynamics

Thermodynamics is the branch of physics concerned with the movement of heat and the energy changes that result. Although much of thermodynamics involves difficult mathematical and theoretical concepts, it has many practical applications, such as predicting or assessing the efficiency of various designs of heat engines.

When a body—a mass of gas, for instance—receives energy, the energy it absorbs increases the motion of its molecules and thereby raises its temperature. Thus the internal energy of the gas, which is the average kinetic energy of its molecules, increases. If, however, the gas expands while being heated, some or all of the heat energy absorbed is used to do external work in pushing back the walls of its container.

The total of the increase in internal energy and the external work done is equal to the amount of heat energy received by the gas. This is the basis of the first law of thermodynamics, which states that in a closed system the total amount of energy is constant. This law is another way of stating the principle of conservation of energy (energy cannot be created or destroyed).

Another fundamental principle is the second law of thermodynamics, which states that, without external assistance, heat cannot flow from a colder to a hotter body.

If heat is to be put to use, it has to flow from a hotter to a colder body. In heat engines, such as gasoline and diesel engines, steam engines, jet engines, and rocket engines, a mass of gas is heated so that it expands and does work. In a gasoline engine, for example, a gasoline-air mixture ignites and expands, forcing a piston to move. A ram jet engine takes in air, which is then heated (by burning fuel) so that it expands and leaves the engine with the exhaust gases in a fast-moving jet. In all types of engines, the overall process is one in which thermal energy is gained by a gas and converted into internal energy and external work. The gas is then returned to the environment at a lower temperature. The higher the initial temperature of the gas, the more heat energy is available for conversion into work; and the lower its final temperature, the less internal energy it possesses and the greater is the work done. Thus the wider the temperature range over which a heat engine operates, the greater is the amount of work it can do. In practice, this range lies between the temperature at which the fuel burns to heat the gas and the temperature of the exhaust gases on leaving the engine.

A rocket engine works by burning fuel to produce exhaust gases, which leave the engine with an enormous amount of energy, and therefore produce a powerful thrust. The amount of work any engine can perform depends on the difference between the combustion and exhaust temperatures. In a rocket engine this difference is very great, so it can do a large amount of work.

Operating cycles and efficiency

To produce continuous power, a heat engine must operate in a cycle. Fuel and air—and water to raise steam if required—must be continually supplied to the engine, and gas must be heated, allowed to expand, and returned to the exterior—either in a continuous operating cycle or, as in a gasoline engine, in repeated cycles made up of a sequence of actions.

The operating cycle consists of a sequence of pressure, temperature, and volume changes undergone by gas passing through the engine. As the cycle begins again, the next mass of gas is brought to the same pressure temperature and volume as the previous mass. Work has to be performed to do this, forcing out the exhaust gases and bringing in the fuel, for example. Throughout the cycle energy is also used in overcoming friction between the engine's moving parts. The amount of work obtained from the engine is, therefore, less than the heat energy that goes into it. The ratio of heat input to work output is called the engine's efficiency; a good heat engine is typically only about 40 per cent efficient, although some types exceed 50 per cent.

It is impossible to construct an engine that is 100 per cent efficient because of unavoidable energy losses due to friction, among other factors. The most efficient engine is an ideal heat engine that performs a totally reversible operating cycle, such as the Carnot cycle, but even this is less than 100 per cent efficient.

In the Carnot cycle (see the diagram above right), a mass of gas is expanded in two stages (an isothermal followed by an adiabatic expansion), and then compressed in two stages (isothermally then adiabatically). Overall, the

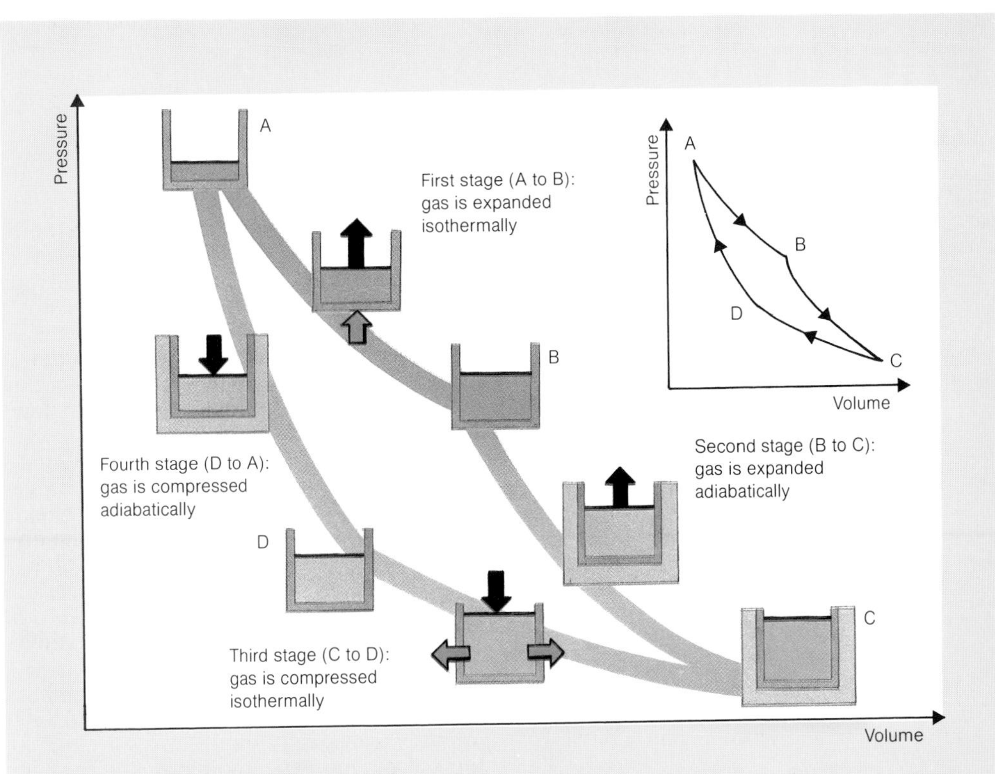

The Carnot cycle consists of four stages. In the first stage (A to B), a mass of gas is heated but allowed to expand freely (an isothermal expansion); thus the heat supplied does only external work (the expansion) and does not increase the gas's internal energy. Hence the gas's temperature remains the same but its volume increases and its pressure decreases. In the second stage (B to C), the gas is expanded adiabatically (without allowing heat to enter or leave the system) so that it does more external work. During this stage the gas's internal energy (and, therefore, temperature) and pressure decrease, and its volume increases. In the third stage (C to D), the gas is compressed isothermally; as a result its pressure increases, its volume decreases, and it releases any remaining heat it possesses. In the final stage (D to A), the gas is brought back to its original temperature, pressure, and volume by compressing it adiabatically.

amount of heat supplied to the gas does not increase its internal energy, but neither is all the heat converted into work; some is lost in the third stage of isothermal compression. Hence, the efficiency of an ideal heat engine performing the Carnot cycle is less than 100 per cent.

Entropy

An important concept in thermodynamics is that of entropy, which can be thought of as the degree of disorder in a system. The more disordered the system, the greater is its entropy. Thus, a crystalline solid, whose molecules are arranged regularly in a lattice, has a lower entropy than the same substance in the molten state, in which the molecules are moving relatively freely and are, therefore, more disordered.

In thermodynamics, a change in entropy can be defined as $dS = dQ/T$, where dS is the entropy change and dQ is the amount of heat taken in at absolute temperature T. Strictly, this applies only to reversible processes; entropy changes for irreversible processes are calculated by postulating equivalent theoretical reversible changes.

In "free" or isolated systems (those that are allowed to distribute their energy without external influences), the total entropy always increases and the available energy decreases. When the system has reached maximum entropy, the matter present is completely disordered and at a uniform temperature, so there is no energy available for doing work. Many scientists believe that the universe will eventually reach this state. Others think that it will collapse in on itself and undergo another "Big Bang" similar to the one that created the present universe.

Organic decay *(left)* and inorganic rusting *(right)* are both examples of increasing entropy. In the case of the decaying tree, the disorder increases as the ordered structure of the wood is broken down into simpler substances (such as sugars) by certain animals and plants. Rusting involves the breaking down (by chemical action) of the regular crystalline arrangement of molecules in steel into less ordered collections of rust (iron oxide) molecules.

Radiant energy

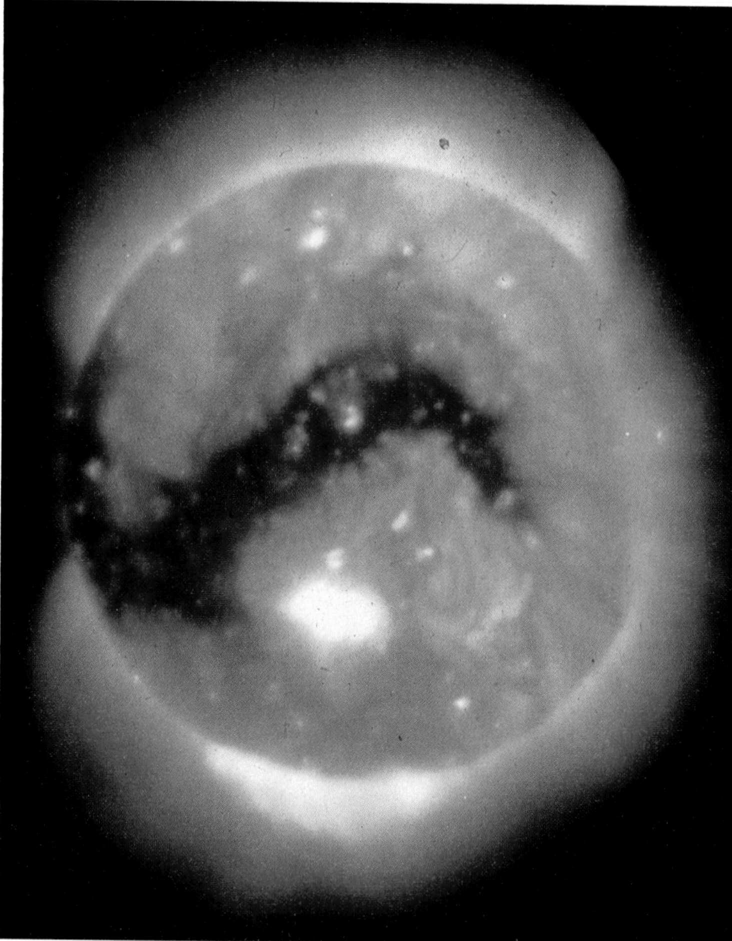

Energy may take various forms—sound, chemical, and electrical energy, for example. One whole range of related forms is known as radiant energy or electromagnetic radiation. Together they make up the electromagnetic spectrum, whose various radiations are characterized by particular ranges of wavelengths and frequencies. At the long-wavelength, low-frequency end of the spectrum are radio waves, followed by microwaves and infrared radiation. Then comes visible light and ultraviolet radiation, and the shortest wavelengths and highest frequencies comprise X rays and gamma rays.

Apart from light, all are invisible to the human eye, although many can be detected by electronic devices and special types of photographic film. Almost all have found practical applications: in radio communications and radar, infrared and microwave heating, photography and spectrographic analysis, and medical and industrial radiology. The most important forms of electromagnetic radiation are visible light and infrared (heat) radiation (both of which are emitted by the sun, our principal natural source of radiant energy); without them, life would be impossible.

Nature and properties of radiant energy

The various forms of radiant energy have many common characteristics, the most fundamental of which concerns their nature. They are all wave motions, consisting of varying

The sun gives off all forms of radiant energy, from gamma rays through to radio waves. The photograph above shows its X-ray emissions. But only visible light, radio waves, and some infrared and ultraviolet radiation can penetrate the earth's atmosphere.

Greenhouses trap energy from the sun. Short-wavelength infrared radiation is absorbed by soil and plants, which become warmer and re-radiate longer wavelengths that cannot penetrate glass. Earth's atmosphere acts in a similar way to the glass, and our entire planet is warmed by a comparable greenhouse effect.

electric and magnetic fields, but they can also be considered as being made up of "particles," or quanta, of energy called photons. Radiant energy therefore has a dual wave/particle nature.

In theory it is possible to explain all the effects and properties exhibited by radiant energy in terms of both wave motion and photons, but in practice it is usually more convenient to employ whichever of the two concepts explains a particular effect most easily.

Another basic characteristic of electromagnetic radiation is the fact that it does not need a medium in which to travel, and so it can traverse the vast emptiness of space. But the various forms of radiation can also travel through other mediums: light travels through air, water, and glass, for example, and X rays and gamma rays can penetrate body tissues and even pass through metal; that is to say, even a metal is "transparent" to X rays and gamma rays.

In a vacuum, all forms of electromagnetic radiation travel at the same constant velocity—called the speed of light. In other transparent mediums, they travel more slowly, the velocity depending on the wavelength of the radiation and, generally, on the density of the medium. The speed of light in air is nearly the same as its speed in a vacuum and for many practical purposes both speeds are regarded as being identical—approximately 3×10^8 m s^{-1} (equal to more than 1,000 million kilometers per second), the experimentally measured value being 186,282 miles (299,792 kilometers) per second.

In other mediums, different wavelengths travel at slightly different velocities (which is why a beam of white light is dispersed into spectral colors when it passes into a prism or lens); the longer the wavelength, the faster the radiation travels. So in the case of visible light, red light travels faster than does blue light in any particular transparent medium.

The various types of radiant energy have different wavelengths and frequencies. These two characteristics multiplied together are equal to the velocity. Because the velocity is constant in any particular medium, wavelength and frequency are inversely proportional to each other; the longer the wavelength, the lower the frequency, and vice versa.

Effects of radiant energy

Radiant energy can be detected only when it is absorbed and brings about an observable effect—usually by causing a physical or chemical change. Gamma rays ionize certain gases when absorbed by their molecules. X rays and visible light cause chemical changes in a photographic emulsion that darken it when it is developed. Ultraviolet radiation causes substances such as fluorite to fluoresce (emit light). Visible light stimulates the nerve cells in the retina of the eye, thereby making vision possible. Atoms and molecules vibrate more energetically, and so substances become hotter when they absorb infrared radiation. And radio waves set up weak alternating electric currents in the metal of receiving aerials.

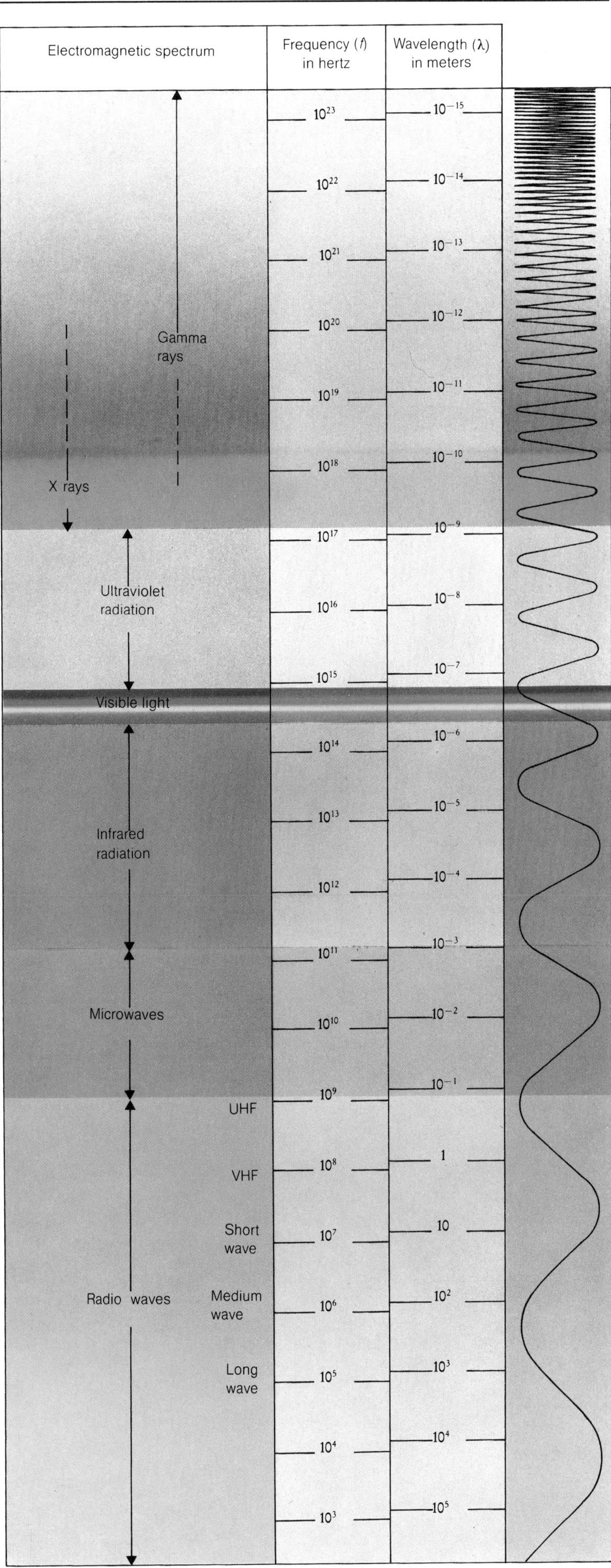

Electromagnetic radiation

Electromagnetic radiation results whenever electrons oscillate, decelerate, or change energy levels in an atom. X rays, for example, can be emitted when fast-moving electrons decelerate rapidly, and radio waves (at the opposite end of the electromagnetic spectrum) are produced by oscillating electrons in the metal of transmitting aerials.

Waves and particles

Electromagnetic radiation consists of a wave motion in both an electric and magnetic field, which oscillate with the same frequency but at right angles to each other. The two fields also oscillate at right angles to the direction of travel, and electromagnetic waves are therefore examples of transverse waves.

Electromagnetic radiation also has properties associated with particles, and its precise nature is something of an enigma. In some circumstances—such as the diffraction of light by a grating—it acts like a wave, whereas in others—for example, the generation of an electric charge when ultraviolet radiation strikes certain metals (a phenomenon called the photoelectric effect)—it acts like a stream of particles. Because of this apparent wave/particle duality, electromagnetic radiation can be considered to behave as a wave motion or as a stream of energy particles. In general, however, wave characteristics tend to predominate at the long-wavelength, low-frequency (radio wave) end of the electromagnetic spectrum, whereas particle characteristics are more pronounced at the short-wavelength, high-frequency (X-ray and gamma ray) end.

Electromagnetic waves are emitted in all directions away from their source. So if the source is a single point, each wavefront can be regarded as an expanding sphere. With increasing distance from the source, the intensity of radiation diminishes as the surface area of the sphere increases. The intensity of radiation is thus inversely proportional to the square of the distance traveled (because the area of a sphere is proportional to the square of its radius).

Another important wave characteristic of electromagnetic radiation is phase. In most sources of radiant energy the atoms emit waves at random intervals. As a result, the vibrations of waves of the same frequency are out of phase with each other—that is, the maximum and minimum strengths of the electric and magnetic fields of one wave do not coincide with those of other waves. Furthermore, although in each single wave the electric and magnetic fields are at right angles to each other, in a collection of waves not all the electric fields (or magnetic ones) lie in the same plane, unless the waves are polarized. Randomly-emitted electromagnetic radiation is described as incoherent, and this is the usual way in which radiant energy is emitted. In lasers, on the other hand, the atoms do not emit randomly and the light produced is coherent—that is, all the waves are in phase.

The particle nature of electromagnetic radiation is explained by the quantum theory, according to which radiant energy is emitted in discrete, individual "packets" of energy called quanta or photons. Furthermore, the amount of energy each photon possesses is proportional to its frequency—the higher the frequency, the greater is the energy. Each photon at the low-frequency (radio) end of the electromagnetic spectrum has an energy of about 10^{-30} J, compared to about 10^{-11} J for high-frequency gamma-ray photons. A gamma-ray photon is therefore about 10^{19} times more en-

Electromagnetic radiation is emitted in all directions away from its source so that the wavefront (an imaginary surface that passes through those parts of the electromagnetic waves that have the same phase) comprises a sphere. As the radiation travels away from its source, the sphere expands and the intensity of radiation diminishes in accordance with the inverse square law. Thus the amount of radiation passing through an area *a* at a distance *d* from the source (S) has to cover an area of *4a* at a distance *2d* and an area of *16a* at a distance *4d*.

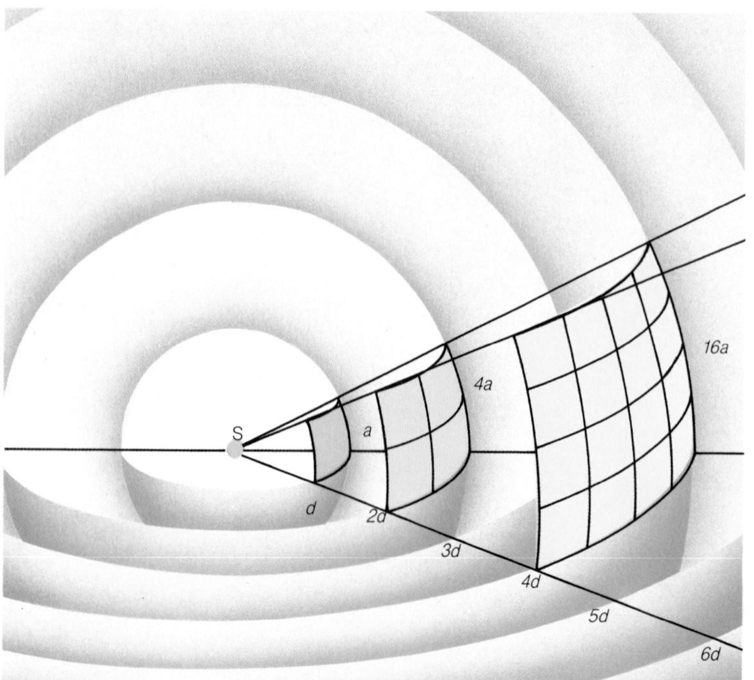

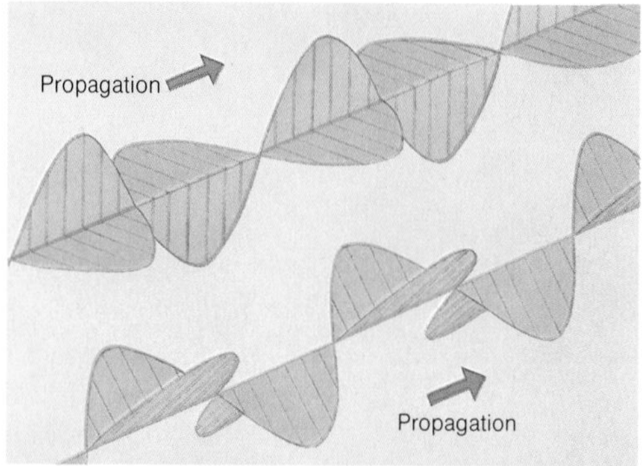

The mutually perpendicular electric and magnetic fields (orange and green respectively in the diagram above) of any one electromagnetic wave have equal energies, and the maximum strengths of the two fields always coincide (as do the minima and other corresponding points). In randomly emitted incoherent radiation, the maxima of different waves do not coincide—as can be seen by comparing the two waves above. In addition, the fields of different waves are oriented at different angles if the waves are unpolarized (which is usually the case).

ergetic than is its radio counterpart—which is why gamma rays can penetrate metal whereas radio waves cannot.

Basic properties

All forms of electromagnetic radiation exhibit several basic properties. One of the most fundamental is velocity: all electromagnetic radiations travel at the same constant velocity through a vacuum. They may also pass through other mediums, although not all mediums are transparent to all types of electromagnetic radiation: X rays, for instance, can pass through metal but light cannot. In transparent mediums the velocity of the radiation is slower than in a vacuum; the exact speed depends on the frequency of radiation and the properties of the medium. Transmission through a medium may be accompanied by refraction (a change in direction) or dispersion (the separation of mixed wavelengths—white light into the colors of the spectrum, for example) or both.

If electromagnetic radiation is not transmitted through a medium, it may be reflected or absorbed. Again, different types of radiant energy are reflected or absorbed by different materials; metals reflect most radiations but not X rays or gamma rays, which pass through thin metal sheets and are absorbed by thick ones. When electromagnetic radiation is absorbed, it interacts with atoms or molecules to increase their vibrational or rotational energies; ultimately, this excitation is converted into heat.

The other principal properties of electromagnetic radiation are scattering, diffraction, interference, and polarization. Scattering is the random deflection of radiation caused by molecules or small objects reflecting or diffracting the radiation. Diffraction occurs when electromagnetic waves encounter a narrow opening (comparable in size to the wavelength of the radiation) or an obstacle and are bent around the edges of the opening or obstacle. This phenomenon may be accompanied by interference—the reinforcement or cancellation of waves that occurs when coherent waves of the same frequency interact. And electromagnetic waves can be polarized so that all their electric fields vibrate in the same plane (which necessarily means that their magnetic fields must also oscillate in one plane).

Radio waves and microwaves, which occupy the long-wavelength end of the electromagnetic spectrum, are extensively used for communications. This transmitting and receiving station has several aerials for radio waves of various wavelengths and a Luneberg lens (the white, almost hemispherical structure on top of the building) for receiving microwaves.

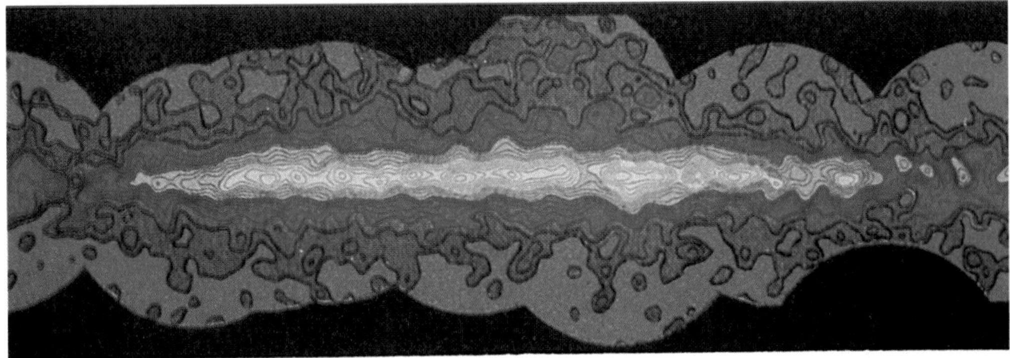

This gamma-ray map of part of the Milky Way reveals that many celestial objects emit this extremely short-wavelength form of electromagnetic radiation. On the map, the intensity of gamma-ray emission is represented by different colors—from yellow (highest intensity) through to blue (lowest intensity).

The surface of still water can act as an almost perfect mirror—as exemplified by the reflection of part of the Sierra Nevada mountains in Shadow Lake, California. The quality of an image reflected in water improves as the angles of incidence and reflection increase because the larger the angles, the less light is refracted into the water and more is reflected.

Light and reflection

Visible light constitutes only a very small part of the electromagnetic spectrum of wavelengths (between 7.5×10^{-7} m and 4×10^{-7} m; the corresponding frequency range extends from 4×10^{14} Hz to 7.5×10^{14} Hz). The longest visible wavelengths (lowest frequencies) are seen as red light, the shortest (highest frequencies) as violet. Various other colors can be identified between these two extremes. Conventionally, the spectral colors are listed as red, orange, yellow, green, blue, and violet.

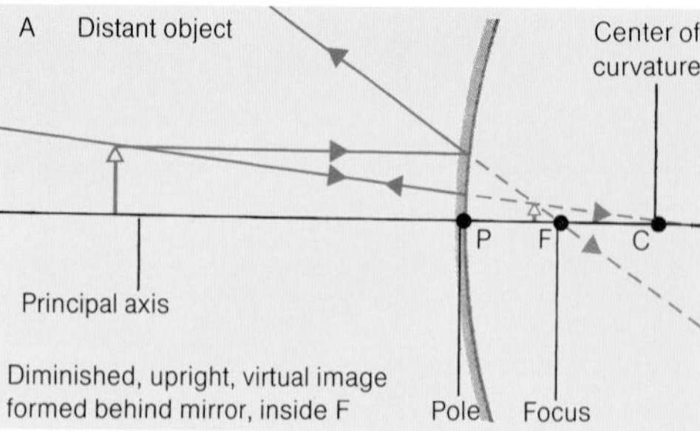

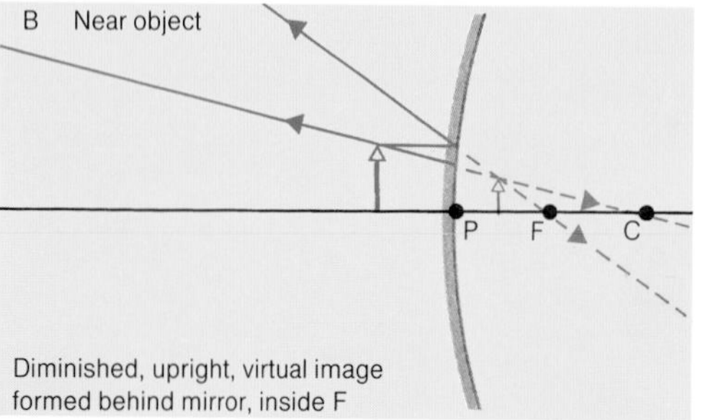

These ray diagrams show image formation by a spherical convex mirror. In A, the object (green) is relatively distant from the mirror, and a diminished, upright, virtual image (red) is formed behind the mirror. In B, the object is nearer the mirror but the same type of image is formed—diminished, upright, and virtual. In fact, the only difference among images formed by a convex mirror is their degree of diminution; the farther the object is from the mirror, the smaller is the image.

Being a form of electromagnetic radiation, visible light exhibits all the properties characteristic of such radiation—for example, wave/particle (photon) duality, reflection, refraction, diffraction, dispersion, interference, and polarization. Reflection is explained in this article; the other properties are dealt with in following articles.

Shadows

Light waves travel outwards in all directions from their source, thereby forming an expanding spherical wavefront. Each individual wave travels in a straight line in a vacuum, or in any isotropic medium (one with uniform properties throughout).

Because light travels in straight lines, shadows are formed behind opaque objects. If the light source is very small, effectively a point source, the entire shadow is equally dark and has well-defined edges. If, on the other hand, the source is relatively large—which is usually the case—the shadow is blurred at the edges and has a dark central region of complete shadow (called the umbra) surrounded by a region of partial shadow (the penumbra), which is illuminated by light from part of the source.

Reflection

All objects and surfaces encountered in everyday life reflect light—which is why they are visible. Only a perfectly matte, black surface absorbs all light, and such a surface is extremely difficult to obtain. Many surfaces reflect only light of certain wavelengths, and so they appear colored when illuminated with light, which contains all visible wavelengths.

The nature of the surface also affects the type of reflection. Irregular surfaces and smooth, matte ones reflect light randomly and so cannot form images. For example, the pages of this book reflect light randomly, so enabling the type to be read rather than reflecting an image of the reader. Very smooth, shiny surfaces on the other hand, reflect light in a regular way and these types of "mirror" surfaces can therefore form well-defined images.

Mirrors and images

There are three main types of mirrors: plane, concave, and convex. Plane mirrors are flat and are the most familiar type; concave mirrors are hollowed—shaving mirrors are an example; and convex mirrors bulge outwards—they are used for some driving mirrors, for instance, because they give a wide angle of view.

Reflection in all three types of mirrors is governed by two principal laws. The first law of reflection states that the incident ray (or beam) striking the surface and the reflected ray (or beam) leaving it are in the same plane as the normal (an imaginary line perpendicular to the surface at the point where the incident ray hits it). The second law of reflection states that the angle of incidence (between the incident ray, or beam, and the normal) is equal to the angle of reflection (between the reflected ray, or beam, and the normal).

Using these two laws it is possible to determine the types of images formed by the different sorts of mirrors. The image formed by a plane mirror is the same size as the object; the same distance behind the mirror as the object is in front; not inverted; and virtual (that is, the image is formed by diverging rays and cannot be formed on a surface such as a screen; it can be seen, however.

The type of image formed by a concave mirror depends on the distance between the object and mirror. In general, the image is inverted and real (that is, formed by converging rays and so able to be formed on a screen) when the object is more distant than the mirror's focus, and upright and virtual when the object is inside the focus. The types of images formed by spherical concave mirrors are shown in the appropriate ray diagrams (right).

Unlike concave mirrors, convex ones always produce upright, virtual images, irrespective of the position of the object. Image formation by a spherical convex mirror is also shown in the appropriate ray diagrams (left).

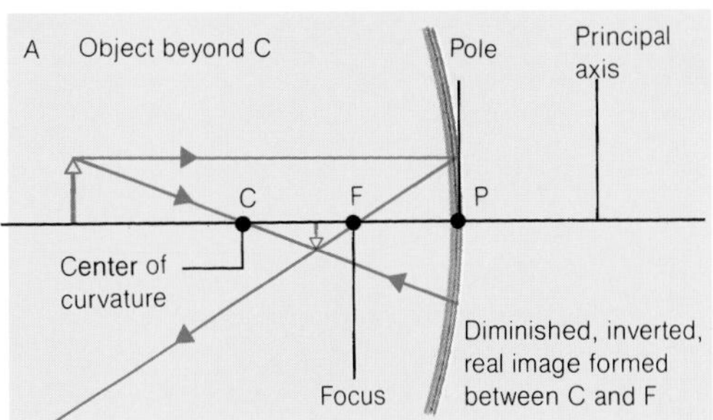

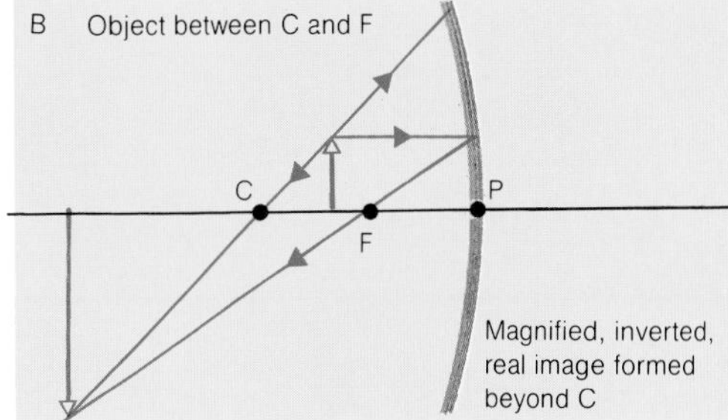

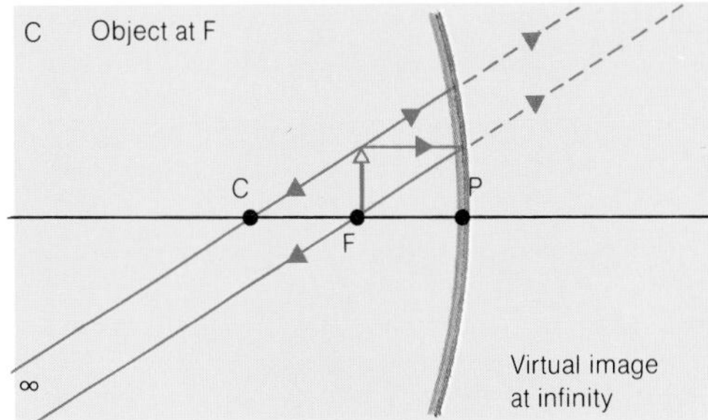

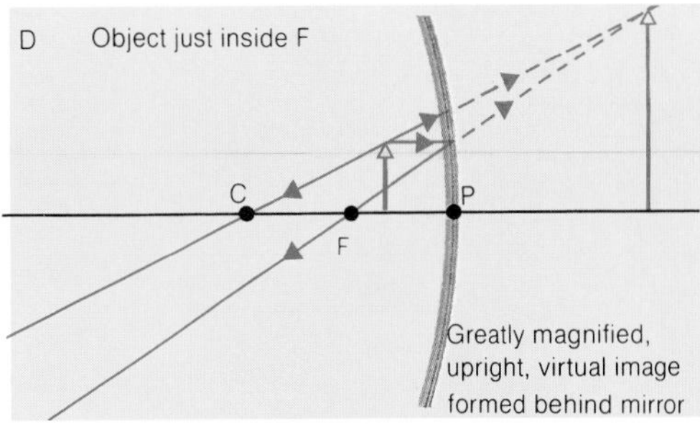

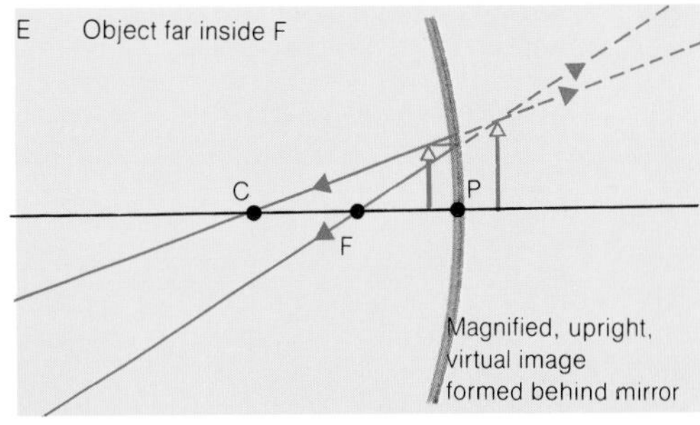

Image formation by a spherical concave mirror is shown in the ray diagrams A to E *(left)*. In diagram A, the object (green) is beyond the mirror's center of curvature, and the image (red) is diminished, inverted, and real. As the object is moved nearer the mirror, the image becomes larger, being the same size as the object when the latter is at the center of curvature and magnified when the object is inside the center of curvature (diagram B). When the object is at the focus, the image is at infinity and is therefore virtual (diagram C). Moving the object just inside the focus produces an upright, much magnified, virtual image (diagram D). And as the object is moved still nearer the mirror (diagram E), the image remains upright and virtual, but its degree of magnification diminishes the nearer the object is to the mirror. Three main classes of rays (or beams) may be used to construct ray diagrams (although any two are sufficient): rays (or beams) passing from the object and through the center of curvature, which are reflected back along their own paths; rays (or beams) running from the object and parallel to the principal axis, which are reflected through the focus; and rays (or beams) running from the object through the focus, which are reflected parallel to the principal axis. These rays (or beams) are used to construct ray diagrams because they are easy to draw and because they accurately represent what happens to light when it is reflected.

Refraction

Refraction is a change in direction that a ray of light (or any other type of electromagnetic radiation) undergoes when it passes from a transparent medium of one density into another medium of a different density. It is one of the most basic and important phenomena that affect light and is fundamental to explaining the action of lenses and prisms. Refraction is also a common phenomenon and can be observed when a straight object is partly immersed in water. The object appears to be bent toward the surface of the water and so its apparent depth is less than its real depth. This effect occurs because light from the object is refracted toward the water surface as it emerges, but the observer automatically assumes that the light has come from the object in a straight line. In this example, the light is traveling from a dense medium (water) to a less dense one (air). If, however, light enters a denser medium, the reverse effects occur and the light is refracted away from the surface.

Although refraction usually occurs when light passes from one medium to another, there is a special case in which the light does not change direction—when it enters or leaves a medium at right angles to the surface.

The multi-image photograph of a flower, taken through a multiple lens arrangement, represents the sort of view an insect might see. Most insects have large compound eyes, each with many individual hexagonal lenses (called facets). The facets refract light so that it is brought to a focus on sensory nerve endings (the arrangement of which is comparatively simple)—just as the human eye focuses light onto the retina (which, in contrast to insects' eyes, has a highly complex nerve arrangement).

Velocity and refractive index

Light travels at different velocities in media of different densities. In general, the denser the medium, the slower light travels in it. The amount by which light is refracted depends on how much its velocity changes—the greater the change, the greater is the refractive power of the medium. The absolute refractive index of a substance is a measure of the extent to which it alters the velocity of light and, therefore, of the substance's refractive power. It is defined as the velocity of light in a vacuum divided by its velocity in the medium concerned. For example, in crown glass, which is a very clear glass used in optical instruments, light travels at about two-thirds of the velocity it has in a vacuum, so the absolute refractive index of the glass is approximately 1.5. Using the same method gives a value of about 1.3 for the absolute refractive index of water. The magnitude of the absolute refractive index varies according to the color of the light. For this reason it is usually specified for yellow light.

Light rarely travels from a vacuum but usually passes between two mediums, each of which refracts light by a different amount. In this situation, the concept of relative refractive index is more useful. It equals the velocity of light in one medium divided by its velocity in the other medium. It is also equal to the ratio of the two absolute refractive indices of the mediums. As an example, for light passing from water into crown glass, the relative refractive index of the glass with respect to the water equals about 1.15 (1.5 divided by 1.3).

Most commonly one of the mediums involved is air, which has an absolute refractive index almost identical to that of a vacuum. For most practical purposes therefore, the refractive index of a substance in air can be taken as its absolute refractive index.

The laws of refraction

There are two main laws governing refraction. The first states that the incident ray, refracted ray, and normal are in the same plane (the normal is an imaginary line drawn perpendicular to the medium's surface at the point where the incident ray enters).

The second law of refraction (also called Snell's law after its discoverer, a seventeenth-century Dutch physicist) states that for a given frequency (color) of light, the sine of the angle of incidence divided by the sine of the angle of refraction is constant for a given pair of mediums. The value of this constant is, in fact, equal to the relative refractive index for the mediums concerned.

Because it is extremely difficult to measure the velocity of light, Snell's law provides a convenient method of determining refractive indices. Such determinations can be carried out using instruments called refractometers to measure angles of incidence and refraction with great precision, and then calculating the refractive index from these measurements. Another way of finding the refractive index of a substance involves measuring the real and apparent depths of an object immersed in the substance; the refractive index equals the real depth divided by the apparent depth.

Lenses and images

The most important application of refraction is the production of images using lenses—of which there are two main types: convex and concave. In general, convex lenses cause light rays to converge and thus produce real images, whereas concave lenses cause rays to diverge and so produce virtual images. Real

A convex lens refracts parallel light rays so that they converge, crossing at the focus of the lens (on the right). If similar parallel rays were incident on the right-hand side of the lens, they would cross on its left-hand side. A convex lens thus has two focal points, each the same distance from the lens's optical center but on different sides.

images can be formed on a screen but virtual images cannot—although they can be seen, as the eye can bring diverging rays of light to a focus on the retina.

One of the principal ways of determining the type of image a given lens/object arrangement will produce is to draw an accurate representation of the set-up—called a ray diagram. By convention, three classes of rays are used to locate the image, although any two rays are sufficient. The first runs from the object parallel to the principal axis (an imaginary line joining the two focal points of the lens) and is then deviated by the lens so that it passes through the focal point on the other side of the lens or diverges from the focal point on the same side. The second runs from the object, through the focus on the same side of the lens as the object and emerges from the other side of the lens parallel to the principal axis. The third ray passes from the object and through the optical center of the lens, without any deviation.

Convex lenses

Lenses that are thicker at the center than at the edge are known as convex lenses (also known as converging, or positive, lenses). There are three main types: biconvex (also called double-convex), with two convex (bulging) surfaces; plano-convex, with one flat and one convex surface; and converging meniscus (also called concavo-convex), with one slightly concave and one convex surface. Despite the difference of these shapes, all convex lenses affect light in the same way; they generally (but not always) refract rays so that they converge and form real images. The precise nature of the images produced by a convex lens is illustrated in the relevant ray diagrams (right).

Concave lenses

Lenses that are thinner at the center than at the edge are called concave lenses (also known as diverging, or negative, lenses). As with convex lenses, they are of three main types: biconcave (or double-concave), with both surfaces concave (hollowed); plano-concave, with one flat and one concave surface; and diverging meniscus (or convexo-concave), with one slightly convex and one concave surface. Again, like convex lenses, all shapes of concave lenses affect light in the same way. Unlike convex

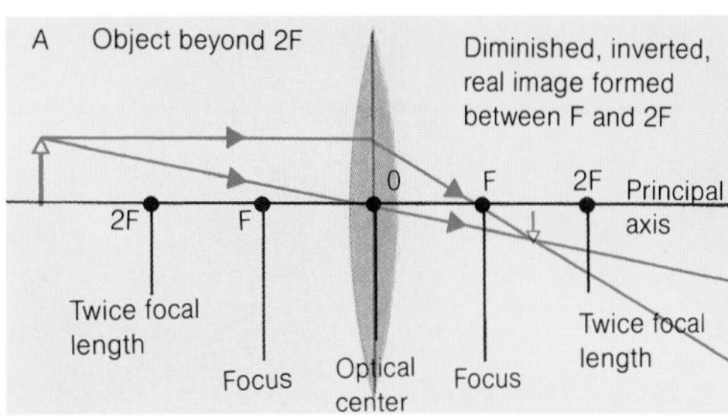

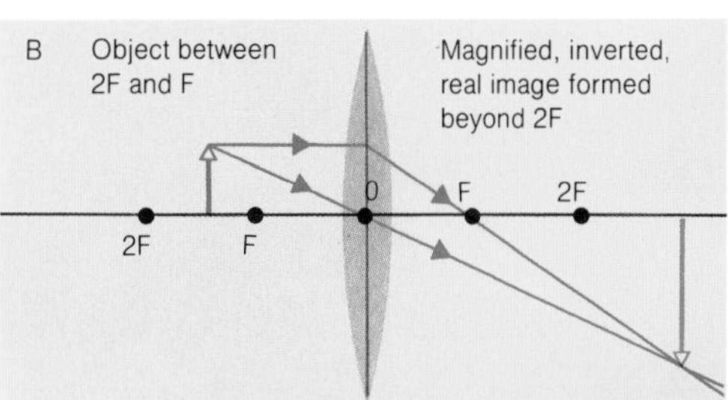

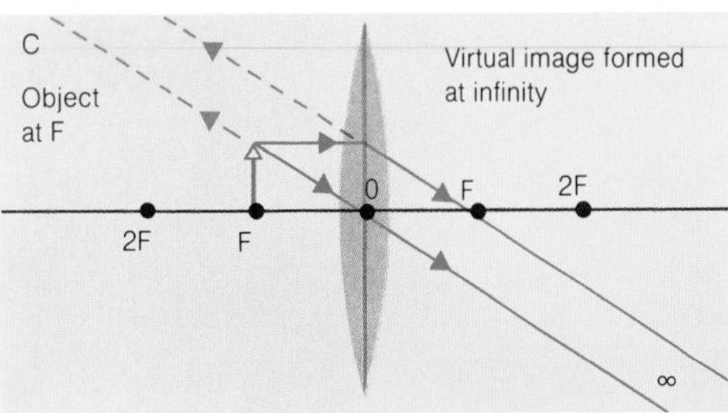

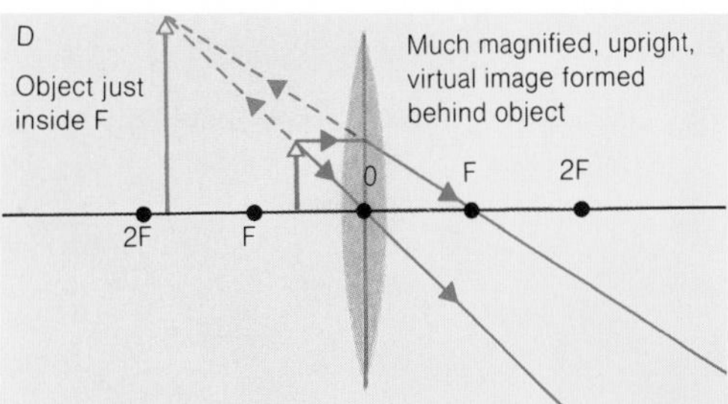

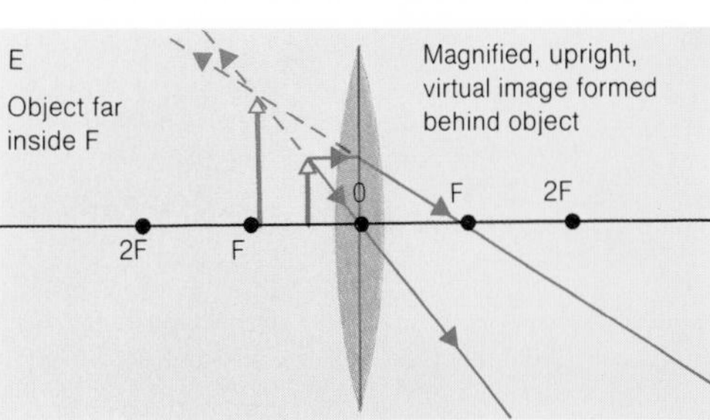

Image formation by a convex lens is shown in the beam diagrams A to E *(left)*. In A, the object (in green) is beyond 2F (that is, more than twice the lens's focal length away from its optical center); the image formed (in red) is inverted, diminished and real, and lies between F and 2F on the other side of the lens. As the object is moved nearer the lens, the image remains inverted and real but becomes larger and is formed farther from the lens. When the object is at 2F, the image is also at 2F (but on the opposite side of the lens) and is the same size as the object. When the object is between 2F and F (diagram B), the image is magnified, inverted, and real and is formed beyond 2F on the opposite side of the lens to the object. When the object is at F (diagram C), the refracted rays emerge parallel, and a virtual image is formed at infinity. Moving the object just inside F (diagram D) produces a much magnified, upright, virtual image behind the object. In diagram E the object is far inside F, and the image is basically the same as that formed in diagram D—magnified, upright, and virtual and behind the object; but the image is enlarged to a lesser degree than in D. In general, the farther the object is inside the lens's focus, the smaller is the amount by which the image is enlarged.

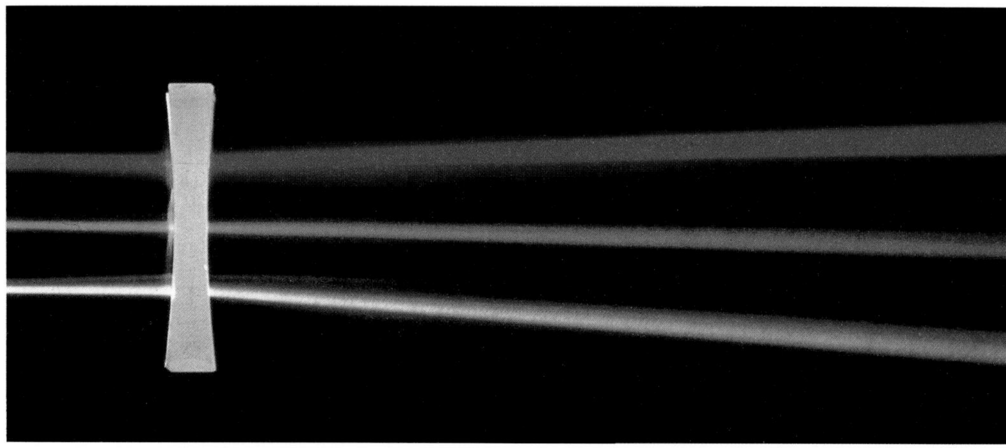

A concave lens refracts parallel light rays so that they diverge. By tracing back the refracted rays it can be seen that they appear to diverge from a single point on the same side of the lens as the incident rays (on the left in this photograph). This point is one of the lens's focuses; the second focus is the same distance from the lens's optical center as the first, but on the opposite side.

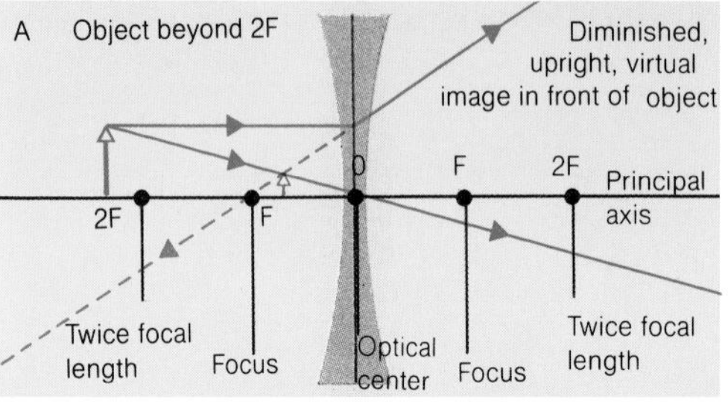

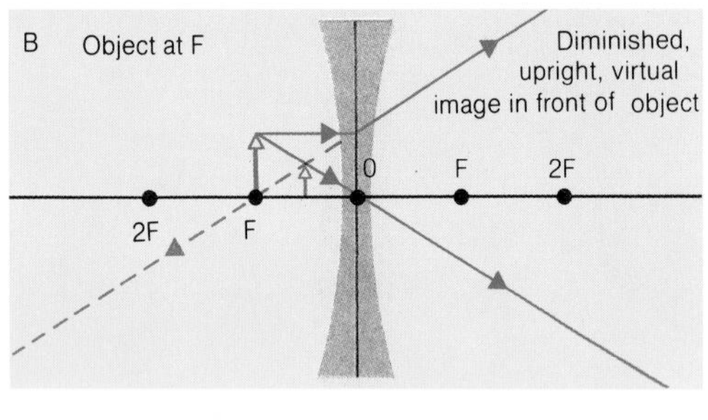

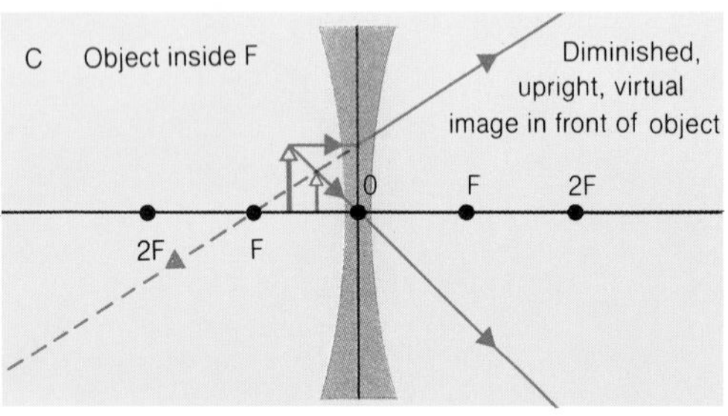

These ray diagrams illustrate image formation by a concave lens. Irrespective of the relative positions of the object and lens, the same type of image is always formed: diminished, upright, and virtual and on the same side of the lens as the object. The only difference among the images is in their degree of diminution—as can be seen by comparing the three diagrams. The closer the object is to the lens, the larger is the image.

lenses, however, concave lenses refract light rays so that they always diverge and, therefore, produce virtual images. Moreover, the image formed is always upright and diminished. The formation of images by a concave lens is illustrated in the relevant diagrams (left).

Aberrations

In optics an aberration is the inability of a lens (or mirror) to form a sharp image of a clearly-defined object. All lenses produce blurred images when they are not focused correctly. This can be remedied simply by moving the object, lens, or viewing screen (or any combination of the three); the characteristic feature of aberrations is that they are present even when a lens is properly focused.

There are two main types of aberrations: spherical and chromatic. With spherical aberration, light rays refracted near the edge of a lens are not brought to a focus at the same point as those passing through the center of the lens. Astigmatism is a similar defect in which, from a point object, a lens forms two short line images at right angles to each other and in different planes of focus. Both defects can be eliminated or minimized by grinding the surfaces of a lens to make them aspherical (that is, so that the curvature varies across the lens surfaces). Decreasing the lens aperture

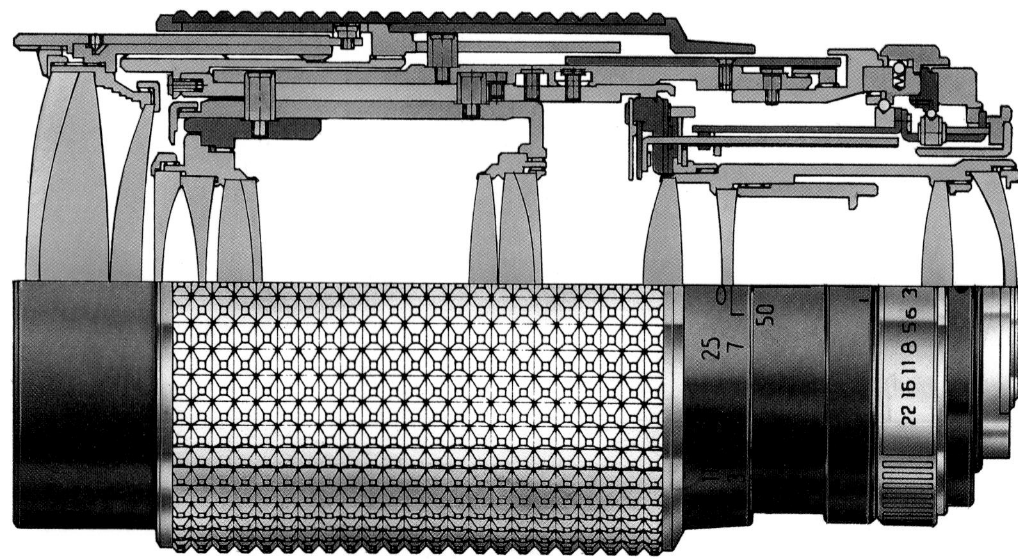

Many optical instruments have complex designs with numerous individual lens elements—for example, the zoom lens for a camera consists of 14 elements, of which 7 are convex (shown in blue) and 7 are concave (shown in turquoise-blue). Such complex designs are necessary because each individual lens element suffers from aberrations, and other lenses are required to correct these and improve image quality.

(by stopping down the diaphragm of a camera lens, for example) so that light passes through only the center of the lens reduces spherical aberration but increases blurring of the image by diffraction at the edge of the diaphragm. For lenses with large apertures relative to their focal lengths (most camera lenses, for example), image deterioration caused by spherical aberration is usually more serious than that caused by diffraction; for lenses with small relative apertures (as in refracting telescopes) the reverse is usually true.

Spherical aberration also affects mirrors. It can be reduced by decreasing the aperture of the mirror or by using a paraboloidal-mirror, which is curved so that all incident rays parallel to the principal axis are reflected through the focus.

The other principal type of aberration, chromatic aberration, results in an image being surrounded by colored fringes. It occurs because a simple convex lens refracts different colors of light by different amounts—a phenomenon called dispersion. (Mirrors do not produce chromatic aberration.) Blue light is refracted more than red light and so is brought to a focus nearer the lens.

Chromatic aberration can be eliminated for two colors (and reduced for all others) by using a special type of lens called an achromatic doublet (or achromatic objective lens), which consists of two different lenses cemented together. One of the lenses is converging (convex) and the other diverging (concave), and they are made of different types of glass (such as crown and flint) so that the dispersion caused by one is canceled by the other.

Total internal reflection

When light is refracted as it passes from one medium to another, some is reflected at the interface—which is why reflections can be seen in glass windows, for example. Usually only a small amount of light is reflected, but this is not always the case when light passes from a dense to a less dense medium. If the light hits the interface at a large enough angle, all of it is reflected back into the denser medium—a phenomenon called total internal reflection. It can be observed when underwater and looking obliquely toward the surface, which if the viewing angle is great enough, acts like a mirror.

The smallest angle (to the normal) at which total internal reflection occurs is called the critical angle, and it depends on the relative refractive index of the mediums—the greater the refractive index, the smaller is the critical angle. For diamond (which has a refractive index of approximately 2.4) it is about 24°, for crown glass (refractive index 1.5) about 42°, and for water (refractive index 1.3) it is about 50°.

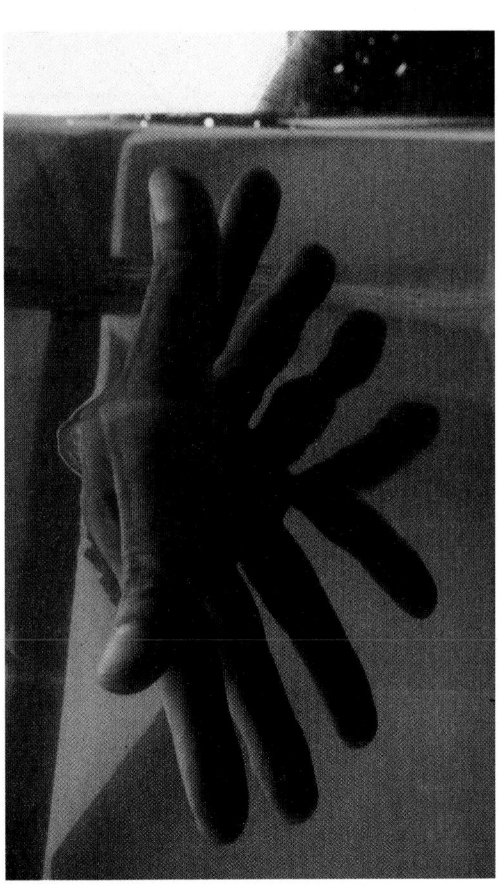

Total internal reflection is vividly demonstrated in this photograph of a hand partly immersed in water and its reflection at the water/air interface. Total internal reflection can occur only when light travels from a dense to a less dense medium (from water to air in this example) and when the angle of incidence is greater than the critical angle for the medium concerned. For light traveling from a dense medium into air, the critical angle can be determined from the equation $\sin c = 1/n$, where c is the critical angle and n is the absolute refractive index of the medium; in the case of water, $\sin c$ equals 0.77 ($\frac{1}{1.3}$), which using sine tables, gives a critical angle of about 50°.

Twin rainbows arc over the bleak landscape of Death Valley in California. A rainbow is a spectrum, containing all the colors that make up white light. In the outer, secondary rainbow, the order of the spectral colors is reversed, for the reason explained in the diagram below. Very rarely a third, tertiary rainbow can also be seen.

Color and spectra

White light is made up of a mixture of different colors—the colors of the visible spectrum—which each have different wavelengths (and frequencies). The spectral colors are usually listed as red (the longest wavelength), orange, yellow, green, blue, indigo, and violet (the shortest wavelength). But there are no definite boundaries between them; they blend together like the colors of the rainbow to form a continuous spectrum.

Dispersion

White light splits into its constituent colors (wavelengths) when, from a medium of one density, it passes at an angle into a medium of another density—from air into glass, for example. This phenomenon is called dispersion. The waves of different wavelengths travel at slightly different speeds in the second medium, and so are refracted by slightly different amounts. As a result, the dispersed white light emerges from the medium with its component colors separated into a spectrum.

A natural example of such dispersion occurs in a rainbow, which is seen when sunlight is refracted (and dispersed) by raindrops. As well as being dispersed inside the drops, the rays of light are also internally reflected toward the ground.

Different colors of light are refracted by different amounts and thus emerge from a drop at different angles, red at 42° through to violet at 41°. An observer sees red at the top of a rainbow and violet at the bottom because when a drop is at the correct angle to reflect

In a primary rainbow, red appears at the top of the arc and violet at the bottom. The colors occur because sunlight entering a raindrop (A) is refracted and dispersed (separating the colors), reflected internally, and refracted again on leaving the drop. Red rays reach the observer at a steeper angle than do violet ones. In a secondary rainbow, there are two internal reflections in each raindrop (B), causing a reversal of the spectral colors.

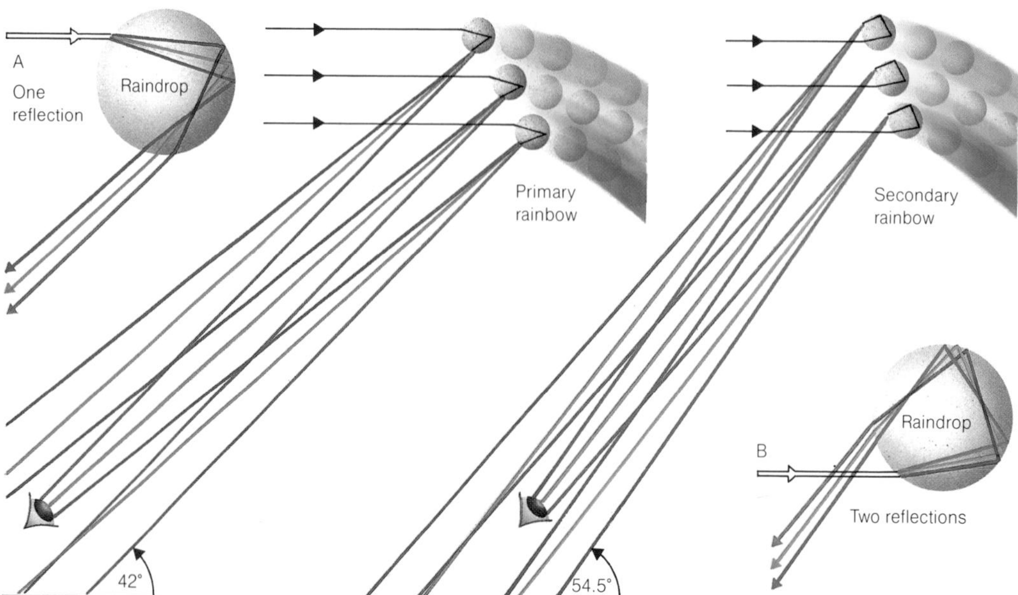

red light to him, the violet light it also returns passes above the observer's head. Similarly, another drop at a lower height is at the angle necessary to return violet light but the red light it reflects at a slightly steeper angle passes below the observer's eye level.

The thickness of a rainbow is a fairly constant 2° of arc, although its apparent location is related to the observer's own. Moreover, a rainbow is always an arc of a circle, but how much of the circle is seen depends on the height of the raindrops, the altitude of the observer, and the angle of the sun above the horizon. Another fainter bow (called a secondary rainbow) can sometimes be seen separated from and outside the first. It results from an extra internal reflection of the light rays within the raindrops, and for this reason the order of the spectral colors is reversed, with violet at the top and red at the bottom.

The ordinary blue of the sky results from a phenomenon called scattering. Molecules of gas in the atmosphere scatter sunlight by diffraction, particularly the shorter wavelengths at the blue end of the spectrum; the result is a blue sky. Higher in the atmosphere there are fewer gas molecules, and so less scattering takes place and the sky looks black. At sunrise and sunset, the light of the sun reaches an observer on earth after passing through a considerably greater depth of atmosphere. The blue component of the light is scattered through large angles, leaving only the longer wavelengths. The result—particularly if the air contains dust particles, which increase scattering—is that the sun and sky around it appear red.

Prisms

One method of dispersing light in the laboratory uses a prism, a triangular block of glass or plastic. More than 300 years ago Isaac Newton demonstrated how a beam of white light angled at the side of a prism is refracted and dispersed to form a continuous spectrum. A second prism and a lens can "collect" the dispersed spectral light and recombine it to form a beam of white light. Spectrometers use prisms in this way.

If a beam of light enters a prism at right angles to one of the faces, it is neither refracted nor dispersed. This effect is employed in various internal reflections in prisms, with the inside of the faces of the prism acting as mirrors. In prismatic binoculars, for instance, prisms reinvert the images already inverted by the objective lenses so that the final images are seen the right way up. The pentaprism in the viewfinder of a single-lens reflex camera performs a similar function.

The spectrometer

In a spectrometer, light passes first through a collimator consisting of a narrow split and a convex lens, which together produce a narrow beam of light. The beam is angled at a prism, which refracts and disperses the beam into its spectral colors. From the prism parallel beams of different colors emerge, which then pass through a second convex lens that brings

A triangular prism disperses white light into a spectrum because the shorter-wavelength colors at the violet end are refracted more than the longer-wavelength red component. The old ornament *(left)* creates a series of spectra as sunlight is split into rainbow colors by the crystal glass pendants.

The sun appears larger and redder at sunrise and sunset than at noon. At noon, it is overhead, and its rays pass directly through the atmosphere; part of the blue light is scattered, and so the sky appears blue and the sun slightly yellow. At sunset the rays have to pass through a much greater thickness of the atmosphere. They are refracted, so that the sun's disk appears to be larger. Nearly all the blue light is scattered, and the sun and the sky around it take on a reddish tinge.

A spectrometer uses a glass prism to split a parallel beam of white light into its component colors, which can be viewed through a low-power telescope or recorded on photographic film. A source of white light produces a continuous spectrum, whereas a single chemical element heated to incandescence gives rise to an emission spectrum consisting of a series of bright lines. The same element in gaseous form in front of a source of white light creates an absorption spectrum of dark lines (Fraunhofer lines) on a continuous background.

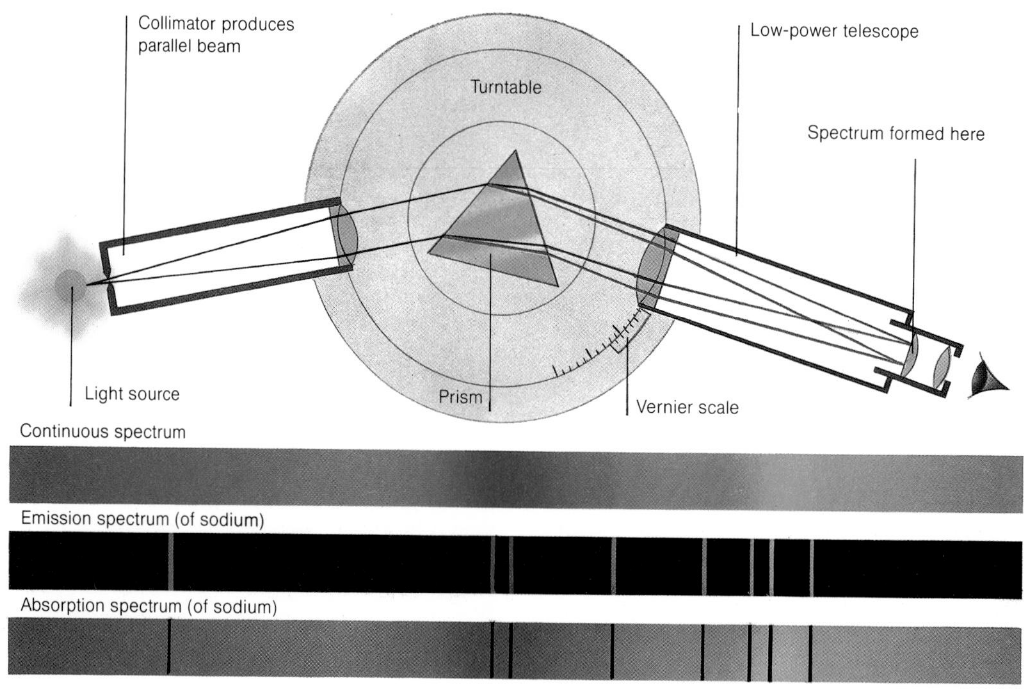

them to a focus. The resultant spectrum can be viewed through a low-power telescope, recorded on photographic film, or registered by some kind of detector. A diffraction grating can also be used to produce spectra, as explained in the following section.

Emission spectra

A spectrometer reveals that many sources of light (unlike the sun) do not produce a continuous spectrum. The light from a sodium lamp, for example, produces a spectrum dominated by a pair of closely-spaced yellow lines. Discontinuous spectra of this type are characteristic of individual chemical elements, which emit light when heated to incandescence. The spectral lines result from the excitation of electrons in the atoms. Excited electrons gain energy and "jump" to higher orbits; when they return to their original (ground-state) orbits they release the extra energy by emitting light of a particular wavelength. The resulting emission spectrum is a unique "fingerprint" of the element concerned, and a mixture of elements can be rapidly analyzed by studying their combined spectral lines using a spectrometer. This technique is particularly useful in chemical analysis and astronomy.

Absorption spectra

Just as light is emitted at specific wavelengths by a substance heated to incandescence, it is also absorbed at the same wavelengths by vapor of the same substance at a lower temperature. The spectrum that corresponds to the wavelengths absorbed, rather than emitted, is called an absorption spectrum. It usually takes the form of a sequence of dark lines or bands within a continuous light spectrum. As with emission spectra, absorption spectra

Color mixing produces different results, depending on whether lights or pigments are used. With lights (A), the three primary colors (red, blue, and green) mix in pairs to produce the secondary colors magenta, cyan, and yellow; all three primaries mix to give white. With pigments (B), the primaries—magenta, cyan, and yellow—combine in pairs to give blue, red, and green as secondaries. Mixing all three primary pigments in the correct proportions theoretically gives black.

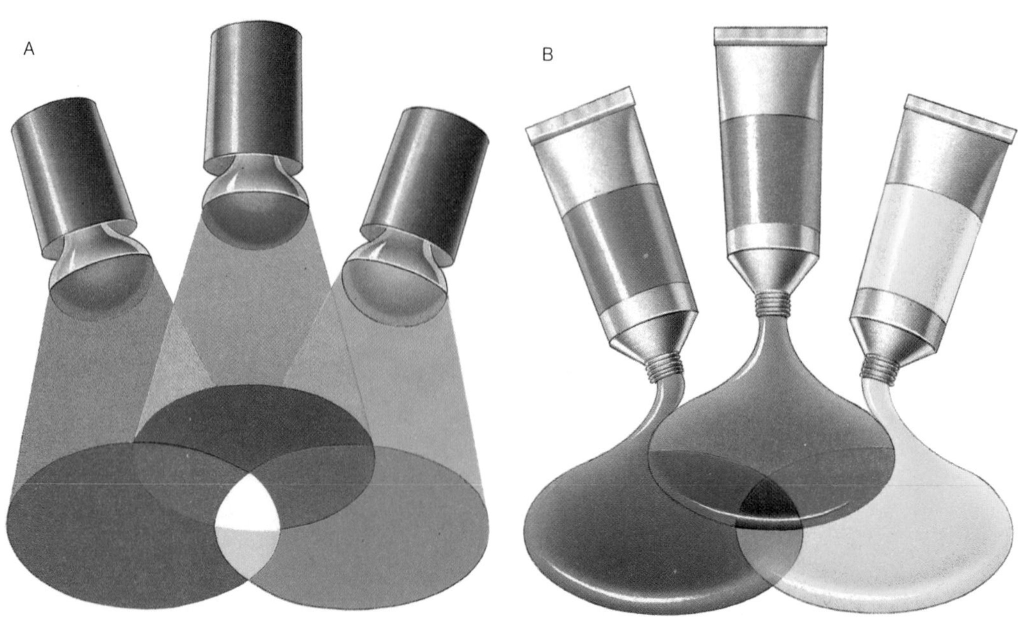

can be used to detect individual elemental constituents in the vapor by means of the dark lines, which are known as Fraunhofer lines after their discoverer Joseph von Fraunhofer.

The sun's absorption spectrum represents the chemical elements present in its outermost layers, which absorb some of the energy radiated from deeper within. (It is thus the hot inner region that produces the continuous emission spectrum.) So, for instance, one pair of dark lines in the yellow part of the absorption spectrum indicates the presence of sodium vapor; other lines show that the gas helium—a product of the energy-generating thermonuclear fusion of hydrogen deep within the sun—is also present. This type of analysis of the gases of the sun led to the initial discovery of the element helium.

Color and coloration

A spectrum formed by the dispersion of white light can be "collected" again to reconstitute white light by using two prisms. But it is not necessary to mix together the entire range of spectral colors in order to obtain white light—red, green, and blue light are a sufficient combination. For this reason they are known as the three primary colors of light. Any other color of light can be made using two or three of these primaries in the correct proportions.

Combining any two primaries in equal amounts results in the following: red and green combine to give yellow; blue and green give a sort of peacock-blue known technically as cyan; and red and blue produce magenta. These are the secondary colors of light. Because each of the secondaries is made up of two primaries, further combination with an equal amount of the third primary again results in white. Thus yellow (a secondary color of light) combines with blue (the third primary color) to give white, and so on.

Such mixing of colors in light is an effect of perception by the individual observer; the various wavelengths can be measured precisely, but the way the human eye and brain "see" a particular color may vary from person to person. We can see colors because the normal eye is sensitive to the wavelengths of the three primary colors, and sensitive also to the amount of each that is received. It is the brain that "mixes" the light accordingly to give the perception of other colors.

With sources of colored lights—stage lighting or a color television screen, for example—the combination of primary colors is obvious, and it is easily demonstrated that red and green light do combine to make yellow. This kind of color mixing is called addition, because the colors add directly together.

Mixing colored pigments, however, does not produce the same results as mixing colored lights. Blue paint and yellow paint, for example, mix to give green. This occurs because the colors of paints and inks are perceived as reflected light, not as direct light from a light source. The perceived color is the only one not absorbed by the surface from which it is reflected.

So when white light illuminates a green surface, for instance, the surface absorbs the red and blue primary colors in white light and reflects only the green. Because the absorbing surface "takes away," or subtracts, all the other colors from the light illuminating it, this process of color formation is known as color subtraction.

With pigments, such as those in paints, inks, and dyes, there are again three colors that are known as primary colors and which can be mixed to give all other colors. But these are not the same as for light. The primary colors of pigments are magenta, cyan, and yellow (the secondary colors of light). The corresponding secondary colors in pigments are produced by mixing the primaries: magenta plus yellow gives red, magenta plus cyan gives blue, and cyan plus yellow gives green. Hence, mixing any two pigment primaries gives a light primary. Mixing all three primary pigment colors subtracts all colors in the light, as a result of which no light is reflected at all and the color produced is black.

Pointillism *(above),* a Neo-Impressionist technique developed by **Georges Seurat,** the nineteenth-century French painter, employed small dots of bright color to create tonal effects. A magnifying glass will reveal that the color photographs in this book are printed using a basically similar principle. Dots of light can also combine to create images, and this is the basis of one of the common color television systems shown in close-up *(left).*

A peacock's feathers shimmer with iridescent hues as the bird performs an elaborate courtship display to attract a female. The colors result from diffraction effects as light waves are reflected and diffracted by the regular array of microscopic barbules along the individual strands of the feathers in the bird's tail and breast.

Diffraction, interference, and polarization

Diffraction and interference are effects that can occur with all wave motions; and with light waves they produce several interesting and important results. The brilliant colors of peacock feathers and some butterflies' wings, for example, result from diffraction and interference. The wave nature of light also gives rise to the phenomenon of polarization, another effect with various important applications in science and technology.

Diffraction

The change in the direction of waves that occurs when they pass the edge of an obstacle or through a narrow opening is called diffraction. The waves bend around the edge by an amount related to their wavelength, with longer waves being diffracted more than shorter ones. The effect is not ordinarily noticeable with light rays because their wavelengths are so small. But it is readily apparent with sound waves, which have much longer wavelengths and will, for example, bend over the top of a wall so that someone speaking out of sight beyond the wall can easily be heard.

Diffraction occurs because waves travel outwards from a source in expanding spherical wavefronts. At a large distance from the source, they act as if they are straight because the radius of the sphere is so large. But if part of a wave is blocked by an obstacle, the wavefront at the edge of it then expands in a sphere, just as if the edge were a new source of waves.

Interference

This effect occurs when two waves meet. If they have exactly the same wavelength and both waves move so that the crests of one coincide with the crests of the other, they reinforce each other and result in a combined wave of greater energy. But if the two waves combine in such a way that the crests of one coincide with the troughs of the other, they cancel out.

With sound waves, interference at slightly different wavelengths causes beats, which are combined low-frequency waves. With light, interference can be made to produce beats by combining light rays with rays reflected from a moving mirror, whose wavelength changes as a result of the Doppler effect. However, light rays, unlike sound, can easily be made to interfere so that complete reinforcement or cancellation occurs, giving rise to patterns of light and dark bands. Spectacular color effects can also be created because interference is related to wavelength. It is essential for the light waves to be coherent—that is, to have exactly the same wavelength and to be exactly in phase, so that the waves are constantly emitted at the same intervals of time. In practice this is normally achieved by using a single source of monochromatic light to produce the interfering waves.

Diffraction gratings

Diffraction and interference act together to produce unusual effects as light passes through small openings or around small obstacles. Diffraction causes the waves to spread out from the edges of the opening or obstacle, and interference occurs where the waves meet. Depending on the distance from the edges, the waves either reinforce or cancel each other, producing patterns of light and dark.

If a small disk is illuminated by a distant source of light, faint circular fringes extend into its shadow as light waves diffract around the edges of the disk and interfere. There is a light spot at the center of the shadow because the interfering rays reaching there from the edges of the disk travel the same distance and reinforce each other.

A similar effect occurs with a narrow slit that is illuminated by a single source of light. A bright strip forms opposite the slit, and parallel light and dark fringes extend into the shadow on each side.

These effects are more noticeable when the dimensions of the slit or any opening are simi-

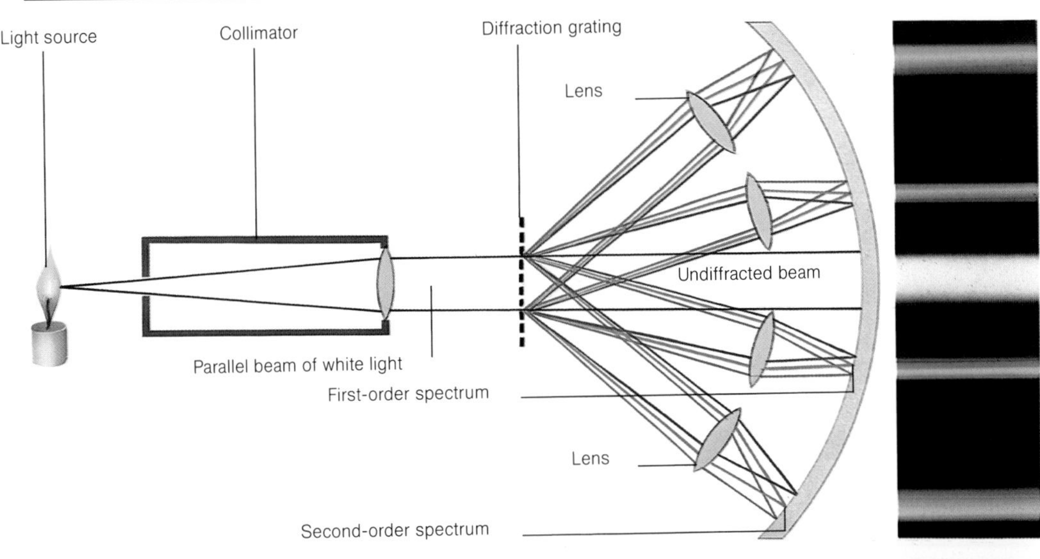

A diffraction grating splits white light into a series of spectra because it diffracts longer-wavelength red light more than shorter-wavelength violet light. The first-order spectra—one on each side of the central, undiffracted beam—are narrowest and brightest. At wider angles there are third-order and even fourth-order spectra, which may overlap. This photograph of a candle flame *(below)*, taken through a grating, clearly shows the first-order spectra.

lar to the wavelength of the light, which is less than one ten-thousandth of an inch or centimeter. They are most marked if the light passes through a series of openings, because the diffraction and interference effects produced by adjacent openings coincide. The phenomenon can be observed by looking at a lamp or candle flame through a fine mesh, such as a handkerchief. Multiple images of the light source can be seen as rays from it are diffracted around the sides of the openings in the mesh, and interference produces regions of light and dark that result in the formation of several images to the sides of the light source.

These effects are utilized in diffraction gratings. They consist of clear glass or plastic plates ruled with a mesh of extremely fine lines, typically 12,000 to 25,000 per inch (5,000 to 10,000 per centimeter). Light rays passing through the gaps between the lines in the grating are diffracted to a precise degree, resulting in the formation of multiple images to both sides of the central image. The angle by which the images are displaced from the center are related to the wavelength of the light, and diffraction gratings provide a very accurate means of determining wavelength.

Furthermore, because the degree of displacement is related to wavelength, a diffraction grating produces several spectra from a non-monochromatic source. Red rays, being of longer wavelength, are diffracted more than blue rays, and for this reason the order of colors is reversed compared with a spectrum produced by a glass prism (in which blue rays are refracted more than red ones).

Diffraction gratings are generally preferred to prisms for spectrographic studies. This is because a grating distributes the wavelengths evenly—unlike a prism—and gives greater resolving power for separating closely-spaced spectral lines.

A reflection grating, which consists of fine lines ruled on a mirror, can be used for wavelengths (such as infrared and ultraviolet) that are absorbed by glass. Such a spectrum can be seen in the fine grooves of a long-playing record or video disk, which act like a reflection grating.

The finely spaced grooves of a video disk *(below)* can act as a reflection grating and generate a series of spectra from reflected white light. In physical terms, the colors are produced in much the same way as those of the peacock's feathers illustrated on the opposite page.

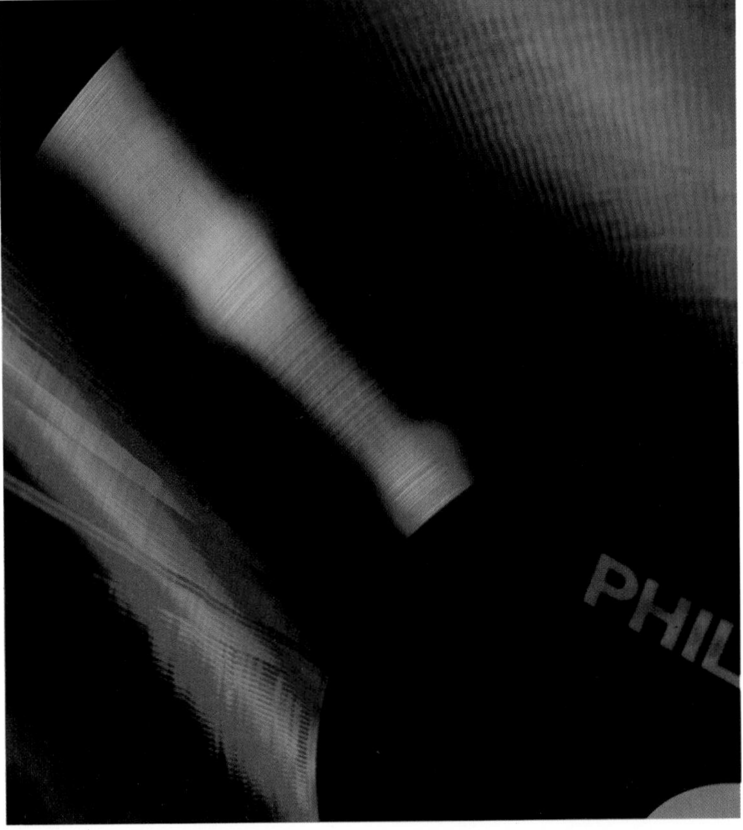

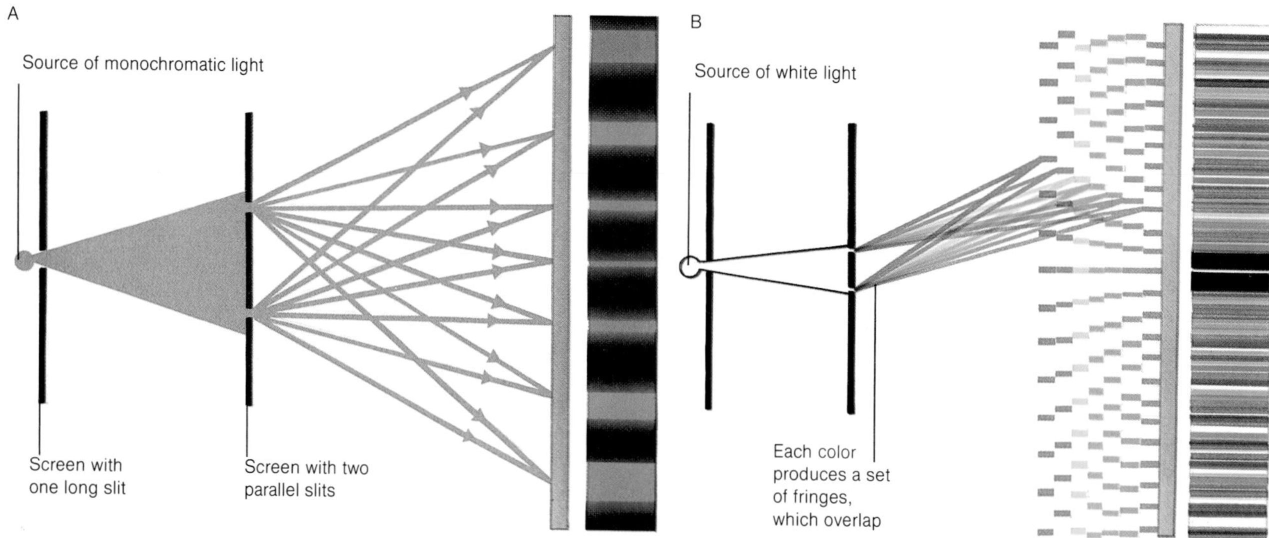

Interference fringes result when monochromatic light passes through two closely-spaced parallel slits (A, *above*). Where the path lengths of two diffracted rays differ by a whole number of wavelengths, the rays reinforce each other, producing a bright fringe; elsewhere, the rays cancel, producing darkness. The various wavelengths of white light are diffracted by different amounts (B), and each color has its own set of fringes that overlap each other.

Interference effects

Interference can easily be demonstrated by illuminating two narrow parallel slits spaced close together. The light waves emerging from the slits interfere and produce a pattern of fringes on a screen. The wavelength of the light can be determined from the spacing of the fringes.

Interference also causes optical effects to occur between two close surfaces. If a convex lens is placed on a flat glass plate and illuminated by a monochromatic light source (such as a sodium lamp), a series of concentric fringes can be seen extending outward from the center of the lens. They are called Newton's rings. What happens is that some of the light is reflected from the lower surface of the lens, while the remainder of it passes through to be reflected from the glass plate. The waves then combine and interfere and reinforce or cancel each other depending on how far the second ray travels between the lens and the plate. This distance in turn varies with the radius of curvature of the lens, and so a series of concentric light and dark interference fringes result. Newton's rings are used to test the quality of lenses and optical instruments.

A similar interference effect occurs in thin films such as films of oil on water, in which shimmering spectral colors can be seen. In this situation, interference involves light rays reflected from the top and from the bottom surfaces of the film, and the different distances traveled by the rays cause different wavelengths to reinforce or cancel, producing the color effects.

Oil films on water give rise to spectacular color fringes. They are caused by an interference effect resulting from differences in path lengths between rays of light reflected from the upper and lower surfaces of the thin oil film. In this respect, the phenomenon is similar to the formation of Newton's rings, explained in the diagram at the top of the opposite page. Colored fringes are also formed in thin soap films and soap bubbles.

Polarized light

Another effect resulting from the wave nature of light is polarization, which occurs because light is a transverse wave motion. The planes of vibration of the electric and magnetic fields that make up a light ray are normally oriented at random. When light is polarized, the electric and magnetic fields are each made to vibrate in one particular plane. The electric plane is considered to be the plane of polarization.

Light can be polarized in various ways. It sometimes happens when light is reflected from a smooth surface—of glass or water, for example. The planes of vibration of the light may be resolved into two components, one parallel to the surface and one at right angles to it. The former vibrations are reflected by the surface as polarized light, whereas the latter vibrations are refracted into the surface or absorbed by it. Polarization by reflection is most marked at an angle of incidence that is related to the refractive index of the surface (for glass this angle is about 57°).

In practical applications, light is polarized using special prisms or other crystalline polarizing materials. In these, the rows of atoms or molecules absorb the light rays that vibrate in some planes and transmit others. A Nicol prism is a single, large crystal of Iceland spar,

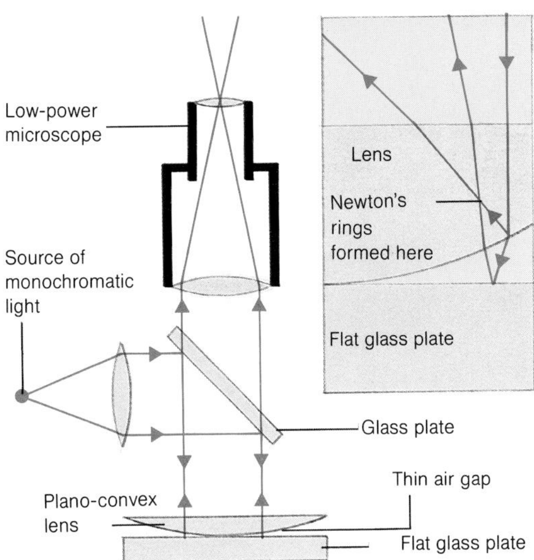

Newton's rings are a series of concentric interference fringes. The diagram *(far left)* shows an experimental arrangement for creating them using a plano-convex lens in contact with a flat glass plate. With monochromatic light, fringes are formed where rays reflected from the lens's lower surface interfere with those reflected from the glass plate beneath it. White light produces colored fringes, as shown in the photograph *(left)*.

a form of calcite (calcium carbonate), which has been cut at an angle and rejoined with Canada balsam cement. Light passing into the crystal is refracted by it into two polarized rays. One ray is internally reflected at the Canada balsam join, while the other polarized ray passes through the junction and emerges from the opposite face of the prism. As a Nicol prism is rotated, so too is the plane of polarization of the transmitted light.

Many crystals, such as quartz, and some liquids, such as sugar solution, rotate the plane of polarized light passing through them. For solutions, the angle of rotation depends on concentration and the length of solution through which the polarized light passes. This effect is put to use in an optical instrument called a polarimeter, which measures the concentrations of sugar solutions and other optically active liquids. When the Nicol prism is rotated so that its plane of polarization is at right angles to that of the solution, all light through the instrument is extinguished. The angle of rotation of the prism is a measure of the concentration of the optically active substance.

Polaroid® sheets are made of a plastic that contains needlelike crystals of iodoquinone sulfate, all aligned in the same direction. These uniformly oriented crystals transmit only light that vibrates in one particular direction and in this way polarize light. One effect of this is to reduce glare, which consists of reflected light that is mainly polarized, and so Polaroid sheet is used in sunglasses and camera filters.

An important application of polarized light is in the stress analysis of transparent materials. Crossed Polaroids (two Polaroids at right angles to each other) extinguish all transmitted light. But when a transparent object under stress is placed between them, it causes some rotation of the plane of polarized light; the Polaroids no longer extinguish each other and some light passes through. The areas of stress in the transparent specimen then show up as complex colored patterns.

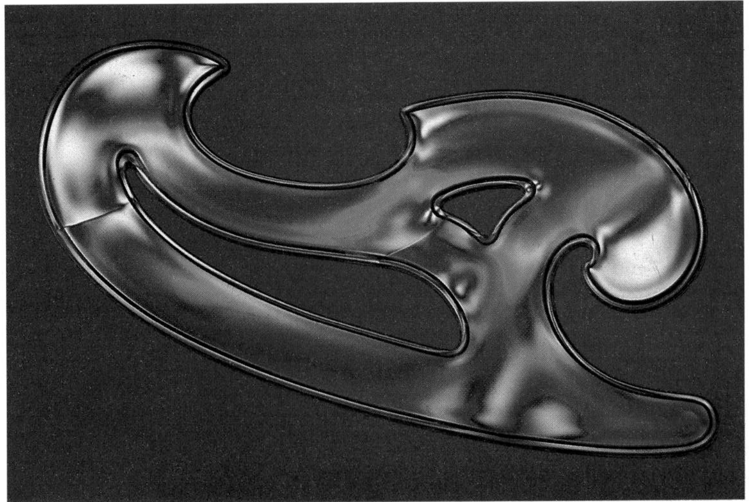

Stress patterns in a plastic drawing template are revealed by placing it between two crossed Polaroids, which normally extinguish all light passing through them. The template has rotated the plane of polarized light, allowing some light to pass through and be photographed as spectral colors.

Luminescence, fluorescence, and lasers

Nearly all light-emitters are extremely hot; examples include the sun and other stars, the filament of an electric lamp, and a candle flame. In physical terms they are all incandescent light sources. But there is another smaller group of "cold" sources that emit light by the phenomenon of luminescence—fluorescent lamps and the luminescent numerals on a clock, which glow in the dark are examples. And a third type of light source—also not incandescent—is the laser.

Many natural substances fluoresce when exposed to ultraviolet radiation. The five objects illustrated below include fluorite crystals *(top left)*, opal *(bottom left)*, amber *(top right)*, ruby, and ivory *(carved into an ornate hollow ball)*. In the atoms of each of them, the energy of ultraviolet radiation excites outer electrons into higher orbits, and as they decay back to their ground-state orbits they emit the excess energy as visible light.

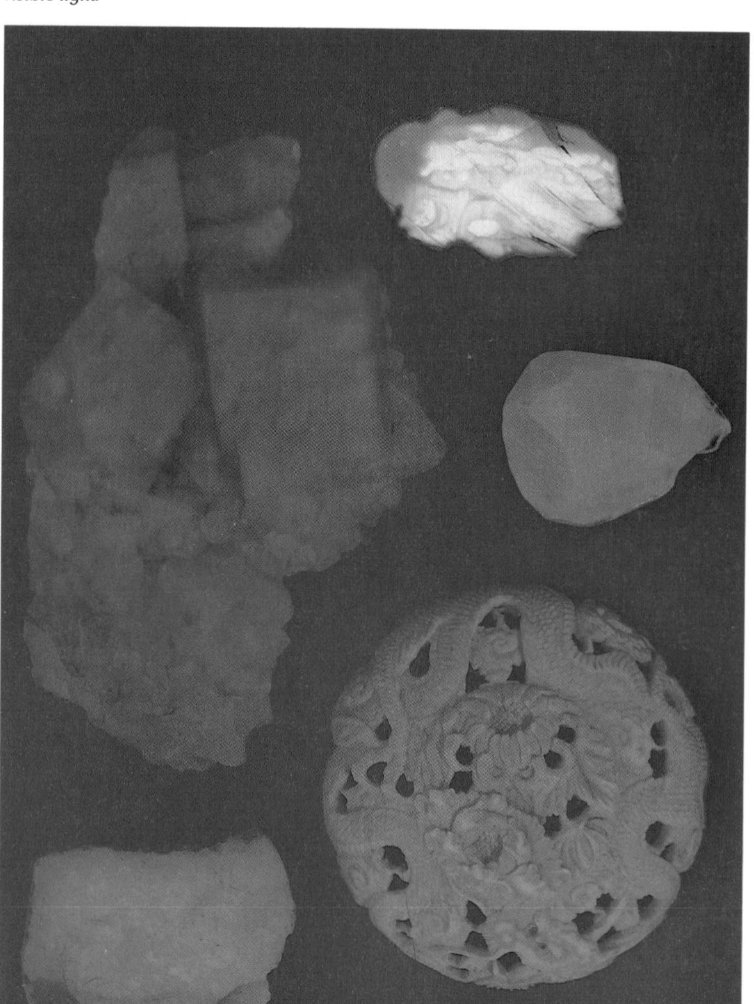

Luminescence

Both incandescence and luminescence have the same fundamental origin, which results from energy changes within the atoms of the radiating object. Electrons within the atoms acquire energy, are "excited" to higher energy levels, and in decaying again to their ground states emit electromagnetic radiation in the visible light range. In an incandescent object, the exciting energy is externally applied heat; in a luminescent object, it derives from within the object itself or from an external source other than heat. There are two kinds of luminescence. The type that ceases when external excitation stops is known as fluorescence; the type that persists even in the absence of external excitation is called phosphorescence.

Sources of energy other than externally applied heat include chemical reactions, which give rise to chemiluminescence. White phosphorus, for example, glows when exposed to air at room temperatures. The exciting energy is provided by an oxidation reaction with atmospheric oxygen that causes the phosphorus to "burn" slowly. Chemical luminescence in living organisms, for example, fireflies, glowworms, and some fungi, is called bioluminescence and results from chemical reactions between complex organic molecules within the organisms.

Fluorescence

Certain bacteria and some minerals and organic chemical compounds that readily glow when irradiated by ultraviolet light are examples of fluorescence. Some washing powders contain fluorescent compounds called optical brighteners that make fabrics appear brighter and cleaner—and also cause freshly laundered white articles to fluoresce a bright violet under ultraviolet light.

Phosphorescence

The cold, steady light given off by certain fungi is an example of phosphorescence: it persists after the removal of the initial stimulus. An inorganic example is the glow of one type of luminous paint, for which the energy source is daylight, which is re-emitted as green light, the glow fading as the "stored" energy diminishes. The light emitted by a television screen is also phosphorescent. Its energy source is a stream of electrons fired at the inside of the screen, which is coated with inorganic substances called phosphors. Because the glow excited in the phosphors persists even only momentarily after the electron beam ceases, it is regarded as phosphorescence—and not as fluorescence, despite its common description as a "fluorescent screen."

Incoherent light

In a fluorescent lamp, atoms of mercury vapor are bombarded by electrons and emit ultraviolet light. This radiation in turn excites a phosphor coating inside the tube that emits visible light. But not all the mercury atoms are hit by electrons at the same instant. Thus, some are in an excited state while others are in the ground state. The overall light output is therefore a randomly discontinuous series of extremely brief pulses. This kind of light, which is also produced by all normal incandescent sources, is said to be incoherent.

Waves of light from two such sources do not stay in phase (in step) with each other for any length of time. Where they meet, they can-

An angler fish, which lives in dark tropical seas at a depth of about 13,000 feet (4,000 meters), attracts its prey with luminescent "lights" at the ends of its dorsal lure and lower barbels. Other bioluminescent creatures include glow-worms and fireflies.

Luminescent numerals on a watch dial *(below)* glow in the dark. The light emitted by phosphorescent paints gradually fades, and they have to be periodically "recharged" by exposure to light. Fluorescent paints contain traces of radioactive elements, such as radium, and glow continuously without fading.

not, therefore, produce the regular pattern of reinforcement and cancellation that is typical of what is called the interference of light waves, even though the wavelengths may be more or less identical and the light thus monochromatic.

Coherent light: the laser

An interference pattern results when two sources emit light of the same wavelength and in phase; such light is called coherent. It is produced, by a process similar to fluorescence, in a laser—a name that denotes *L*ight *A*mplification by *S*timulated *E*mission of *R*adiation. Stimulated emission can be best understood by considering light as a stream of electromagnetic radiation particles (photons)

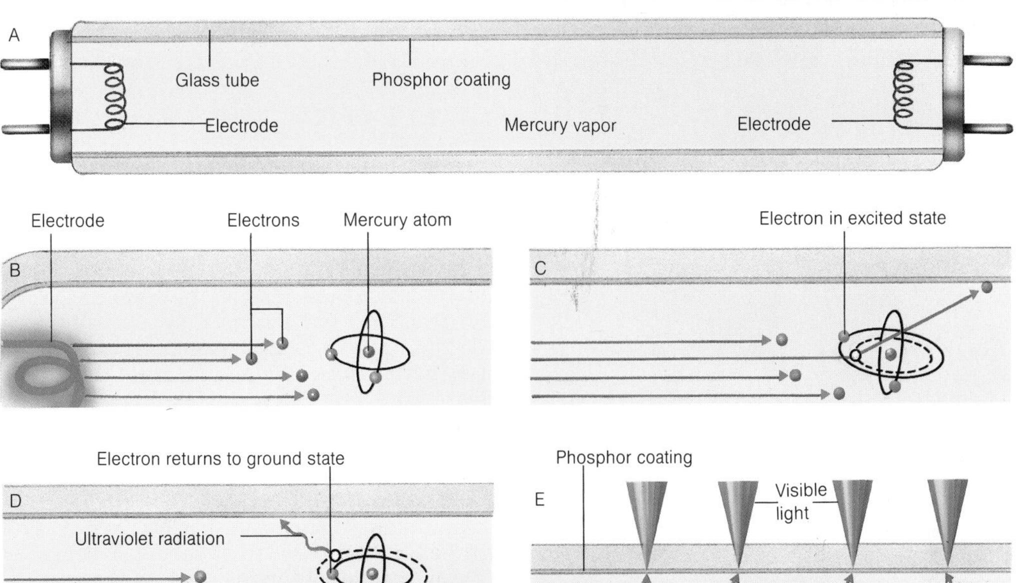

A fluorescent lamp (A) contains mercury vapor at low pressure. Electric current heats the electrode (B) so that it emits electrons, which stream through the near-vacuum of the tube. When an electron interacts with a mercury atom (C), it makes an outer "mercury" electron jump to a higher orbit. On returning to its ground-state orbit (D), this electron emits its excess energy as ultraviolet radiation, which stimulates the phosphor coating inside the tube to emit visible light (E).

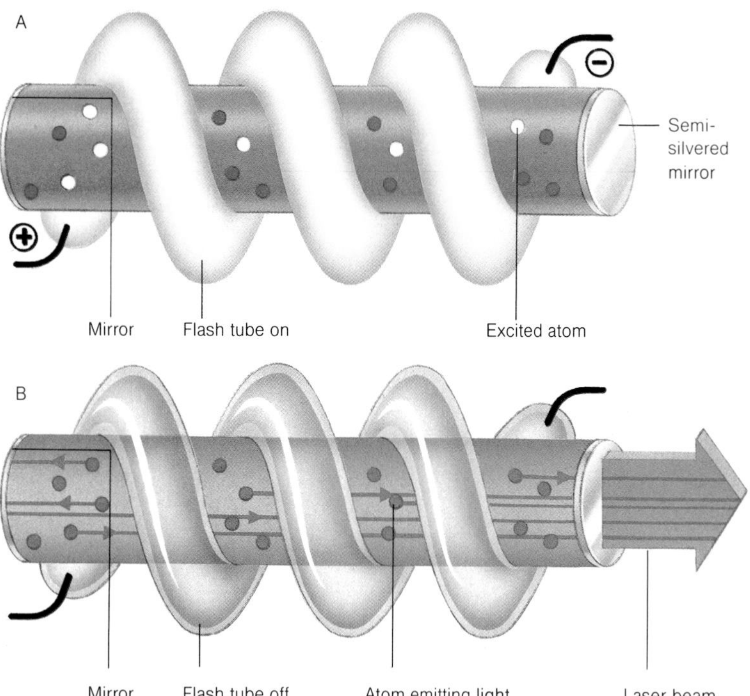

A ruby laser consists of a rod of ruby crystal with a mirror at one end and a semisilvered mirror at the other. When a flash tube coiled around it lights up (A), some atoms in the ruby are excited into a higher energy state. With the flash tube off (B), the excited atoms return to their ground state and emit light, some of which excites other atoms; some light is reflected within the crystal, and some emerges as a laser beam.

rather than waves.

All the photons in a laser beam have the same amount of energy—the light has a single frequency because it is emitted by atoms excited to an identical energy level above the ground (unexcited) state. In addition, all the photons are emitted by the atoms at the same instant, so they are in phase.

The first type of laser invented—the ruby laser—has as its energy source a flash tube; as the active medium (from which light is emitted), it has a man-made cylindrical ruby with its sides and one end silvered, and the other end is semisilvered. Light from the flash tube coiled around the ruby excites ("pumps") outer electrons in atoms in the ruby to a higher and unstable energy state, from which they then rapidly and spontaneously decay to a lower, slightly more stable state that is still well above the ground state. When an atom in this condition is struck by a photon, two things happen. The atom returns to the ground state and simultaneously emits another photon of the same frequency (it "lases"). This photon strikes another atom in the raised, metastable state, causing it to emit yet another identical photon, and so on. The pumping and lasing actions take place in milliseconds.

A cascade of stimulated emissions of photons then occurs, enhanced ("amplified") as the photons are reflected by the silvered surfaces at the ends of the crystal. The resulting photons travel together and are in phase, resulting in short pulses of red light emitted from the semisilvered end of the ruby crystal as a narrow beam.

Most lasers in use today are gas lasers. Unlike a ruby laser, a gas laser produces a continuous beam of light. The active medium is a tube of gas, such as carbon dioxide or a helium-neon mixture, with semisilvered mirrors at each end. A powerful source of radio waves is used to pump the gas in the tube (which may lase in the infrared as well as in the visible spectrum). The main advantage of gas as an active medium is that there are no minor imperfections to cause slight divergence of the beam and a consequent spread of wavelength as there may be in a crystal.

The maser

A closely related device (invented several years before the laser) is the maser—a name denoting *M*icrowave *A*mplification by *S*timulated *E*mission of *R*adiation. Instead of coherent light as the radiated product, a beam of coherent microwave radiation is produced.

Only two energy states are involved (unlike the ruby laser's three), so the method of pumping is somewhat different. After pumping, the stimulated emission is amplified in an environment that resonates at the frequency of transition between the two energy states; a beam of coherent microwaves is emitted.

Properties and uses of lasers

The best-known use of lasers is in medicine, where it is now established as an important surgical tool. A laser beam can be used as a concentrated fine ray, much in the way of an ordinary surgical scalpel—and has the additional qualities of being totally aseptic and able to cauterize minor blood vessels instantly.

Other major applications include the cutting or shaping of a great variety of materials, particularly metals; the accurate marking or measurement of straight lines; experimentation in wind-tunnel projects; and, again because of the laser beam's property of coherence, holography.

The energy concentrated in a thin laser beam can be used to cut hard materials such as metals and, as illustrated here, glass. Lower-power laser beams are used in medicine as a form of scalpel—which simultaneously cauterizes the tissues it cuts—and for "welding" a detached retina back into place to cure a common form of partial blindness.

Holograms

A hologram is a special type of three-dimensional photograph. It is produced on a photographic plate or film using a laser beam, and provides a three-dimensional image when viewed by laser light. By changing the viewpoint, the viewer can see around the objects

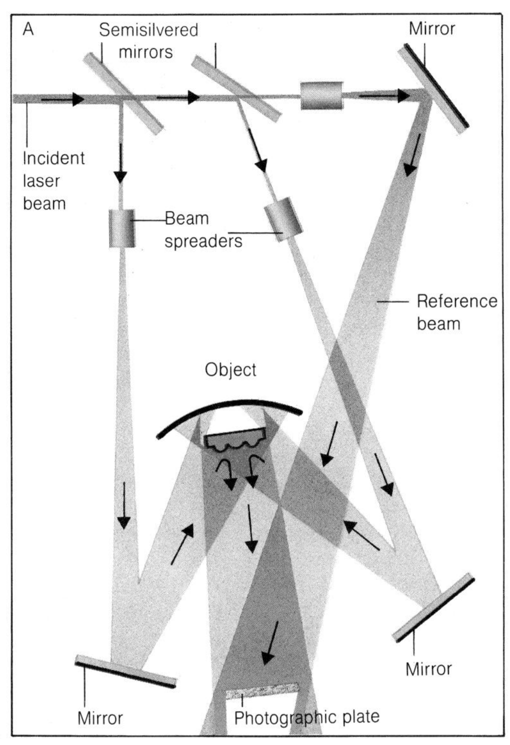

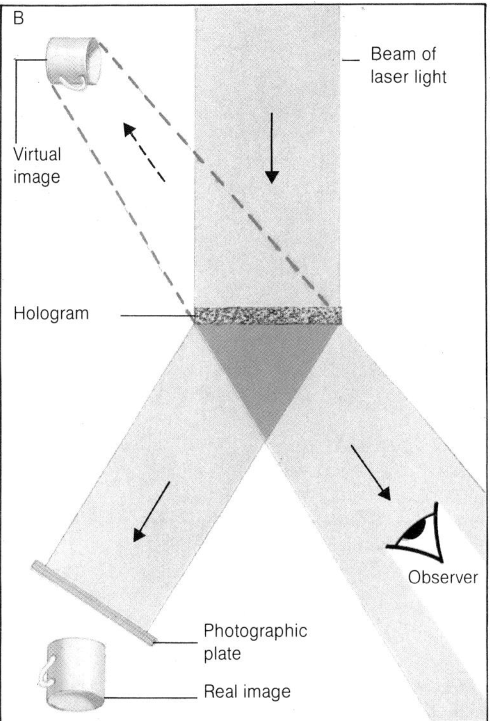

A hologram is a photographic record of the patterns produced when laser light reflected from an object interferes with a reference beam derived from the same source. In making a hologram (A), two semisilvered mirrors split a laser beam, which is reflected to illuminate the object. Light from the reference beam is directed onto a photographic plate, where it interferes with the laser light reflected by the object. To view a hologram (B), it is lit with laser light. An observer sees a three-dimensional virtual image. A real image is also produced, and this can be recorded directly on a photographic plate.

in the foreground and objects at the edge may disappear out of the field of view.

Holographic images are produced using reflected coherent light, which contains information about the surface of the object from which it is reflected. A hologram is generated by splitting a laser beam into one or more object beams and a reference beam—by means of semisilvered mirrors. The object beams, after further reflection by mirrors if necessary, are directed to illuminate the object. This then reflects the laser light onto a photographic plate placed in the path of the reference beam, reflected directly off another mirror.

The two beams interfere at the surface of the plate, producing an interference pattern that is then photographically developed on the plate to give the hologram. The amount of interference in the pattern at any one point on the plate depends on the difference in phasing between the beams, and therefore on the difference in the distances traveled by the object beams (after reflection from various parts of the object) and the reference beam (after reflection from its mirror).

To view the holographic image, the hologram is generally illuminated by a laser. The hologram acts rather like a diffraction grating, reconstructing the original light rays that were reflected from the object to all parts of the plate. On looking through the hologram from the side opposite the laser, the viewer sees a three-dimensional virtual image (apparently located behind the hologram) because the hologram completely reconstructs the original light rays from the object, and each eye therefore sees a slightly different image—just as happens when looking at a solid object under normal conditions. In addition, a real image is formed on the side opposite the laser, and this image can be photographed by placing a photographic plate at its point of formation.

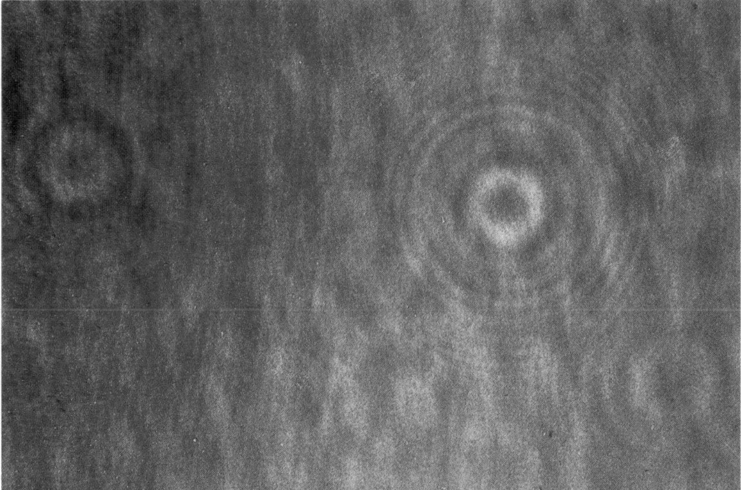

The interference patterns recorded on a hologram *(above)* consist of an apparently meaningless jumble of dots and circular whorls. But when viewed using laser light in the manner described at the top of the page, a three-dimensional image is formed. A holographic image can be photographed, producing the result shown *(left)*.

Radio waves and microwaves

Radio waves and microwaves are two of the more familiar types of electromagnetic radiation. Radio waves are used for communications, as are microwaves, which are also used in radar and for cooking food in microwave ovens.

Both radio waves and microwaves have long wavelengths (always measured in metric units) and low frequencies. The microwave band lies immediately beyond the infrared region and comprises waves with frequencies between 3×10^{11} and 10^9 Hz and wavelengths between 1 mm and 30 cm. Beyond the microwave band is the radio part of the electromagnetic spectrum. Radio waves have frequencies lower than 10^9 Hz and wavelengths greater than 3×10^{-1} m. In general, those used in commercial radio and television broadcasting have frequencies between about 5×10^5 and 8×10^8 Hz, and wavelengths varying from hundreds of meters to less than one meter.

Tycho's supernova remnant is all that remains of a star that exploded violently in the sixteenth century; it was so bright that it was visible in daylight; it faded rapidly, however, and is now invisible. Nevertheless, the supernova remnant emits radio waves—as do many other celestial objects—and these can be detected on earth using radio telescopes. Moreover, the intensity of emissions can be represented by different colors to give a false-color image of the remnant. In the photograph, for example, the brighter the color, the more intense is the emission; hence it can be seen that the strongest emissions emanate from the edges of the supernova remnant.

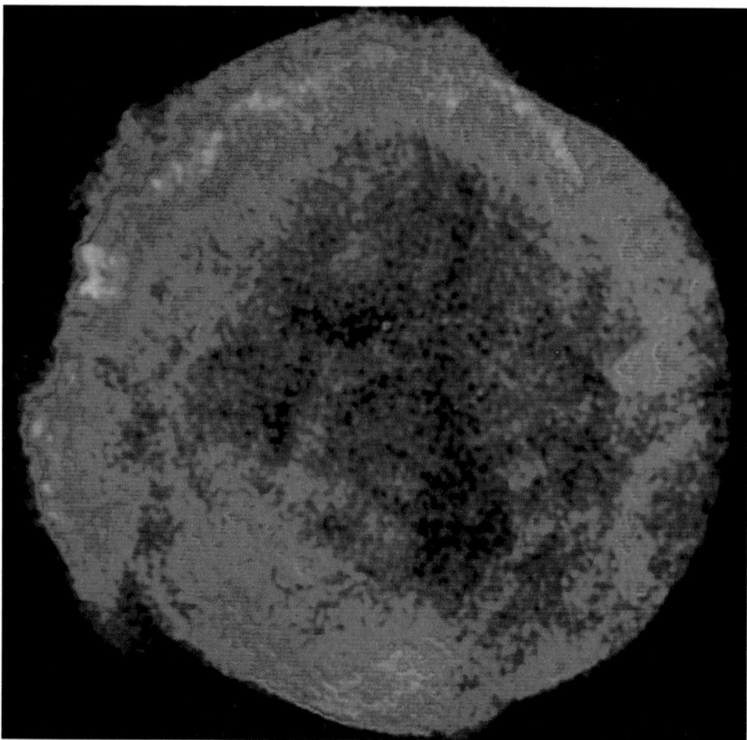

Radio waves

Like other forms of electromagnetic radiation, radio waves are vibrations in mutually perpendicular electric and magnetic fields. They are produced naturally by some stars, and their emissions can be detected with radio telescope, which are basically very large receiving antennas. For radio and television transmissions—the most common sources of radio waves on earth—they are artificially generated by oscillating electrons in metal transmitters.

In its simplest form a radio transmitter consists of a long metal wire connected to a source of electric current so that the wire is positively charged at one end and negatively charged at the other. As a result of this separation of electric charges, an electric field is produced around the transmitter. When it is connected to a rapidly alternating electric current, the charges in it reverse as the current reverses (that is, the charge at each end alternates between positive and negative in step with the changing current); this, in turn, causes repeated reversals of the electric field around the transmitter. The reversals constitute the electric field component of the radio wave; they also give rise to the magnetic field that is an essential part of all electromagnetic radiation.

The radio wave thus generated travels at the speed of light; its frequency (and therefore wavelength) is determined by that of the alternating current in the transmitter—the greater the rate of reversal of the current, the higher the frequency of the radio wave.

The radio waves used in radio and television transmissions are, however, much more complex than this, for they carry the information necessary for radio and television programs. This information is coded in the form of variations, called modulations, in either the amplitude or frequency of the simple wave. To produce such modulations, electrical signals from microphones or television cameras are superimposed on a simple wave (known as a carrier wave), producing variations in its amplitude (in *a*mplitude *m*odulation, or AM, transmissions) or frequency (in *f*requency *m*odulation, or FM, broadcasting). The modulated wave is then amplified before being sent to the transmitter.

When the modulated signal encounters a suitable receiving antenna, an electric current is set up in it. This current, which varies as the modulated signal varies, is passed to a tuning device. The tuner can be altered to select one particular signal (now in the form of a modulated electric current) from the many received by the antenna, and the selected signal is then amplified to increase its strength. After amplification, the modulated current is decoded by a demodulator and then passes to a loudspeaker or television tube.

Microwaves

In many ways, microwaves resemble radio waves but are more difficult to generate, requiring special electronic devices, such as the klystron (like a radio valve). Masers also generate microwaves, but they are used only when a coherent beam of microwaves is required. Once generated, microwaves are sent to a transmitter along waveguides, which are simply hollow pipes.

Microwaves can be concentrated into powerful, highly directional beams and are widely used for satellite communications. Radio waves are generally unsuitable for this application because they tend to be reflected by the ionosphere—a layer of earth's atmosphere in which there is a high concentration of charged particles.

Microwaves can be specially coded to carry many separate transmissions—several thou-

Microwave communications dishes, such as the one shown left, are typically about 100 feet (30 meters) across and are used for both transmitting microwave signals to, and receiving retransmitted signals from, communications satellites. Two principal frequency bands are used for this purpose: 5.9 to 6.4 $\times 10^9$ Hz for earth-to-satellite transmissions; and 3.7 to 4.2 $\times 10^9$ Hz for satellite-to-earth retransmissions.

sand telephone conversations, for example—at the same time (a technique called multiplexing), and so are very valuable for communications on the earth's surface. Extensive microwave transmission networks have been set up in some countries, such networks comprising a series of towers—recognizable by their horn- or dish-shaped antennas—to detect, amplify, and retransmit microwave signals. The towers are usually about 30 miles (48 kilometers) apart on flat terrain because microwaves can be transmitted only over line-of-sight distances.

Radar basically involves transmitting a microwave signal and then detecting its echo after reflection by a distant object. The distance to the object is calculated from the time elapsed between the transmission of the signal and the detection of its echo. As well as in navigation, radar is employed in meteorology to detect rain storms.

Microwave cooking relies on the fact that water and organic molecules vibrate very energetically (an example of forced vibration) when subjected to microwaves of a certain frequency (about 2.4 $\times 10^9$ Hz). Friction between the vibrating molecules rapidly generates heat and so cooks the food.

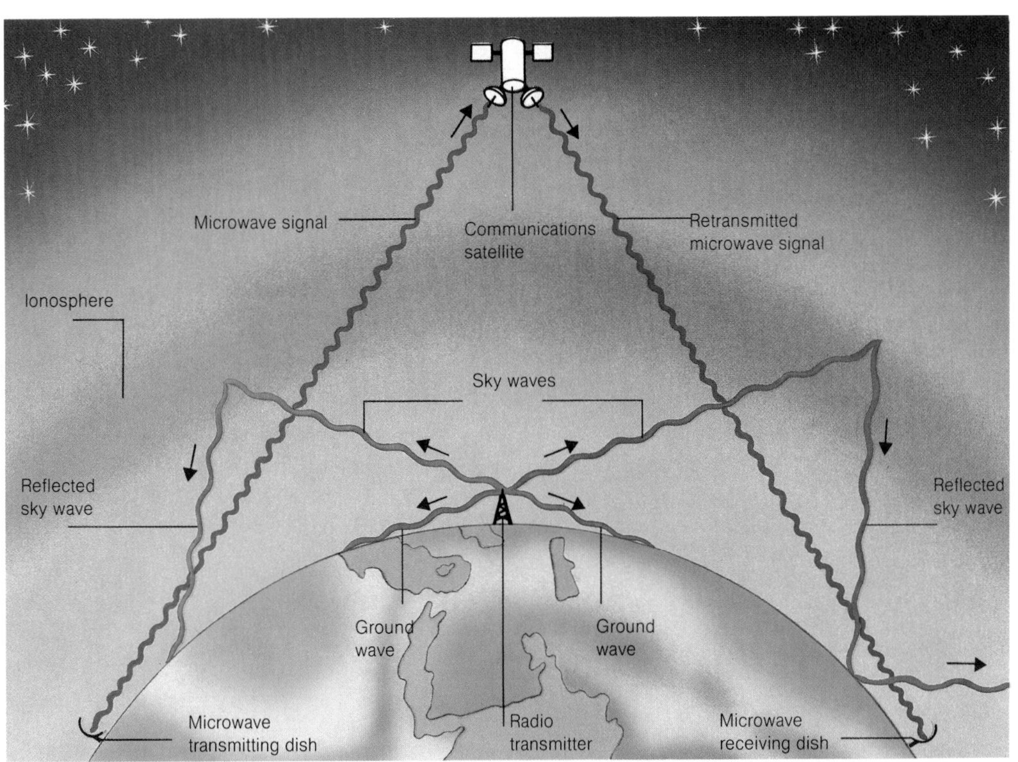

A radio transmitter radiates waves in all directions. Sky waves travel upward and are reflected by the ionosphere back to earth; hence they can be received from beyond the horizon. Some reflected sky waves may hit the ground and be reflected again. Ground waves travel horizontally or toward the ground and so cannot be received much beyond the horizon. Microwaves are not reflected by the ionosphere and can therefore be transmitted to satellites—either for communication with the satellite or for retransmission to a receiving station on earth.

Infrared and ultraviolet radiation

Wavelengths
Metric conversions:
10^{-9} m
= 4 hundred-millionths of an inch
10^{-7} m
= 4 millionths of an inch
10^{-3} m = 1/25 inch

Lying astride the visible part of the electromagnetic spectrum are infrared rays and ultraviolet rays. Infrared radiation is familiar as radiant heat, and ultraviolet radiation from the sun is largely responsible for causing a suntan. Both have important practical applications in various branches of science and technology.

In scientific terms, ultraviolet radiation occupies the part of the electromagnetic spectrum between visible light and X rays. Its frequencies range from about 7.5×10^{14} Hz (at the visible light end) to 3×10^{17} Hz (at the X-ray end). The corresponding wavelength range extends from about 4×10^{-7} m to 10^{-9} m. Infrared radiation, on the other hand, lies on the other side of the visible spectrum, between visible light and microwaves. Its frequencies range from approximately 4×10^{14} Hz (the visible light limit) to 3×10^{11} Hz (the microwave limit); wavelengths range from about 7.5×10^{-7} m to 10^{-3} m.

Ultraviolet radiation

Ultraviolet radiation is emitted as a result of energy changes in the orbital electrons of atoms in substances that are extremely hot. The sun is the principal natural source of this type of radiation, and it emits a wide range of ultraviolet frequencies. Most of the high-frequency, energetic ("hard") and potentially dangerous ultraviolet radiation does not reach earth's surface but is absorbed by the upper atmosphere, where it ionizes the gas and also supplies the energy needed to create and maintain the ozone layer. Lower frequency, "soft" ultraviolet radiation from the sun does penetrate the atmosphere; it helps to tan and make vitamin D in the human skin and is also used by green plants in photosynthesis.

The eyes of hornets *(above)*, bees, and some other insects can detect ultraviolet radiation (which is invisible to the human eye). As a result, the insects are attracted to flowers that may appear relatively dull to us, but which display striking markings in ultraviolet light. For example, the evening primrose (*Oenothera* sp.) appears as a plain yellow flower in daylight *(top right)* but reveals a prominent dark pattern when photographed under ultraviolet illumination *(bottom right)*. This effect occurs because there are pigments around the center of the flower that absorb ultraviolet radiation and therefore appear dark in the bottom photograph.

Ultraviolet radiation can be produced artificially by passing an electric current through mercury vapor. This method of production, together with the fact that ultraviolet radiation makes certain substances (notably phosphors and fluorite) fluoresce, is put to use in fluorescent lamps. In these lamps, a phosphor coating on the inside of a glass tube, which contains mercury vapor, absorbs the invisible ultraviolet radiation and converts it into visible light. Various colors of light can be produced, depending on the composition of the coating.

Although ultraviolet radiation is invisible to the human eye, it registers on photographic film; in fact, ordinary film is more sensitive to certain frequencies of ultraviolet radiation than to visible light. Ultraviolet rays can also be detected with fluorescent screens and with some types of photoelectric cells.

Hard ultraviolet rays can destroy living cells and so are used to sterilize some foods. Soft ultraviolet radiation is sometimes employed to treat vitamin D deficiency. The fluorescent effect of ultraviolet is used for such purposes as identifying ores and testing materials, and some washing powders contain substances called optical brighteners that fluoresce to enhance the brilliance of white clothes.

Another use of ultraviolet radiation is in microscopy. Ultraviolet microscopes are basically the same as conventional light microscopes, except that they use an ultraviolet "light" source and record the image on photographic film; they also use quartz or fluorite lenses, which are more transparent to ultraviolet radiation than is normal glass. Their main advantage is that they can resolve finer details than can light microscopes.

Infrared radiation

This results from changes in the energy states of the outer electrons of atoms and from changes in the vibrational and rotational energies of molecules. Because atoms and molecules and their associated electrons are continually moving and changing in energy, all objects constantly emit some infrared radiation—except at absolute zero, when all movement theoretically ceases.

The amount and frequency of the radiation an object emits depends on its temperature; the higher the temperature, the greater the amount of radiation and the higher its frequency. Extremely hot objects emit visible

Vegetation appears very bright and falsely-colored in infrared color photographs *(left)* because chlorophyll—the pigment that gives plants their green color and also plays a major part in photosynthesis—is a good reflector of infrared radiation. In addition, plants generate heat as a result of their metabolic processes and therefore emit infrared radiations as well as reflecting them.

light in addition to infrared radiation—which is why they glow.

Infrared radiation is readily absorbed by many substances. In general infrared frequencies of about 10^{14} Hz tend to be absorbed most strongly because this is the frequency at which atoms and molecules vibrate naturally. But because all objects also give off infrared radiation, in practice an exchange of radiant energy takes place so that two objects adjacent to each other eventually absorb and radiate at the same rate and reach the same temperature.

Some substances are transparent to certain frequencies of infrared radiations. For example, high-frequency radiation given off by the sun readily penetrates both the atmosphere and glass—a property put to practical use in greenhouses. The high-frequency radiation passes through the glass and is absorbed by the soil and plants, which being relatively cool, emit infrared radiation of a much lower frequency. Glass is opaque to the lower frequency radiation, and heat therefore accumulates in the greenhouse. This phenomenon—called the greenhouse effect—occurs naturally and warms our entire planet. In this case, the radiant energy from the sun is absorbed by the ground and re-radiated at lower frequencies that cannot penetrate the atmosphere.

Being radiant heat, infrared radiation can be detected with thermometers and, more fundamentally, by sensory nerve endings in the skin. Also, a special type of photographic film is sensitive to infrared radiation, making it possible to photograph objects in the dark. In daylight, infrared film depicts colors falsely—foliage and heat sources such as factories appear red, for example—but this effect is put to practical use in infrared satellite photography of the earth's surface to assess agricultural and industrial resources. Infrared sensitive photoelectric detectors can produce a video image to provide night vision—useful in studying the nocturnal behavior of animals, for example.

In medicine, infrared radiation is used in heat treatment (diathermy) and in a special type of diagnostic technique called thermography, in which a heat picture of the body is used to reveal areas that are abnormally hot or cold and may, therefore, be diseased. Another important application is infrared spectroscopy, which is employed in chemistry and biology to determine molecular structures.

High-frequency infrared radiation from the sun penetrates glass and so can pass into greenhouses *(below left)* where it is absorbed by plants; as a result, the plants become warmer and reradiate low-frequency infrared, which cannot pass through glass and so heats the greenhouse. This heating effect also occurs on a global scale *(below right)* because the atmosphere acts like glass and traps infrared reradiated by the ground. The effect is greater in the polluted air above cities.

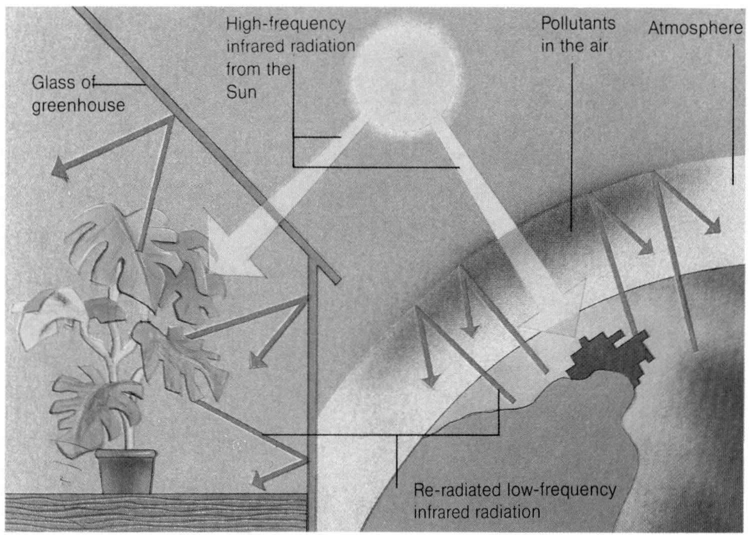

Wavelengths
Metric conversions:
10^{-9} m = 4 hundred-millionths of an inch
10^{-10} m = 4 billionths of an inch
10^{-12} m = 4 hundred-billionths of an inch
10^{-15} m = 4 hundred-trillionths of an inch.

X rays and gamma rays

The part of the electromagnetic spectrum with very short wavelengths (between about 10^{-9} m and 10^{-15} m) and very high frequencies corresponds to X rays and gamma rays. Both are high-energy, penetrating types of radiation that can ionize gases and cause other chemical and physical changes, such as fluorescence. The main difference between the two types of radiation is in their origins. X rays are produced by the excitation of atomic electrons, whereas gamma rays derive from the excitation of the atomic nucleus itself. In both cases, characteristic radiation is emitted when the excited electrons or nucleons return to their original energy levels.

X rays

Spanning the wavelength range 10^{-9} m to about 10^{-12} m—corresponding to frequencies of the order of 3×10^{17} Hz and upward—X rays are one of the types of radiation emitted by some stars and galaxies. Studies of heavenly bodies from artificial satellites have revealed various X-ray sources, among which are possible locations of black holes. The observations have to be made from space because X rays do not pass through the atmosphere (although highly energetic, X rays do not penetrate the atmosphere because they interact with—and in so doing, ionize—molecules in the air). Here on earth, X rays are generated artificially.

X rays are produced using thermionic vacuum tubes that operate at potentials of up to 2 million volts. A stream of electrons, fired from the cathode of the tube, strikes a metal anode (the target), which emits X rays. The anode of an X-ray tube gets very hot and is often water-cooled or rapidly rotated and made of the high-melting point metal tungsten. The X-ray beam is emitted at right angles to the electron beam. Because of the high voltage necessary and the penetrating power of the radiation itself, commercial X-ray machines are heavily insulated and shielded with lead, which absorbs X rays. Even infrequent exposure to X rays has been known to cause cancer, so any exposure to ionizing radiation should be kept as low as possible.

The best-known uses of X rays are in medicine: in radiology to "photograph" the internal structures of the body and in radiotherapy for destroying cancerous tumors. X rays are also used extensively for internal examination of industrial materials, for example in the detection of flaws and other weaknesses in metal castings and welds. (Although if the metal is very thick, X rays may be insufficiently penetrating and gamma rays may be used instead.) In both types of application, the X rays are usually made to expose photographic film, although fluorescent screens are also employed in "instant" X rays and in cinematography in medicine.

The penetrating power of X rays is controlled by varying the voltage between the electrodes of the X-ray tube: the higher the voltage, the more penetrating are the rays. It

In a hot-cathode X-ray tube a stream of electrons from an electrically heated filament strikes a target of tungsten mounted on the end-face of a copper anode. The anode is cooled by water circulating within it. In another type of X-ray tube, the anode takes the form of a rapidly rotating tungsten disk.

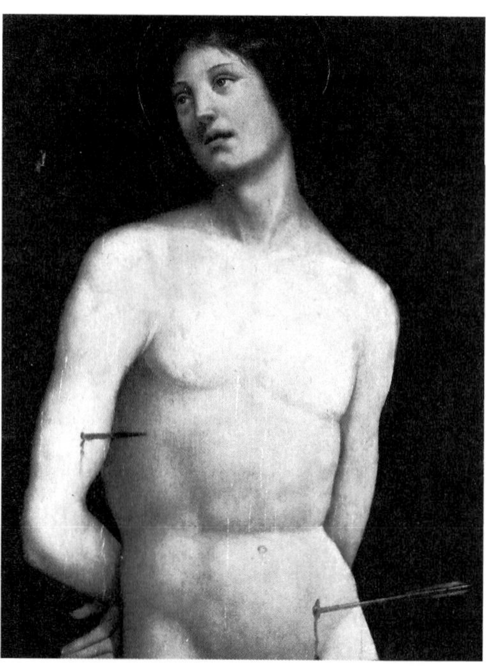

This portrait of Saint Sebastian by the fifteenth-century Italian painter Francia is painted on a wooden panel, now in Hampton Court Palace, near London. The X-ray photograph *(far right)* reveals that the artist changed his mind and repainted the head, inclining it to the left of the picture instead of the right. Such X-ray techniques allow paintings and other inanimate objects to be examined internally without damaging them.

also varies according to the composition of the substance being irradiated. Absorption of X rays results from the Compton effect (in which electrons in atoms scatter X-ray photons), the photoelectric effect (in which X-ray photons strike atoms and cause the emission of electrons) and pair production (electron-positron pairs produced by X-ray photons). These effects all depend to some extent on the number of electrons in the atoms of the absorbing material; as a result, elements of high atomic number (such as lead) are better absorbers of X rays than are elements of low atomic number. In medical radiography, for example, X rays penetrate flesh (which consists largely of low-atomic number elements) more readily than they penetrate bone (which contains elements of higher atomic numbers). And a mixture containing the heavy element barium is swallowed to make the esophagus and stomach opaque to X rays so that they show up on an X-ray photograph.

X-ray diffraction and spectroscopy

Just as light is diffracted on passing through the finely ruled lines of a diffraction grating, X rays are diffracted by crystals. The spacings between the regular array of atoms in a crystal approximate to X-ray wavelengths, and the resulting diffraction patterns can be interpreted to provide information about crystal structures. The technique also provides a means of measuring X-ray wavelengths.

When bombarded with high-energy electrons, different metals emit X rays of different, but characteristic, wavelengths. X-ray spectra can therefore be used in analysis to identify metals—in fact, a given X-ray frequency is directly related to the atomic number of the metal producing it. This results from the fact that the energetic electrons bombarding the target—the anode in the X-ray tube—cause discrete energy changes that involve the atom as a whole.

Gamma rays

During the radioactive decay of unstable isotopes of elements, such as uranium, thorium, and radium, gamma rays are emitted. These constitute electromagnetic radiation of extremely short wavelength, from about 10^{-10} m down to 10^{-15} m or less (frequencies of 3×10^{18} Hz and upward), overlapping to some extent with those of "hard" X rays.

The properties of gamma rays resemble those of X rays. They are even more penetrating and can be contained only by thick shielding, typically several inches of lead or several yards of concrete. Like X rays, gamma rays can be used to reveal flaws in solid materials; a suitable source for such applications is cobalt-60 or some other long-lived radioisotope. In research into particle physics, gamma rays often occur as product of collisions between high-energy subatomic particles. They are, therefore, important in the study of the ultimate structure of matter.

The gamma-ray telescope at Mount Hopkins Observatory in Arizona has a parabolic "mirror" 32.8 feet (10 meters) across consisting of an array of hexagonal plates. It detects the visible-light photons produced when gamma rays interact with gas atoms on entering the earth's atmosphere.

Gamma radiation from the radioactive isotope iridium-192 was used in taking this photograph that reveals an electric drill's internal construction. (X rays are insufficiently energetic to penetrate great thicknesses of metal and so would not have produced such a well-defined image.) A similar technique is used in industry to examine metal objects for flaws.

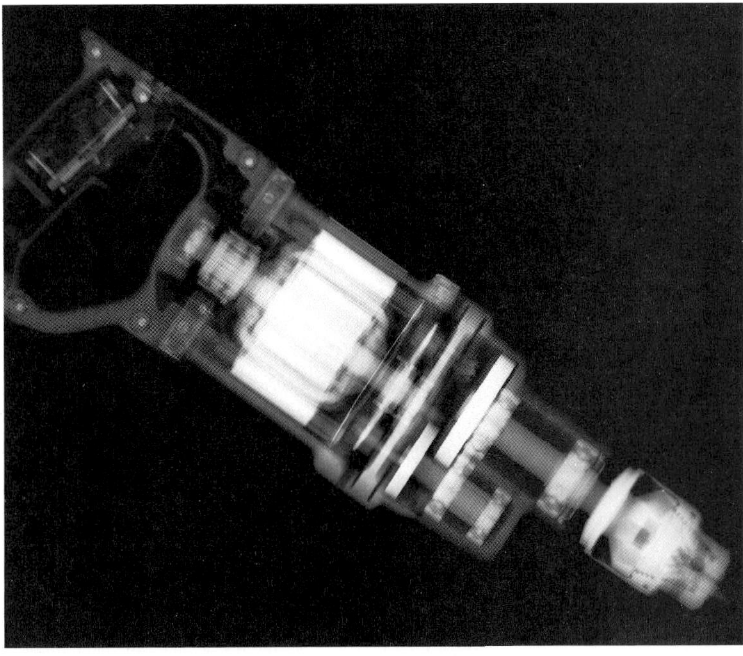

Sound

Terms, units, and abbreviations
wavelength in meters (m)
time in seconds (s)
velocity in meters per second (m s^{-1})
frequency in hertz (Hz) (equals cycles per second)
sound intensity in watts per square meter (W m^{-2})
loudness in decibels (dB)

Sound consists of waves of alternate compression and rarefaction that transmit kinetic energy through a medium. If there is no medium, there is no sound; sound waves cannot pass through a vacuum. All sounds originate from vibrating objects, such as the vocal cords in the human throat, the skin of a drum, or the strings of a violin. The vibrations are then passed on through a medium—most commonly air—until they strike the ear drum; the ear converts the vibrations into nerve impulses, which pass to the brain where they are interpreted as sounds. Microphones are the other most common instruments of sound detection, and they work by converting sound waves into electrical signals.

Waves and sound

There are two types of waves in physics: longitudinal and transverse. Sound waves are of the longitudinal type—that is, their vibrations take place in the same direction as they travel. The phenomenon of longitudinal wave propagation can be demonstrated with a long, large-coiled spring. When the spring is fixed at one end and the other end is moved backwards and forwards, alternating regions of compression (where the coils are bunched together) and rarefaction (where the coils are farther apart than normal) move along the spring.

Transverse waves are those in which the vibrations occur at right angles to the direction in which the waves travel—as happens, for example, when a length of rope is moved regularly up and down to give it a wavelike appearance. Ocean waves and light waves also travel by transverse propagation.

As with longitudinal waves along a spring, sound waves consist of a series of compressions and rarefactions. Consider sound waves generated by a vibrating tuning fork and passing through air. When each prong of the fork moves outward, it compresses the air; when it moves back, it allows the air to "stretch," or become rarefied. Hence each complete wave cycle is made up of one compression and one rarefaction. The wavelength is the distance between two adjacent points (or particles) that are in the identical phase of the wave's vibration—two points of maximum compression (or rarefaction), for example.

A sound wave consists of a series of alternate compressions and rarefactions of the medium through which it passes. The diagram below is a simplified representation of what happens to the air molecules (in the colored bands) near a vibrating tuning fork, with graphs of the associated pressure variations.

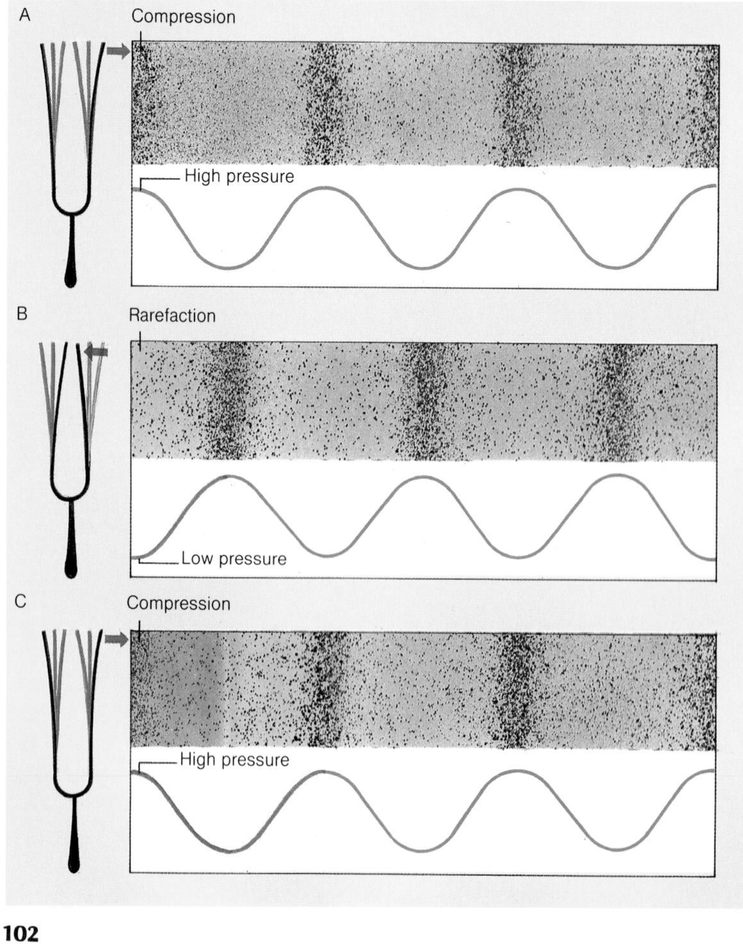

Frequency and intensity

Two obvious ways of distinguishing one "pure" sound (that is, a pure tone) from another are by their pitch and loudness. Neither property is a scientifically accurate notion, however. Pitch is a word used by musicians and is closely related to frequency; but it is a subjective term, whereas frequency can be measured physically. Similarly loudness, closely related to a sound's intensity, differs according to the sensitivity of the hearer's ears, whereas intensity can be measured in exact physical quantities. (Real sounds are mixtures of many different pure sounds and also have a characteristic "quality," which can be assessed by detailed examination of the shape of a real sound's waveform.)

The frequency of a sound wave is the number of wavelengths that are completed in a given period of time. The universal unit of measurement is the hertz (Hz), which represents one complete wavelength, or cycle, per second. The human ear is sensitive to sound in the frequency range from about 20 Hz (a very low hum) to about 20,000 Hz (an extremely high whine). In musical terms, middle C has a frequency of about 262 Hz, although in physics it is sometimes assigned the "scientific" frequency of 256 Hz. As people grow older, their ability to perceive high frequencies diminishes.

The intensity of sound is measured in terms of the amount of power that passes each second through a given area perpendicular to the direction of the sound wave. The intensity unit is the decibel (dB). There is, however, no absolute decibel scale; it is a relative scale, set against an arbitrarily chosen level, which is approximately the minimum audible intensity of a sound of frequency 3,000 Hz. (This minimum intensity, 0 dB, is defined as 10^{-12} watts per square meter—watts are units of power.) The intensity of a sound is related to the amplitude and frequency of the sound wave and to the density of the medium through which it passes. The greater the amplitude, the higher the frequency; and the denser the medium, the greater is the intensity.

Velocity

Sound can be transmitted by any medium—gas, liquid, or solid—and all sound waves travel at a constant velocity through any given medium at a constant temperature. For example, the speed of sound through dry air at 68° F. (20° C) is about 1,096 feet (334 meters) per second. But depending on their elasticity and density, some mediums transmit waves faster than others. Sound travels through a solid, such as glass or metal, about 14 times faster than it does through dry air. It is extremely difficult for sound waves to pass from a medium of one density to another of a much higher or lower density. Thus cavity walls, in which the waves must pass through brick then air then brick again, provide better soundproofing than do solid walls.

Anything that changes the density of a medium also alters the speed at which sound travels through it. Hence the velocity of sound is affected by temperature; as the temperature increases, the medium expands. Its density decreases, with the result that the velocity of sound increases.

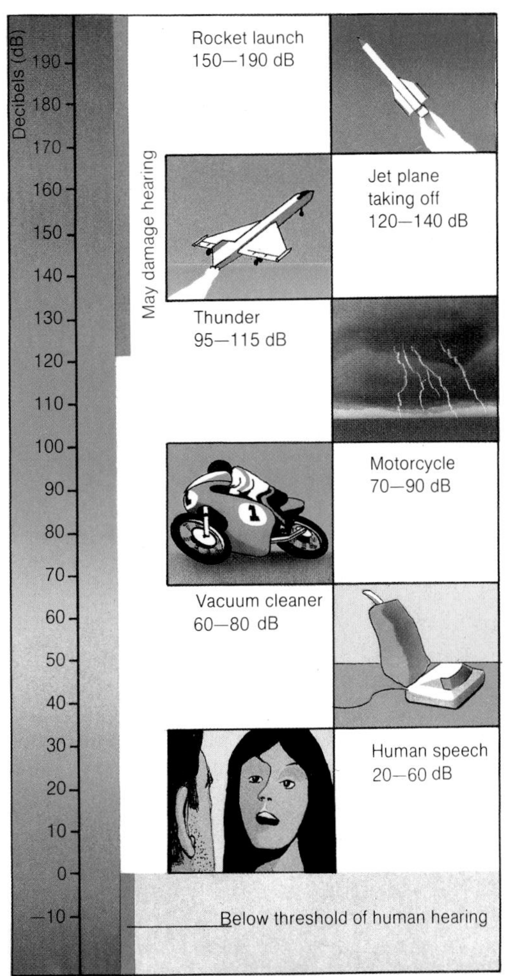

The power of a sound, perceived as loudness, is measured in decibels (dB). The decibel scale is a relative one, with an arbitrary zero point (usually taken as the lower threshold of human hearing). It is also logarithmic, so that a sound of 50 dB is 10 times as loud as one of 40 dB, but 10,000 times as loud as a sound of 10 dB.

A thunderstorm exemplifies the great velocity difference between light and sound. A clap of thunder is the shock wave produced by a lightning flash, so both occur virtually simultaneously; but lightning is seen before thunder is heard (the interval between them being greater the farther away the storm is) because light travels at about 186,000 miles (300,000 kilometers) per second whereas sound travels much more slowly—at only about 1,096 feet (334 meters) per second. Thus, if a lightning flash is seen four or five seconds before the thunder is heard, then the lightning occurred about one mile away.

Generation and properties

Sound from a point source spreads out in all directions—upward, downward, and sideways. The pressure waves—a series of compressions and rarefactions—are not flat, but take the form of concentric spheres with the sound source at their center. Each sphere moves away from the source at a constant velocity, so that the distance between one sphere and the next is equal to one wavelength. As each wave moves outward, its surface area increases. This spreads the amount of energy in each compression ever more thinly, thus reducing the intensity of the wave and its perceived loudness. Only if sound waves in a gaseous medium (such as air) are confined to a straight tube do they travel along a straight path.

Reflection, refraction, and diffraction

The reception of sound waves by the human ear depends on other factors as well as distance. Two of the most important are the presence of an obstacle between the hearer and the sound source, and a change in the distance between the hearer and the sound source while the sound is being emitted. The first gives rise to the reflection, diffraction, and refraction of sound waves. The second gives rise to what is known as the Doppler effect (named for the Austrian physicist, Christian Doppler, who described it in 1842), which is a change in a sound wave's frequency (and wavelength) that results when the hearer and sound source move relative to each other.

The reflection of sound is a common phenomenon, giving rise to echoes and reverberations. Like all other wave motions (including light), sound obeys the law of reflection—the angle of incidence (that is, the angle at which a sound wave hits a reflecting surface) equals the angle of reflection. There are, however, differences between the reflection of sound and that of light. Only a highly polished surface gives a regular reflection of light, whereas very rough surfaces can reflect sound waves. In addition, sound waves require a large surface (of at least several square yards or square meters) to be reflected with little distortion, whereas light waves can be reflected by a tiny surface area. The reason for this difference is that light waves have very short wavelengths (a few millionths of an inch or centimeter); thus complete waves can be reflected by a minute area. Sound waves, however, have long wavelengths (often several yards or meters) and so require much larger areas for reflection of entire waves.

When a sound wave encounters a relatively small obstacle (smaller than the wavelength of the wave), very little of it is reflected: instead, the wave is scattered and passes around the obstacle. This phenomenon, in which waves are bent into regions not directly exposed to the source of the waves, is called diffraction. It also occurs when sound waves encounter an edge or an opening.

Just as sound—like other types of wave motion—undergoes reflection and diffraction, so it is also refracted when it passes between two mediums of different densities. When this occurs, a considerable amount of the sound is reflected at the boundary of the two mediums, so that its intensity is reduced and refraction cannot be detected easily. Nevertheless, sound refraction can be demonstrated using a balloon filled with a dense gas (such as carbon dioxide) which acts as a converging (convex) "sound lens." If a faint sound source is placed a few yards or meters from the balloon, there is a point on the other side at which the sound is loudest. This point is where the refracted sound waves are brought to a focus—exactly like the focusing of light by a glass lens.

The spherical shape of a sound's wavefront can be seen clearly in this photograph of an explosion, in which the wavefront looks like a bubble. This extremely rare photograph, which is a single frame from a high-speed cine film, shows the explosion only milliseconds after the detonation. A special technique called Schlieren photography (which makes visible differences in air density) was also used to enable the wavefront to be seen.

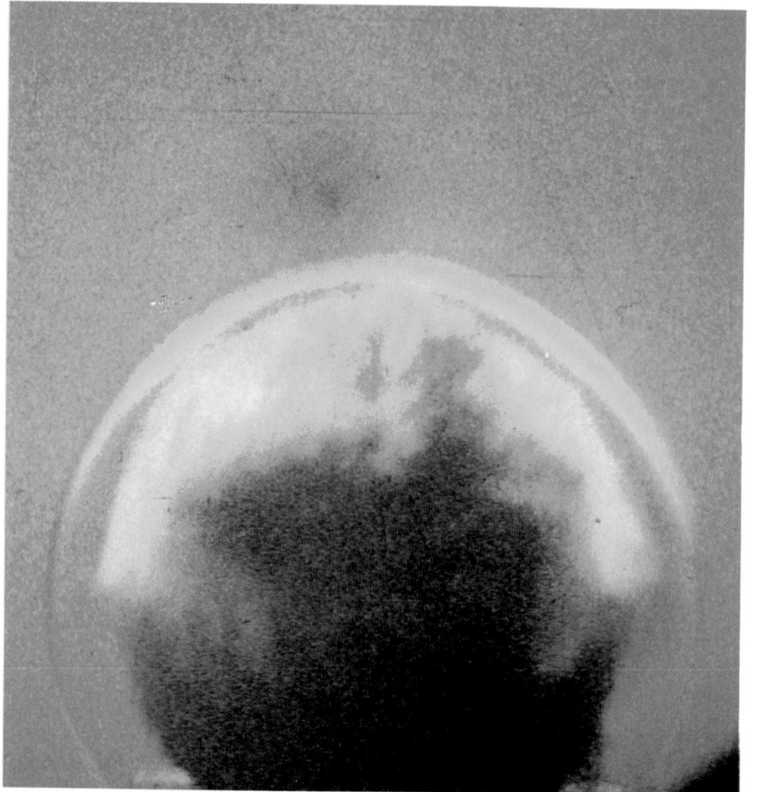

The Doppler effect

The Doppler effect (probably best known in the contexts of astronomy and radar) is the effect that relative velocity has on the observed frequency of waves. An example is the fall in pitch (frequency) of a note from the whistle of a fast-moving train as it passes the hearer. This phenomenon of a frequency change occurs whenever a sound source and hearer (or receiver) move relative to each other. If the source and hearer move toward each other, the time interval between successive compressions or rarefactions arriving at the hearer is reduced. In other words, the frequency is increased, and the note sounds higher in pitch. The reverse effect occurs when the hearer and sound source move apart.

Interference and beats

Like other wave motions, sound exhibits interference effects. Interference occurs when two

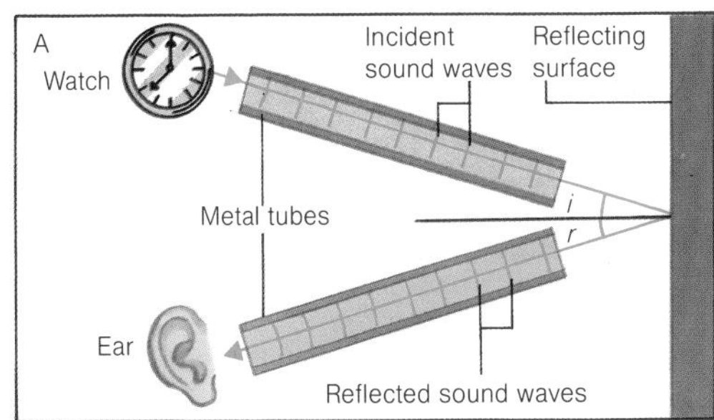

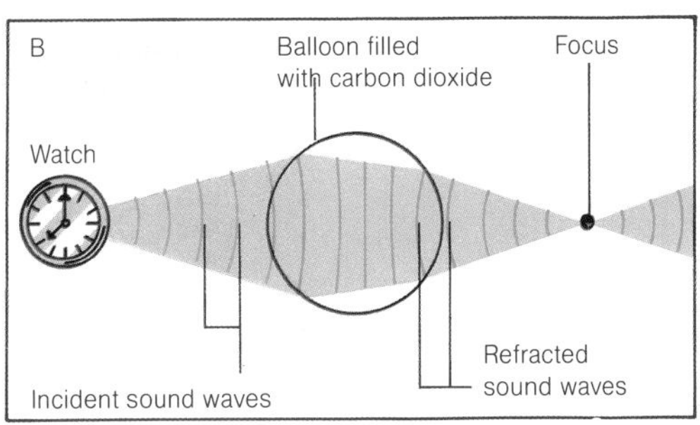

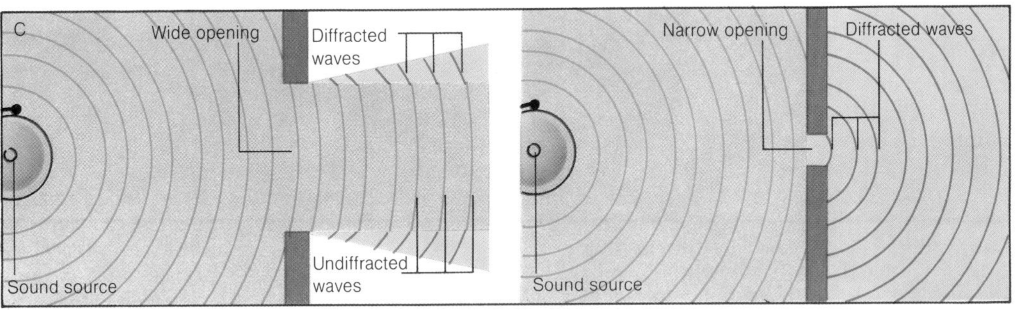

Sound can be reflected (A), refracted (B), and diffracted (C). In A, sound from a watch is directed along a tube to a reflecting surface, and the reflected sound is heard through another tube. By adjusting the angles of the tubes it can be proved that the angle of incidence *(i)* equals the angle of reflection *(r)*. B shows how sound from a watch is refracted and brought to a focus by carbon dioxide gas in a balloon. C demonstrates that sound is diffracted minimally by a wide opening but by a large amount when passing through a narrow opening.

sound waves pass through the same point. If the displacements of each wave reinforce each other—that is, if the crest of one coincides with the crest of the other (and the two waves are in phase), so that the particles of the "carrying" medium vibrate with extra force—there is an increase in amplitude (and hence loudness). If, on the other hand, the displacements cancel out each other—that is, if a crest coincides with a trough (and the two waves are out of phase)—the particles vibrate with less force and the sound decreases in intensity.

Sound interference can give rise to what are called beats. When two waves of nearly equal amplitude and frequency come together, regular fluctuations in intensity occur—known as beats. They are caused because their combined amplitude varies between a maximum (when the two waves are momentarily in phase and reinforce each other) and a minimum (when they are out of phase and cancel out each other). The number of beats heard per second (called the beat frequency) equals the difference between the frequencies of the two waves. Thus, if the waves have frequencies of 140 and 145 Hz, for example, the crests of the two waves are in phase only once every one-fifth of a second and the beat frequency is 5 Hz. Beats can be useful in the tuning of musical instruments. When an instrument is sounded at the same pitch as a tuning tone and no beats occur, then the instrument is in tune with the tuning tone.

Standing waves

Of special significance for musical sound is the interference phenomenon that produces what are called standing (or stationary) waves, which occur only in vibrating bodies generating sound—strings and pipes, for example. A standing wave is produced when two waves of the same frequency and amplitude are reflected from the ends of the vibrating body and interfere with each other. When this occurs, certain points at intervals of half a wavelength never move from their equilibrium positions throughout the interference. These stationary points are called nodes. They result from the fact that a displacement caused by one wave is always accompanied by interference in the form of an equal but opposite displacement caused by the second wave. Between the nodes are antinodes, where the displacements reinforce each other to give the maximum amplitude of the standing wave.

The sound produced by string and wind instruments is the result of standing wave motion. A freely-vibrating string of a guitar, for example, vibrates simultaneously in a number of different modes. These modes are standing waves with frequencies related in a harmonic

The Whispering Gallery in St. Paul's Cathedral, London, is famous for its reflection of sound. A sound whispered against the circular wall is reflected around the gallery so that it can be heard clearly on the other side, more than 105 feet (32 meters) away.

The sounds produced by songbirds, such as the chaffinch, are well-known for their melodiousness. The sound-producing organ of birds is the syrinx (located at the base of the windpipe), which contains vibratory membranes that generate different tones according to how tense they are—just as a string produces different tones at different tensions.

(whole number) series 1,2,3,4,5 . . . The relative amplitudes of each mode—which would have frequencies of, say, 440 Hz, 800 Hz, 1,320 Hz, 1,760 Hz, 2,200 Hz, and so on (that is, a single pitch with a first harmonic frequency of 440 Hz)—determine the sound quality. Lightly touching the center of the string tends to extinguish the odd-numbered modes, which have antinodes there. Thus only even modes (2,4,6,8 . . .) remain, and in our example, they have frequencies of 880 Hz, 1,760 Hz, 2,640 Hz . . . This combination of modes is a new pitch with a first harmonic frequency of 880 Hz. The new tone or harmonic, as it is called by musicians, is one octave higher in pitch and radically different in quality. Similarly, touching the string one-third of the way along its length tends to extinguish all modes except those that are multiples of three. Hence the modes 3,6,9,12 . . . predominate, producing a new tone with a first harmonic frequency 3 times that of the original tone—1,320 Hz in the example.

Musical sounds

The ability to produce sounds of set frequencies is one of the chief factors distinguishing what we call noises from musical tones. Sounds are distinguished from each other not simply by pitch (frequency) and loudness (intensity), but also by their quality.

The quality of a sound is difficult to define scientifically. Musicians refer to the predominant or characteristic quality of the sound made by a particular instrument as its timbre. The timbre of a musical instrument depends on a number of factors, the most important of which is the relative intensity of the harmonics (or overtones) produced. The quality of a tone is also influenced by its "attack," the short time taken for the tone to reach its full volume.

It is normal for many of an instrument's modes of vibration to occur, with widely differing amplitudes, at the same time. The mixture of these various modes is a major determinant of the instrument's sound quality. The violin has a full, vibrant sound because its tones are characteristically rich in overtones. A mellow flute tone, on the other hand, approximates more to a pure sound—that is, a sound produced by a fundamental vibration without overtones. The piano produces a deliberate mixture of overtones. The strings that are of different lengths are struck by a hammer at a point about one-seventh of the way along their length, thus producing enough of the higher overtones to yield a brilliant sound, while eliminating overtones that lend harshness to the sound.

Resonance

Sometimes a particular tone causes a nearby object to vibrate or buzz. This effect—called resonance—reveals that the object has a natural standing wave frequency in common with the tone. It occurs whenever one body or system is made to vibrate at its own natural frequency by impulses received from another body vibrating at the same frequency. This is exploited in the construction of musical instruments to add richness and volume to the sound; for example, the piano has a sounding board that resonates with the tones produced by the strings.

Stationary waves can be produced in pipes that are open at only one end or open at both ends. A *(below left)* shows the first three natural modes of vibration in a pipe open at one end. In all three cases, there is an antinode (a point of maximum vibration, shown in red) at the open end, and a node (a point where the air molecules do not move, shown in black) at the closed end. Closed pipes can produce only odd-numbered harmonics. B *(below right)* shows the first three natural modes of vibration of an open pipe, which can generate odd- and even-numbered harmonics. The solid blue line is a graphical representation of the amount of vibration of the air molecules along the length of the pipe.

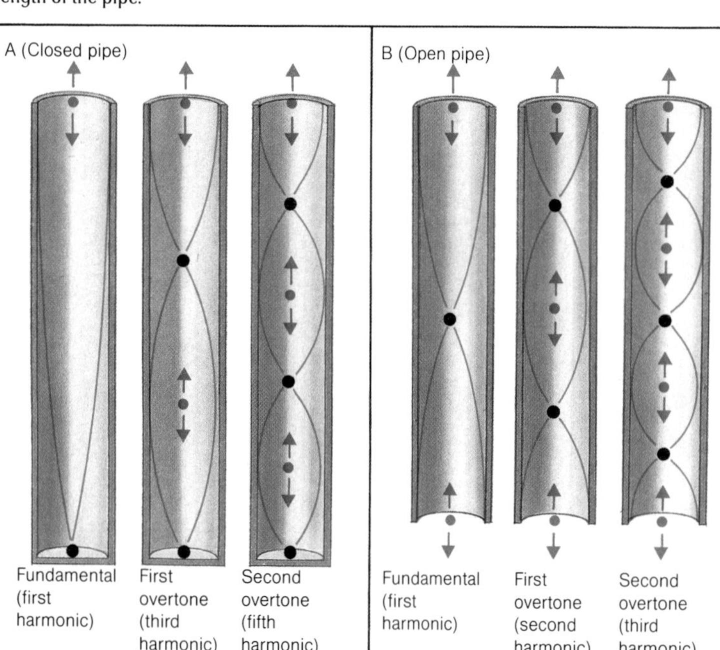

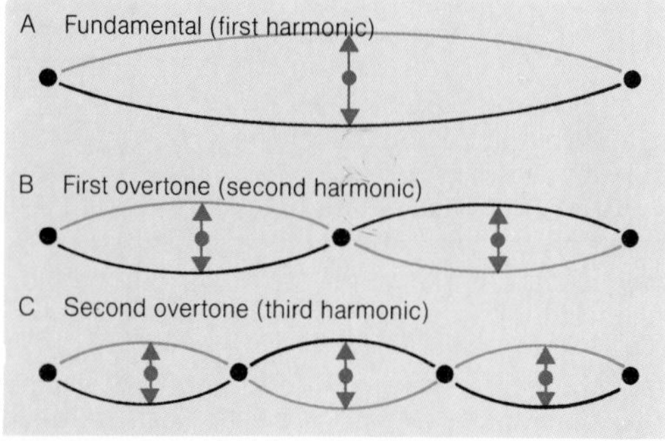

The first three modes of a vibrating string are the fundamental, or first harmonic (A), and the second (B), and third (C) harmonics. Whatever the mode, there is a node (in black), a point of zero displacement, at each end of the string where motion is impossible. In the second harmonic there is also a node halfway along the string. In the third harmonic the intermediate nodes are one-third of the way along the string. Midway between each node is an antinode (in red), a point of maximum displacement.

The human voice

When air passes from the lungs to the mouth, it may vibrate the vocal cords. All vowel sounds are produced by the vocal cords, which are the human equivalent of the double-reed mouthpiece of an oboe or bassoon. Many consonants are produced without the aid of the vocal cords. A *b*, for example, is produced by vibrations of the lips. It has been demonstrated that the pitch of a human vowel sound depends on the length and tension of the vocal cords (hence the different ranges of voices) and that its quality depends largely on the shape of the resonant cavities, the mouth and the throat, in which it is produced.

Noise

The concept of noise (like sound quality) is difficult to define because it is, to some extent, a subjective phenomenon—for example, to many people the unfamiliar sound of some modern or electronic music seems to be little more than a series of noises. In the scientific sense, noise is a sound consisting of all frequencies. White noise is a sound in which all frequencies have equal intensities; it sounds like a hiss.

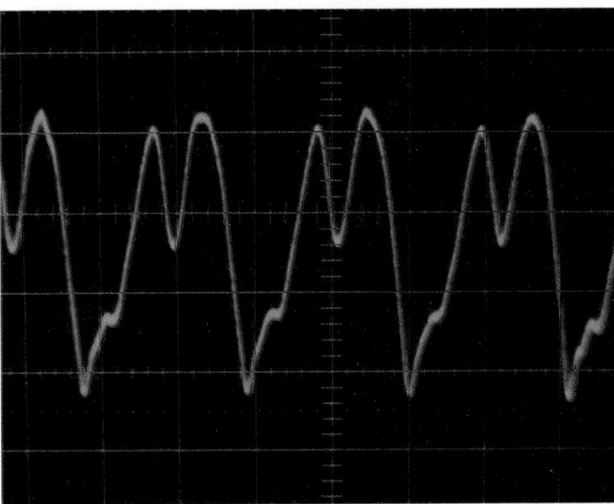

The flute *(top left)* produces sounds of a characteristically mellow quality. Its tones are nearly pure sounds and have few overtones—as can be seen from the audiogram *(below left)* of an *A* played on a flute; the trace of the note is very smooth, with few peaks and troughs, and approximates to a sine curve.

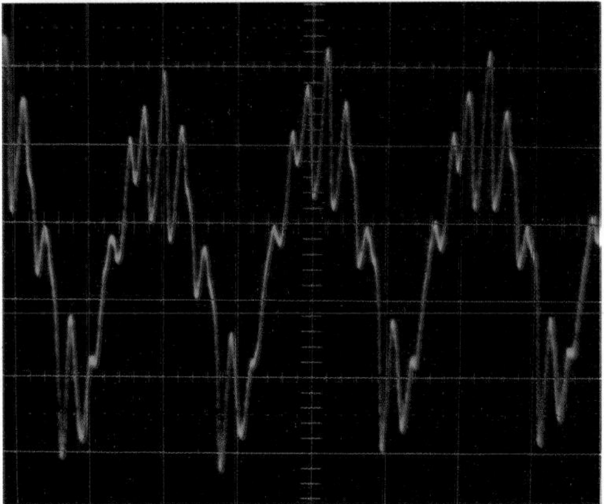

The guitar *(far left)*, in contrast to the flute, produces tones with many overtones—as is evident from the audiogram *(left)* of an *A* played on a guitar. Compared with the flute's audiogram *(above)*, that of the guitar is much more uneven, which shows that the sound has many overtones.

Acoustics and ultrasound

Acoustics is strictly the science of sound in its entirety, but the term is often used in a more limited sense to refer to the sounds that people hear in enclosed spaces, especially in public buildings such as theaters and concert halls. Acoustics in this narrow sense concerns mainly the reflection and absorption of sound.

Echoes and reverberations

When a sound wave is reflected, it produces either an echo or a reverberation, depending on the length of the interval between the emission of the original sound and the return of its reflection. Because the sensation of sound persists in the human ear for about one-twentieth of a second after the sound ceases, a sound that is reflected back to the hearer in less than that time is not perceived as a separate echo because there is no apparent interval of silence between the hearing of the original sound and the hearing of its reflection; instead the reflected sound prolongs the sensation of the original sound—an effect called reverberation. The speed of sound in dry air at 68° F. (20° C) is about 1,096 feet (334 meters) per second, so echoes are not heard unless the hearer is more than approximately 30 feet (9 meters) from the reflecting surface (assuming that the hearer is next to the sound source or between it and the reflecting surface). If he or she is closer than 33 feet (10 meters) to the surface, reverberation occurs.

The submerged wreck of a sailing ship is clearly outlined in black in this false color, computer-enhanced photograph of an ultrasonic sonar scan. (The colored areas are created by the computer processing.) Sonar, which relies on the reflection of high-frequency sound waves, is also widely used to detect fish shoals and to measure the depth of water.

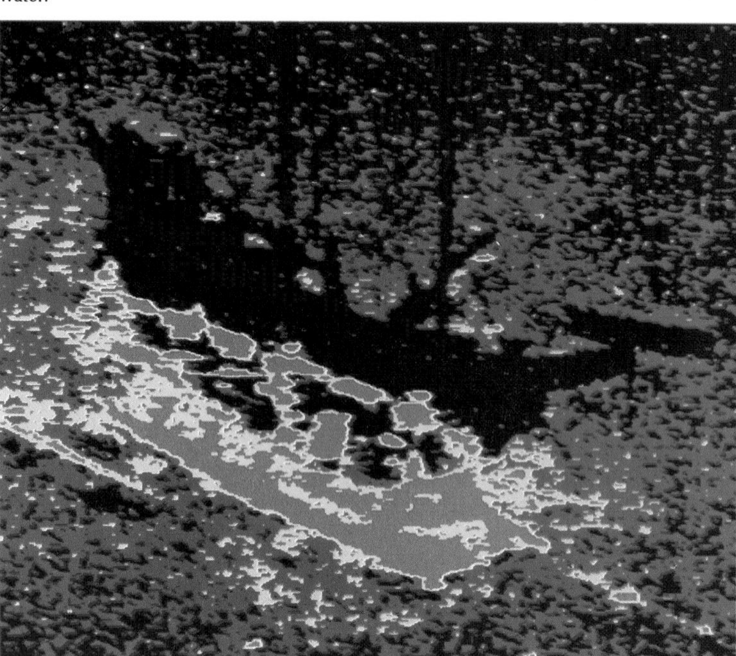

Acoustical control

Because the interior surfaces of buildings reflect sound, setting up echoes or reverberations or both, sounds do not fade as quickly inside buildings as they do in open spaces. The time taken for a sound's intensity to diminish to one-millionth of its initial value is called the time of reverberation. This time can vary from about 10 seconds in a cathedral to less than one-half second in a room crowded with people. Acoustical control in building design is largely concerned with achieving the optimum reverberation period. A relatively long reverberation period, which adds richness to music, is a characteristic of most concert halls. In lecture rooms, a short reverberation time gives clarity to a speaker's voice.

To study the way sound waves might behave in a proposed building, architects sometimes make cross-sectional models of the building and then use the models as ripple tanks. Analyzing how water waves are reflected or absorbed by the various structures in the model gives a good indication of what will happen to sound waves in the building itself. Such experiments indicate that concave surfaces such as domes inside buildings should generally be avoided because they tend to focus sound waves in unwanted

Some concert halls have poor acoustics because of their design. The Royal Albert Hall, London, for example, is oval (exaggerated by the extremely wide angle of view in the photograph) and has a concave ceiling—undesirable features for good acoustics. To improve the acoustics, convex disks (visible at the top of the photograph) have been suspended from the ceiling.

places, though this can sometimes be overcome by erecting false ceilings and walls with convex surfaces. In concert halls it is common to control the sound by placing large, angled reflecting surfaces behind the performers and absorbent surfaces behind the audience. As yet, however, acoustical control is an imperfect science, and it is not always possible to predict exactly how sound waves will behave in a building.

Insect-eating bats navigate and locate their prey by echolocation—in principle similar to man-made sonar. The bats emit ultrasonic squeaks and, from the echoes received by their large ears, sense the positions of and distances to obstacles. The frequency of the squeaks varies with the species, but most are between 30,000 and 80,000 Hz, well above the upper limit of human hearing.

Ultrasonics and infrasonics

Pressure waves ("sounds") with frequencies lower than the lowest limit of human hearing (about 20 Hz) are called infrasonic waves; those beyond the upper audible limit (approximately 18,000 Hz) are called ultrasonic waves. Although inaudible to humans, infrasonic and ultrasonic waves can be heard by some animals. For example, many species of bats (generally those that feed on insects) navigate by using ultrasound; they emit ultrasonic squeaks and can sense the distances to obstacles from the time that elapses between emitting a squeak and receiving its echo. At the other end of the frequency scale, some species of spiders can detect infrasonic waves (the main natural source of which is earthquakes).

Ultrasound has several practical uses, one of the most common of which is measuring the depth of and mapping the ocean floor using the technique called sonar (an acronym of *SO*und *N*avigation *A*nd *R*anging). Unlike electromagnetic waves such as light, ultrasonic waves are not significantly absorbed by water. Moreover, because of their very high frequency, they can be concentrated into narrow beams that can travel deep into the ocean without losing much energy by diffraction; they are therefore well suited to measuring depths underwater. Sonar works in exactly the same way as does the echolocation system of bats—by using the time difference between the transmission of an ultrasonic pulse and the reception of its echo to determine distance.

Ultrasound is also valuable in medicine, particularly in fetal examinations during pregnancy. Again, the basic technique is similar to that used in sonar—using ultrasonic waves reflected by the fetus to "map" the unborn child and check that he or she is not suffering from any abnormalities. This type of examination has several advantages: it does not involve surgery; it is relatively easy to perform; and it is completely safe for the mother and her unborn child. Ultrasonic waves are also used to examine the brain and, by making use of the Doppler effect, to monitor heart movements and the circulation of blood.

Other applications of ultrasound include detecting the positions of flaws in metal, determining the thicknesses of lean and fat meat in livestock while the animals are still alive, and welding together pieces of metal without the metal melting.

In contrast to ultrasound, infrasound has few uses. Its main application is in seismographic surveying, in which the behavior of infrasonic waves generated by underground explosions is analyzed to determine the types and properties of surrounding rocks.

Infrasonic waves, with frequencies below 20 Hz, can be "heard" by some species of spiders, which probably use this ability to detect the approach of potential prey or predators. In addition to detecting very low frequencies, spiders can also respond to sounds with frequencies as high as 45,000 Hz.

Electricity

Electricity and all its observable effects result from properties of stationary or moving electric charges. There are two types of charges, known as positive charges and negative charges, which exist in all substances and in all states of matter—solid, liquid, and gas.

All substances are made up of atoms, which consist of a nucleus (containing protons and neutrons) and orbiting electrons. Protons are positively charged, and electrons are negatively charged. Electrons can be removed from atoms and made to take part in various electrical phenomena. Protons, on the other hand, are usually found as part of a cluster of protons, neutrons, and electrons known as an ion. Ions may be positively or negatively charged, depending on whether they have an excess of protons or electrons.

In an ordinary piece of material the numbers of protons and electrons within each molecule, and therefore within the material as a whole, are equal. The positive and negative charges cancel each other to a large extent, and no electrical effects are observed. But if extra charges of one type are added, the material is left with a net charge and electrical effects result.

Charges can be transferred from one material to another by rubbing them together. The study of these net stationary charges and their effects on each other constitutes the subject of static electricity or electrostatics.

In certain solid materials—metals in particular—some electrons are sufficiently "free" to be able to move through the material in a stream, forming an electric current. This is the basis of current electricity.

Solids, liquids, and gases can all conduct electric current if there are enough free electrons or ions. Electrons and positively charged ions move in opposite directions through the material, but in general both contribute to the current in a conducting medium. In an insulator, the electrons are too firmly fixed by chemical bonding to be able to move, and the material does not normally conduct electricity.

Sparks dance between someone's fingertips as they ground an electrostatically charged glass plate, photographed from below.

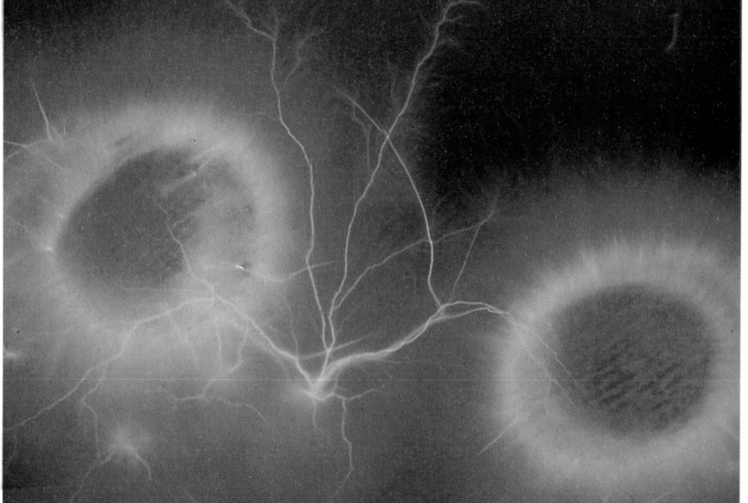

Static electricity

Static electricity can be produced by rubbing an insulating material such as a polyethylene rod with a dry cloth. Charges are transferred so that the polyethylene is left with a net charge of one type and the cloth with a net charge of the other type. The rod and cloth then attract each other.

On the other hand, two rubbed rods of the same material repel, and the two cloths repel. Unlike charges thus experience a force of attraction, whereas like charges repel each other.

It is impossible to find a material that is repelled by both polyethylene and cloth. There are therefore only two types of charges, called positive and negative. By convention, the charge on polyethylene is taken to be positive.

Electrostatic induction

Although nothing is repelled by both polyethylene and cloth, uncharged pieces of material (for example, small scraps of paper) are attracted to both. This results from a phenomenon called electrostatic induction. The paper is neutral overall, but the influence of the charged polyethylene redistributes the charges in the paper. The polyethylene is positively charged and exerts an attractive force on the paper's negative charges (which outweighs the repulsive force on the paper's more distant positive charges). The net result is mutual attraction.

A gold-leaf electroscope is an instrument that uses the repulsion of like charges as a means of detecting small amounts of charge. It has a strip of gold foil hinged to a metal plate and lying against it. When the electroscope is charged (by touching it with a charged rod), both the leaf and the plate acquire a similar charge and therefore repel each other. The leaf diverges from the plate, giving a visual indication that charge is present.

An electroscope can also be charged by induction. If a charged polyethylene rod is brought close to it, the leaf diverges from the plate, even though the rod is not touching the electroscope. Overall the instrument still has equal numbers of positive and negative charges. But the negative charges are attracted toward the top of the instrument by the polyethylene, and so the positive charges are repelled downward to the leaf and plate, causing them to diverge.

The phenomenon of electrostatic induction is important in a lightning conductor, which usually consists of a metal rod at the top of a tall building, connected to the ground by a thick metal wire.

When a negatively-charged thundercloud passes over a building, positive charges are induced on the roof. There is then a danger that an electric discharge or flash will pass between the cloud and the building. However, if there is a lightning conductor on the roof, the positive charges concentrate there. And if the conductor is sharply pointed, the charge concentration and hence the local electric field are very high.

Air molecules in the region may break down into positive and negative charges: they are then said to be forming an "ion wind." A stream of positive charges may flow upward toward the thunder cloud and neutralize its negative charges; but if an electric discharge does occur, it passes harmlessly to earth down the conductor.

Coulomb's law and electric field

The magnitude of the force between two charges depends on their size, distance apart, and the substance they are in. The force is proportional to the product of the charges, and inversely proportional to the square of the distance between them—a relationship known as Coulomb's law.

The force also decreases if the charges are placed in a material that undergoes electrical polarization, which separates the charges and has the effect of partly shielding them from each other. The amount of shielding is quantified by a property called the permittivity of a substance.

A vacuum, which cannot undergo polarization, has the lowest permittivity, called the permittivity of free space. The permittivity of other materials depends on their structure; ionic substances such as water, for example, generally have higher permittivities than non-ionic substances.

The concept of electric field is useful in considering the forces between charges. A charge creates an electric field in the region surrounding it, and another charge experiences a force if it is in this field. The direction

Artificial lightning flashes over a section of a 1,300 kV transmission system under test on a high-voltage rig. With an applied potential of 3.5 million volts, the air has become ionized and conducting, resulting in a spark nearly 16 feet (5 meters) long.

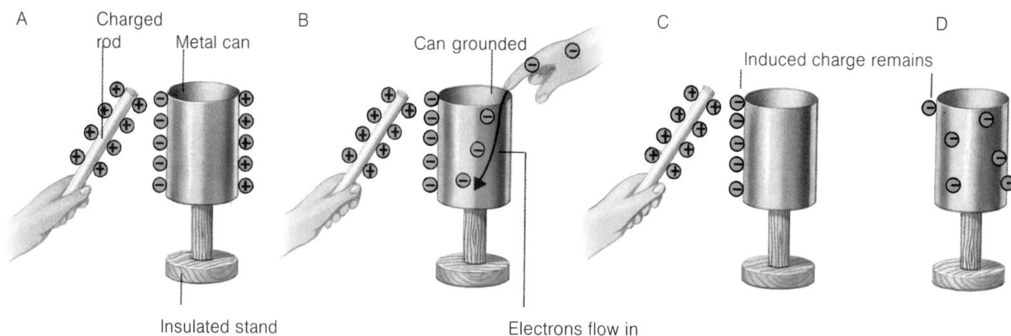

In charging by induction, a positively charged polyethylene rod is brought up to a metal can on an insulated stand (A), inducing negative charges on the side near the rod. The can is grounded (B) by touching it, when electrons flow in and neutralize the positive charges. When the finger is removed (C), only the induced negative charge is left. Finally, the rod is removed (D), and the induced charge spreads out evenly over the can.

of a field is taken to be the direction of the force on a positive charge placed in it. The field is strong where this force is large and weak where it is small. The strength of the field around a point charge decreases with distance, corresponding to the decrease in force between two charges as their separation increases. A visual representation of an electric field is made by drawing electric field lines.

If an insulator is placed in a gradually increasing electric field, it eventually breaks down and begins to conduct. The forces exerted by the electric field on charges in the material of the insulator become sufficiently great to free them from their fixed positions. Lightning flashes, for example, are caused by a breakdown in the insulation of the air when a field of more than 3 million volts per yard or meter is applied. The very large current that results consists mainly of electrons, which collide with air molecules. The collision of electrons with air molecules causes heating and the emission of light.

The electric field inside a hollow conductor is always zero, no matter how highly charged it is. The inside of the conductor is completely shielded from electrical discharges outside it. Thus the inside of an aircraft or a car is generally unaffected if it is struck by lightning.

Electrical potential

When a charged conductor is connected by a metal wire to an uncharged conductor, some of the charges flow from the former to the latter. The charges on the first conductor have potential energy, which they reduce by transferring to the second conductor. Initially, the first conductor is said to be at a higher potential, or voltage, than the second. When charged objects are connected together, positive charges flow from high to low potential, and negative charges from low to high poten-

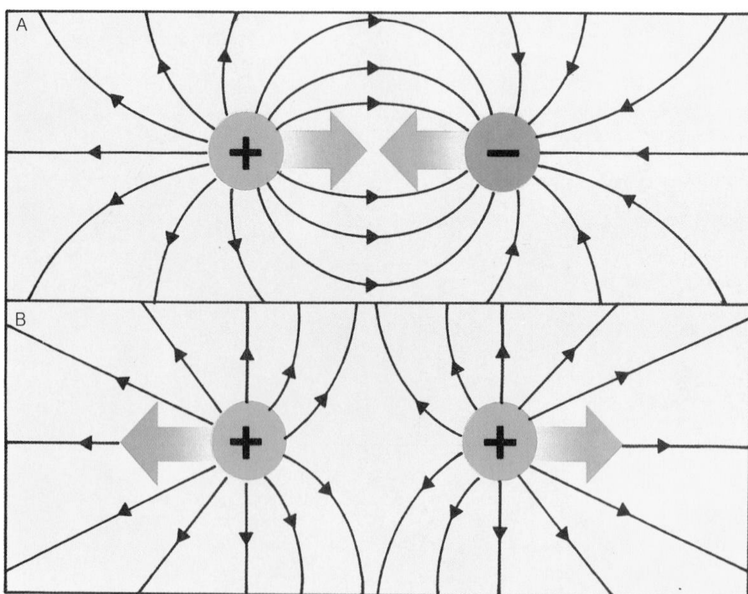

Unlike charges attract (A), and electric field lines join them (rather like the magnetic field lines that join the unlike poles of a bar magnet). Like charges—two positive charges or two negative ones—repel (B), as shown by their field lines.

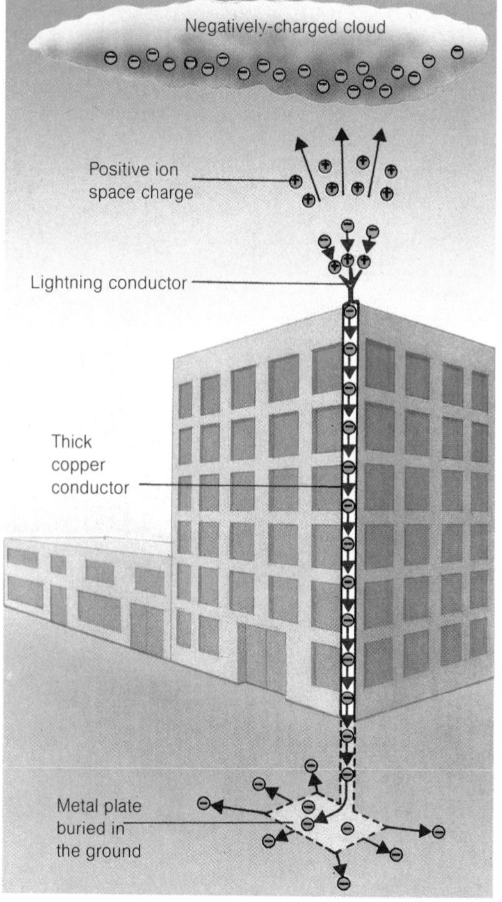

A lightning rod *(right)* has a pronged tip joined to a thick copper conductor that runs down the outside of the building to a metal plate buried in the ground. A negatively charged thundercloud overhead induces a positive charge on the rod's prongs (and a negative charge on the buried plate). An upward stream of positive ions above the rod constitutes a space charge, reducing the electrical forces between the cloud and the building.

tial, thus altering the potentials of the objects themselves. Charge continues to flow until they are all at the same potential.

In general, the greater the potential difference between two objects, the more readily charge flows between them. As a reference, the earth is taken as being at zero potential, so that all objects connected to earth by a conductor also reach zero potential (whether they are charged or not).

When polyethylene is rubbed with cloth, it acquires a positive potential; and the cloth acquires a negative potential compared with the earth. The potential difference is, however, only small. To create large potential differences, various kinds of electrostatic generators are used. One particularly important and useful type is the Van de Graaff generator, which is extensively used in research laboratories as an ion and electron accelerator.

Capacitance and capacitors

In a Van de Graaff generator, charge is transferred to a large hollow conductor at the top of the apparatus. A large conductor holds more electric charge at a given potential than does a small one and is said to have a larger electrical capacity, or capacitance.

The amount of charge stored depends on the capacitance of the conductor and the potential it is charged to, much as the amount of gas in a container depends on its size and the pressure of the gas. The capacitance can be significantly increased by bringing up a second, earthed conductor close to the first. Such an arrangement is called a capacitor, or condenser.

Capacitors used as charge-storing devices may have two parallel metal plates as the two conductors. Positive charge is stored on one plate and an equal amount of negative charge on the other. The capacitance is increased if the space between the plates is filled with a substance that can be polarized. Such a substance is called a dielectric, and it should have as large a permittivity as possible. It must also be an insulator, otherwise the charge stored on one plate would merely leak across the gap and neutralize the charge on the other plate.

The ratio of the capacitance of a capacitor with such a medium between its plates to that of the same capacitor with a vacuum between them is known as the relative permittivity, or dielectric constant, of the medium.

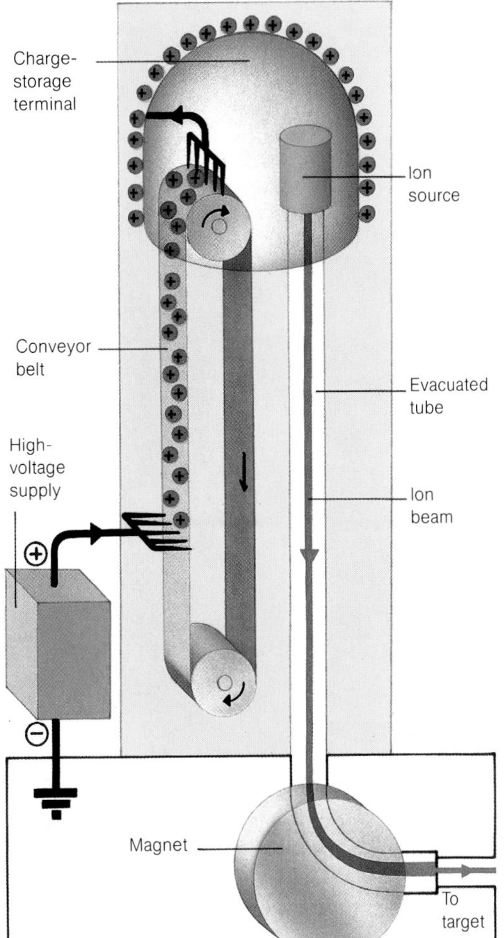

In a Van de Graaff accelerator, a high-voltage (50,000 volts) supply "sprays" positive charges onto a conveyor belt, which carries them to a domed terminal where they are stored at up to 6 million volts. The huge potential difference is used to accelerate ions along an evacuated tube. A magnetic field turns the ion beam through a right angle, and directs it at a target.

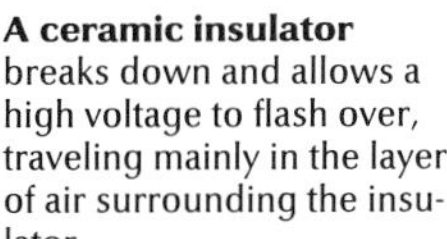

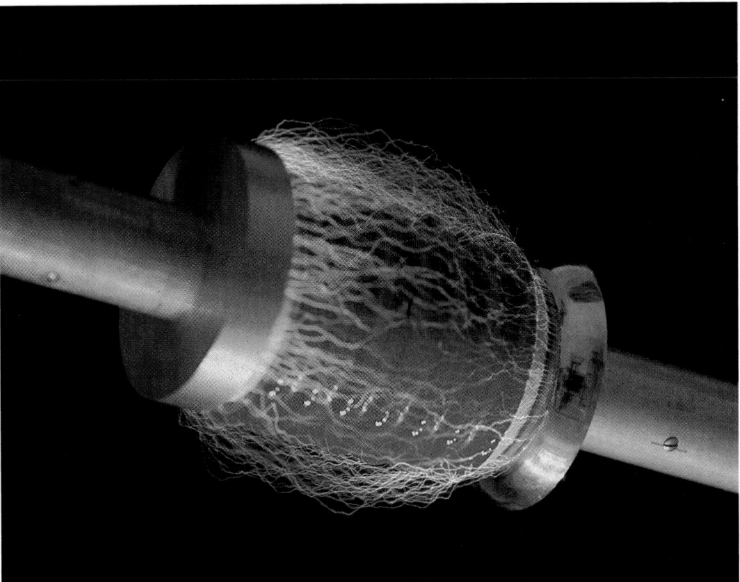

A ceramic insulator breaks down and allows a high voltage to flash over, traveling mainly in the layer of air surrounding the insulator.

Current electricity

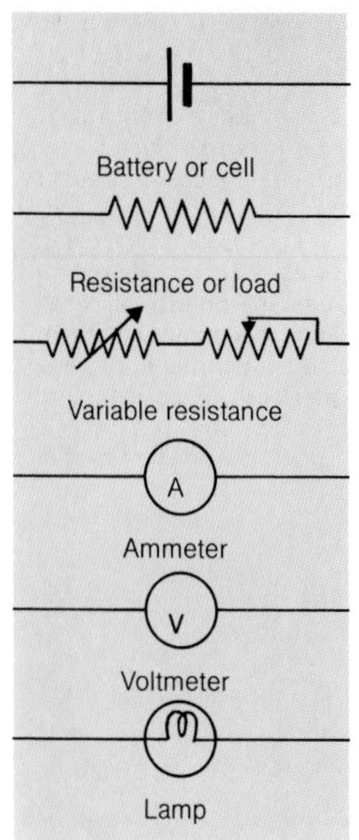

Conventional symbols are used to represent various circuit components.

Current electricity is the flow of charges—usually electrons or ions—through a conductor. (Static electricity, or electrostatics, on the other hand involves stationary charges.) An electric current can produce certain effects directly, such as heat, light, magnetic fields, and chemical reactions. Indirectly, through various types of electrical machines and devices, it can produce rotary motion and sound.

The most important concepts in current electricity are current, voltage, and resistance. Consider a pump driving water through a pipe in which there is a constriction at one point. The amount of water flowing in the pipe is measured in, say, liters per second, referring to the quantity of water passing a fixed point in the pipe in a given time. There is water in the pipe all the time, whether or not the pump is switched on; the purpose of the pump is to provide a pressure difference between the ends of the pipe to make the water flow. Furthermore, the greater the pressure difference, the greater is the flow.

The other factor that affects the flow is the resistance offered by the constriction in the pipe—it is more difficult for the water to flow through the constriction than through the rest of the pipe. If the constriction narrows and its resistance increases, the amount of water that flows throughout the whole pipe system decreases.

In many ways the flow of an electric current in a circuit is similar. A battery or cell provides a potential difference (analogous to the pressure difference), which forces the current along the wires. The size of the current measured in coulombs (charge per second, analogous to liters per second) depends on the size of this potential difference and on the resistance to current flow caused by the various components.

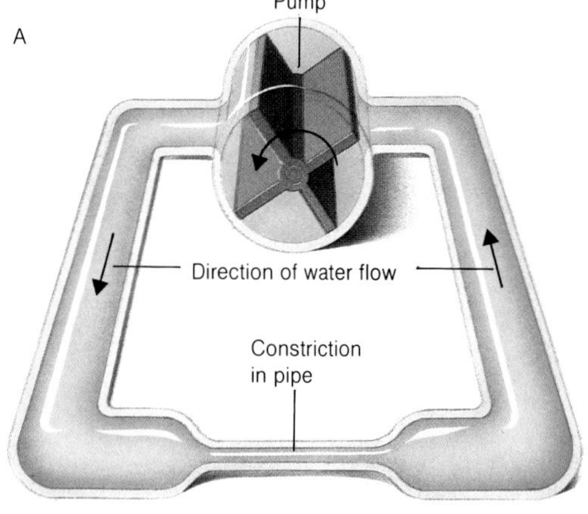

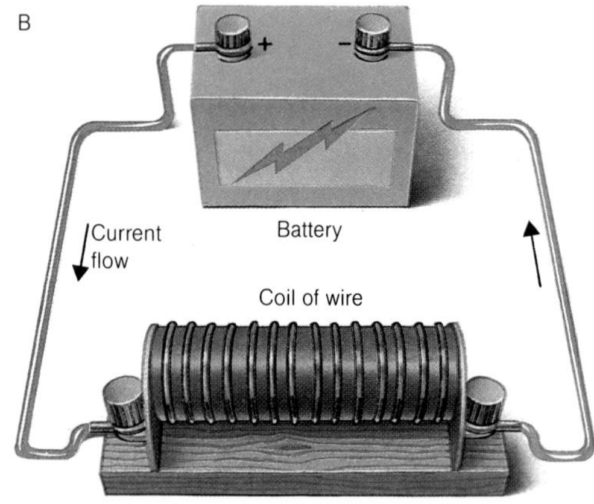

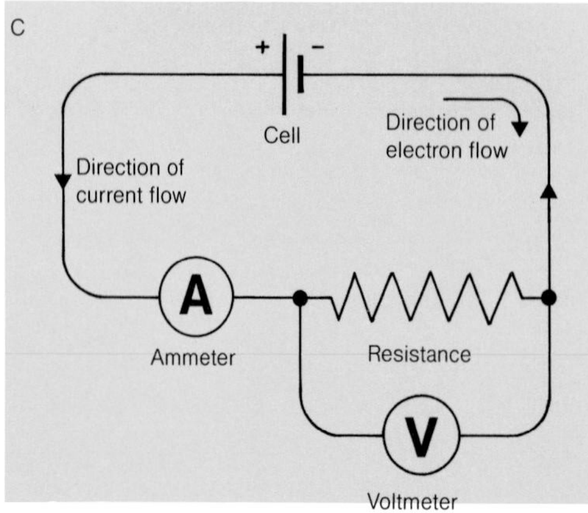

A simple electric circuit, with a battery and a resistive load, can be likened to a pump driving water around a closed pipe with a constriction in it (A). The water current is constant, and there is a pressure difference across the constriction. In the equivalent electric circuit (B), the battery acts like the pump, and the coil of wire—a resistance—is analogous to the narrow part of the pipe in diagram A; there is a potential difference across it. Current flows from the positive terminal of the battery, through the coil, and back to the negative terminal. The electric current is the same in all parts of the circuit. Diagram C uses conventional symbols to represent the same circuit, with the addition of an ammeter (which measures the current flowing around the circuit) and a voltmeter (which measures the potential difference across the resistance). By convention, current is taken to flow from positive to negative, although the actual flow of electrons is in the opposite direction.

A simple electrical circuit

If a resistor is connected to the terminals of a battery using copper wires, a current flows. The current consists of negatively charged electrons flowing through the wires and the resistor from the negative to the positive terminal of the battery, and then through the battery itself from the positive terminal to the negative. It is equivalent in its effects to a flow of positive charges in the opposite direction—that is, from the positive terminal through the wires and the resistor to the negative terminal. This is the convention that is used: current is said to flow from positive to negative outside the battery and from negative to positive inside it.

Current electricity consists then of a movement of electric charges. The amount of charge passing a fixed point in the circuit per second is called the current and is measured in amperes, or amps. The potential difference provided by the battery is called its electromotive force, or EMF, measured in volts. The larger the EMF, the greater is the current that flows. There is a potential difference or voltage across the resistor (which resists the current flow) and its resistance is measured in ohms. The electric potential is higher where the current enters the resistor and lower where it leaves.

Resistance and Ohm's law

The resistance of the resistor is defined as the potential difference across it divided by the current flowing through it. If the resistance is constant, then the current through the resistor is proportional to the voltage across it. This relationship is known as Ohm's law.

All conductors have at least some resistance, but they do not all obey Ohm's law. For example, an ordinary light bulb does not. The reason is that as the current passing through the filament of the bulb increases, it gets hotter; and in metal (of which lamp filaments are made) heating causes an increase in resistance.

Overhead power lines criss-cross the landscape, distributing high-voltage electricity for industrial and domestic use. The electrical engineer faces a dilemma: to use thin cables to keep their weight to a minimum, or to use thick cables to keep their electrical resistance to a minimum. As in much of engineering, the practical solution is a compromise between the two conflicting requirements.

The speed of electric current

Even when a conductor is not carrying an electric current, the heat energy it inevitably possesses causes the free electrons to move about randomly at speeds up to 620 miles (1,000 kilometers) per second. When a potential difference is applied, the free charges acquire a drift velocity, which is superimposed on their random movements. The drift velocity is low, typically less than an inch or a few millimeters per second, so it would take several minutes for a charge to pass right along a wire a yard or meter in length. But an electric signal as a whole is extremely fast—it can travel up to the speed of light. This is because it is passed directly from one electron to another when they collide. Such collisions are extremely frequent in metals because their "conducting" electrons are free to move, and there are many of them.

Direct current and alternating current

A *d*irect *c*urrent (DC) is one that flows around a circuit in the same direction all the time. An *a*lternating *c*urrent (AC) is one that reverses its direction at regular intervals, usually many times a second. The domestic electricity supply in America is 60 Hz AC, meaning that the current changes its direction of flow 120 times every second.

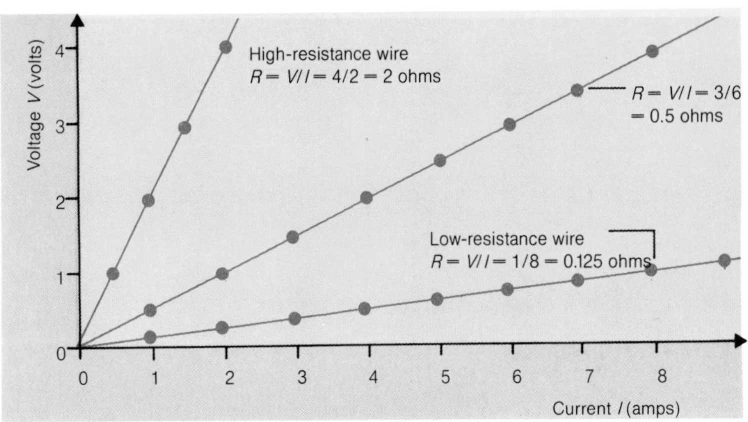

Ohm's law states that resistance equals voltage divided by current. A graph of voltage against current *(above)* is a straight line, whose slope equals resistance.

Christmas tree lamps are often wired in series and connected to the main electricity supply, so that their resistances "drop" the voltage. For example, each of the 60 lamps in series on a 240-volt supply passes only 4 volts.

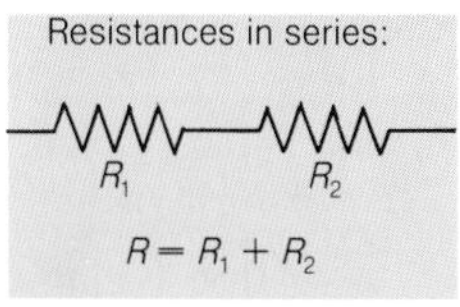

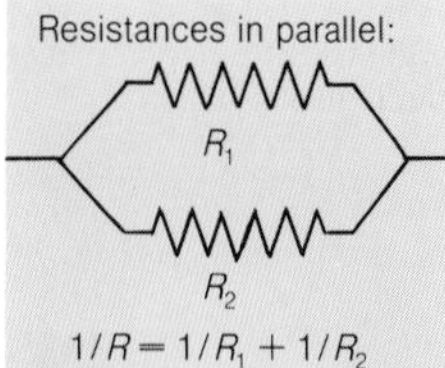

The combined resistance of a combination of resistors depends on how they are connected: in series or in parallel.

Direct current circuits

The resistance to current flow in a DC circuit is known as the load. Usually a load consists of not one but several different resistances, which are equivalent in their effect to a single resistor. But the size of this equivalent resistor depends on how the component resistances are connected together.

Resistors in series and parallel

If a set of resistors are joined end to end they are said to be connected in series. The resistance of the equivalent resistor is found by simply adding together the component resistances. Whatever current flows out of one must flow into the next—the current cannot alter or disappear unless the wire branches; the current in each resistor is thus the same.

If the connecting wire in a circuit branches, however, and the current flows along two separate paths through two different resistors before recombining, the resistors are said to be connected in parallel. When the current arrives at the branching point it divides and some flows down each branch. This is "easier" than forcing all the current through only one of the resistors, and as a result the equivalent resistance of the circuit is less than either of the component resistances. (The reciprocal of the equivalent resistance is equal to the sum of the reciprocals of the component resistance.) More complicated resistive circuits are a combination of resistances in series and resistances in parallel. And in all circuits the resistance of the connecting wires themselves contributes to the overall resistance of the circuit.

The potentiometer

A potentiometer is a variable resistor. An example of this is the volume control on a radio receiver. One of the uses of a potentiometer in the laboratory is for finding—or, more strictly speaking, comparing—the EMFs of cells or other DC voltage sources. It consists of a battery driving a constant current around a circuit that includes a straight length of bare wire of high resistance. The potential (or voltage) across the wire decreases uniformly from one end to the other. Therefore the potential difference between one end of the wire and a movable contact can be altered by moving the contact to various points along the wire. If the contact is positioned so that the potential difference is the same as the EMF of another cell, this cell can be connected so that the two voltage sources oppose each other. As a result, there is no effective current flow and a galvanometer connected to the cells reads zero. If the second cell is then removed and replaced by another, a different balance length (the length of wire needed for the galvanometer to read zero) is required. The ratio of the balance lengths is the ratio of the EMFs of the two cells. If one of the EMFs is known (for example, by using a standard cell), the other can be calculated.

The Wheatstone Bridge

Bridges are circuits used for finding (or comparing) resistances. The one improved upon by Charles Wheatstone in 1843 is particularly useful in physics experiments. A galvanometer forms a "bridge" between two parallel circuits, each containing two resistors. If the four resistances are chosen at random, a current flows through the galvanometer. But if their values are related in a particular way, no current flows and a null reading is obtained on the galvanometer. The required relationship is that the ratio of the two resistances in each arm of the bridge is the same.

Suppose only one of the resistances in one arm of the bridge is known. The ratio of the

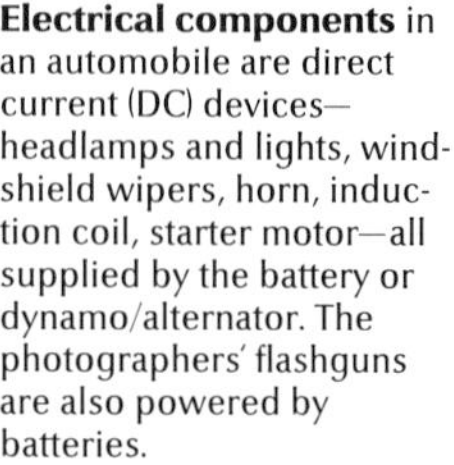

Electrical components in an automobile are direct current (DC) devices—headlamps and lights, windshield wipers, horn, induction coil, starter motor—all supplied by the battery or dynamo/alternator. The photographers' flashguns are also powered by batteries.

Most underground railways use DC electricity, which is collected through a third rail (and returned through the running rails), called the three-rail system. In the system illustrated here, there are two extra rails, one supplying current and one returning it to complete the circuit.

two resistances in the other arm of the bridge is altered until a null reading is obtained. This ratio is then the same as the ratio of the known resistance to the remaining, fourth resistance, whose value can thus be calculated. In summary, in order to find one of the resistances we need to know one of the other three plus the ratio (but not the actual resistances) of the remaining two. A form of the Wheatstone Bridge circuit that satisfies these criteria is the Meter Bridge, which has a meter of high-resistance wire and movable contact (like a potentiometer) that forms two of the resistances.

Resistivity and conductivity

The resistance of a piece of material depends on its composition and its size. Consider a cylindrical conductor, with a current entering one end and leaving at the other. The wider the cylinder, the "easier" it is for the current to flow along it, and the lower is its resistance. (In the same way, more water flows through a wide pipe than through a narrow one.) Also, the longer the cylinder, the greater is the potential difference needed to make a given current flow, and the higher the resistance.

Electrical wires are essentially of cylindrical shape. A thick, short wire has a lower resistance (and carries more current for a given potential difference) than a long thin wire. But a copper wire of a given size and shape has a lower resistance than a similar tungsten wire. Copper is the better conductor; it is said to have lower resistivity than tungsten—in fact, only one-third as much. The reciprocal of resistivity is called the conductivity of the material. Copper has a conductivity about three times that of tungsten.

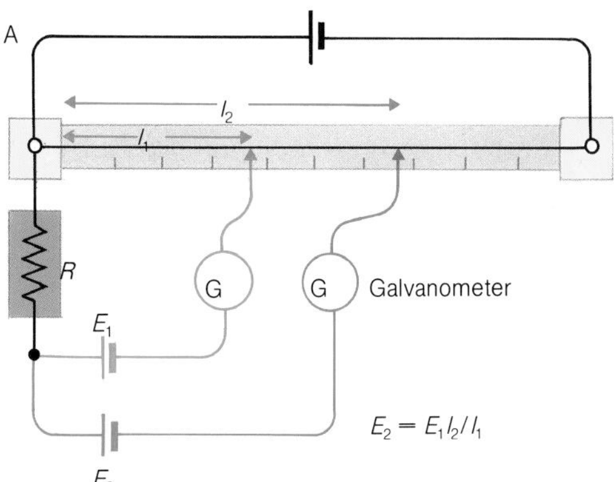

A laboratory potentiometer (A) can be used to find the EMF of a cell. First a cell of known EMF (E_1) gives a balance point l_1 (no deflection of the galvanometer). Then the cell of unknown EMF (E_2) is connected and balanced at l_2. A Meter Bridge (B) can be employed to find an unknown resistance (R_1) using a known resistance (R_2) and a potentiometer.

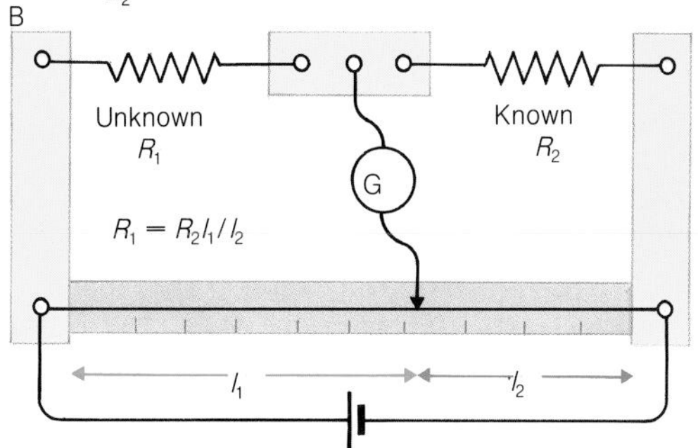

The resistance of wire depends on its composition, length, and thickness. The diagram below compares two thicknesses of 100 meters of copper, aluminum (both low resistance), and Nichrome (high resistance) wires. In customary units, the wires are 328 feet (100 meters) long with thicknesses of $\frac{1}{25}$ inch and $\frac{1}{12}$ inch.

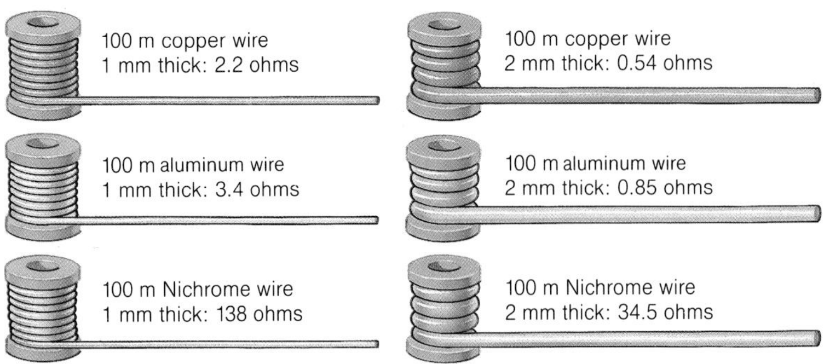

Alternating current circuits

AC source

Resistor

Capacitor

Inductor

Diode

Symbols commonly used in AC circuits are shown above.

When an alternating voltage is applied to an electrical circuit, an alternating current (AC) of the same frequency results. The size of the current is, however, different from that which results from using a direct current (DC) voltage source, and the behavior of the circuit is also fundamentally different.

When an alternating current flows, resistors are not the only circuit components that "resist" the flow of current. Others, such as capacitors and inductors (with a property called reactance), also resist the flow. The chief difference is that whereas current flowing in a resistance generates heat, current flowing in a component with reactance does not. Resistances and reactances have different uses. A space heater must have resistance; on the other hand, in many electronics applications the generation of heat is undesirable and reactive components are very important.

In general, the various components in an AC circuit have both resistance and reactance. The total resistance to current flow is described by what is called the impedance of the circuit. The greater the impedance, the lower is the current; the lower the impedance, the greater the current.

Capacitors and inductors

A capacitor cannot pass direct current—the dielectric between the plates of a capacitor is an insulator—but it can pass alternating current. If positive charges "flow" into one plate through one connection, negative charges are electrostatically attracted to the other plate through the other connection. In effect, the current passes through the device, although a different set of charges is involved on each side. During the flow of AC the capacitor is continually charged and then discharged. When the frequency of the AC is high, and the charging and discharging process occurs very quickly, a large current can flow. As a result, the reactance of a capacitor is low if the frequency is high. Unlike a resistor, its resisting effect alters as the frequency of the AC alters. When the frequency is zero, the reactance is theoretically infinite (corresponding to the fact that a capacitor cannot pass DC).

A common type of inductor is a solenoid (a cylindrical coil of insulated wire), usually with an iron core to increase its inductance. When the current flowing in an inductor changes, the magnetic field it produces changes and induces an EMF in the circuit. The induced EMF in turn affects the flow of current. If the current is increasing, the induced EMF opposes it; if the current is decreasing it tries to maintain it. An inductor therefore reacts to changing currents, but is unaffected by a constant current. The reactance of an inductor varies with frequency. It is zero when the frequency is zero (no opposition to DC), but increases as the frequency increases. Essentially, this is the reverse of the behavior of a capacitor.

An ideal capacitor has infinite resistance for DC current. However, this is never quite so in practice. An ideal inductor has zero resistance for DC current. Again, in practice this is not possible, because the wire from which it is made inevitably has some resistance, and so a certain amount of energy is lost in the inductor as heat.

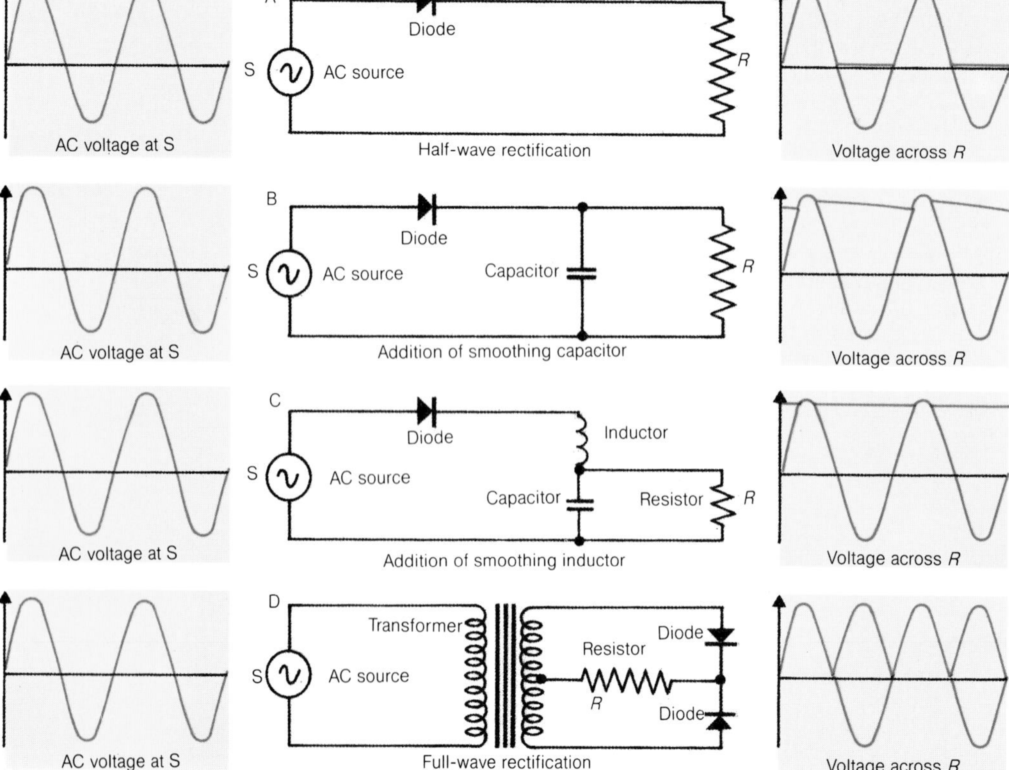

Most electric power is generated as alternating current (AC), so that for many devices and electronic circuits, which require direct current (DC), the AC has to be rectified. The simplest form of rectifier is a single diode (A), which produces a series of half-wave pulses. The addition of a capacitor (B) produces a better DC, which can be further "smoothed" by including an inductor in the circuit (C). Two diodes can be employed to provide fullwave rectification (D). In each case, the input and output voltages are shown here as red lines.

Modern street lighting and advertising signs use alternating current, usually from the main electricity supply; so do most domestic appliances and most of the machines employed in industry and commerce.

Phase differences in AC circuits

In a DC circuit, the current and voltage are always in phase with each other. As a result, if the voltage rises, the current rises at the same time; if the voltage falls, the current also falls. This is not always the case with AC. The current and voltage can be out of phase, usually by different amounts in different components. In a capacitor the current leads the voltage—an increase in voltage is preceded by an increase in current. In an inductor the current lags behind the voltage. The amount of lead or lag depends on how closely the actual components approach the ideal. In a resistor, AC current and voltage are always in phase with each other, as in a DC circuit.

Resonance in AC circuits

When a resistor, an inductor, and a capacitor are connected in series, the resulting circuit has an impedance that depends on the resistances and reactances of the separate components. The impedance varies with frequency and reaches a minimum at one particular frequency called the resonant frequency. At this point the effective impedance of the circuit is a minimum and, as a result, the current is a maximum. Thus a circuit of this type can be "tuned" to one particular frequency, giving a large current at that frequency but only a small current at all other frequencies. This phenomenon is called electrical resonance.

Electrical resonance forms the basis of tuning circuits in radio and television receivers, which may have a capacitor and an inductor in series or—more commonly—in parallel. The circuit selects just one of the many signals that arrive at the antenna by passing a high current at that frequency, and ignores the others. The frequency selected can be varied by altering the size of the capacitance in the circuit, which for this reason usually takes the form of a variable capacitor.

Energy in AC circuits

In an AC circuit, energy is provided by a voltage source and dissipated in the various circuit components, as in a DC circuit. The source is an AC generator or alternator, working on the principles of electromagnetic induction. (Cells and batteries are sources of DC only.) But only components that have resistance dissipate any energy. Reactances absorb energy in parts of the AC cycle, but then return it to the circuit in the other parts. The overall result is that pure capacitors and inductors do not "use" any electrical energy.

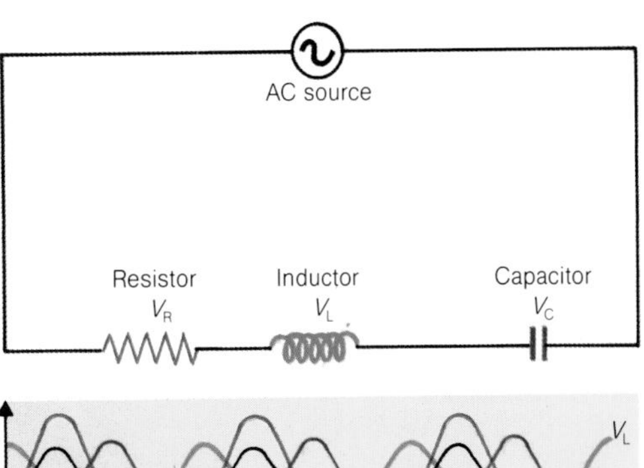

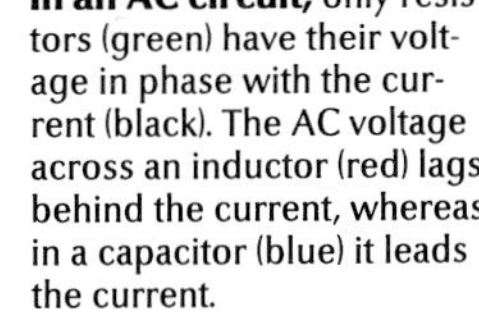

In an AC circuit, only resistors (green) have their voltage in phase with the current (black). The AC voltage across an inductor (red) lags behind the current, whereas in a capacitor (blue) it leads the current.

Cells and batteries

Cells are DC voltage sources in which chemical energy is converted into electrical energy—a process that is essentially the reversal of electrolysis. The term battery originally described an assembly of several cells, although today it is often used merely as another word for a cell.

Primary cells

A primary cell gradually uses up the chemical fuel it contains and cannot be recharged. The original Daniell cell, for instance, consists of a plate of copper and another of zinc immersed in an electrolyte of dilute sulfuric acid. The zinc slowly dissolves in the acid, forming zinc ions and leaving behind electrons on the zinc plate, which therefore becomes negatively charged. At the copper plate, positively-charged hydrogen ions (present in the acid) are neutralized by gaining electrons from the copper. The hydrogen is liberated as bubbles of hydrogen gas, and the copper is left positively charged. When a length of wire is connected between the electrodes, electrons flow from the cathode to the anode.

Such an arrangement, called a simple cell, has various drawbacks (such as polarization, which stops the current flow after a short time), and more complex designs have been developed. An example is the Leclanché cell, which has been further developed into the Leclanché dry cell—e.g., a common flashlight battery. It has the advantage of using only dry components or ones in the form of paste and is, therefore, easier to transport and store.

Other types include the Weston cadmium cell, important for standardizing other voltage sources, and mercury and silver oxide cells. They all work on the same principle. At one electrode, chemical oxidation takes place and as a result a negative charge builds up. At the other electrode, chemical reduction takes place, which uses up negative charges and leaves the electrode with a net positive charge. An *e*lectro*m*otive *f*orce (EMF) develops between the electrodes, and if they are connected by a wire a current flows. All such cells have a limited lifetime because the oxidation or reduction process reaches completion. In the simple cell, for example, the reaction must stop when all the zinc has dissolved. The total energy the cell is able to deliver—its capacity—is therefore limited. Standard cells are never used as sources of current; their voltage is "standard" only at low current drain.

The EMF of a cell depends on the chemical composition of the electrodes and electrolyte, not on how large it is. The size of a cell does, however, affect its capacity. Large electrodes and a large volume of electrolyte give a cell a large capacity—it will continue to produce current for a longer time. The current it is able to deliver depends partly on its internal resistance, which in turn depends crucially on the size and separation of the electrodes. A cell with large electrodes close together has a low internal resistance and is consequently capable of delivering a larger current.

Fuel cells

In the simple cell, the zinc electrode dissolves away, zinc ions appear in the electrolyte, and hydrogen is lost from the cell in the form of gas (at the copper electrode). If it were possible to make good the loss of hydrogen, to remove the zinc ions from solution, and to replace the zinc electrode as the cell operates, the cell would not become exhausted. This is the principle of a fuel cell—it is a primary cell that is continuously supplied with a chemical fuel as it is being used.

In a fuel cell, hydrogen or a liquid fuel is fed to one electrode, and oxygen or air is supplied to the other. In the hydrogen-oxygen type (which has electrodes of porous nickel) hydrogen enters the electrolyte as positive hydrogen ions, leaving behind negatively-charged electrons on the electrode—which therefore becomes the cathode. The ions move through the electrolyte to the other electrode, which is supplied with oxygen, where they combine with the oxygen to produce water. In doing so, they take electrons from this electrode, leaving it with a net posi-

A simple cell (A, *below*) has electrodes of copper (Cu) and zinc (Zn) in an electrolyte of sulfuric acid (H_2SO_4). Hydrogen (H_2) is evolved at the copper. In a Daniell cell (B), a copper can containing copper sulfate ($CuSO_4$) solution is one electrode; the other is a zinc rod in a porous pot containing sulfuric acid. The original Leclanché cell (C) has electrodes of zinc and carbon in manganese dioxide (MnO_2), and ammonium chloride as electrolyte. In a dry cell (D) the electrolyte is in the form of a paste. A typical lead-acid accumulator (E) has electrodes of lead oxides and lead (Pb) in sulfuric acid electrolyte.

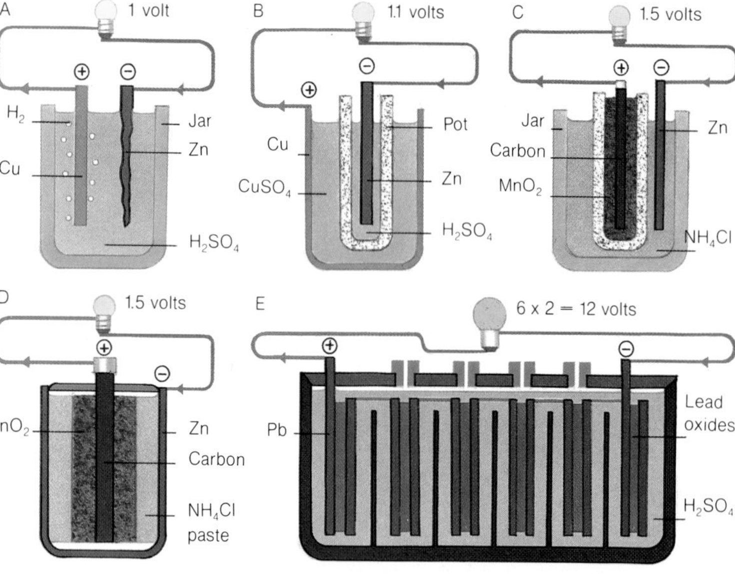

An accumulator can be recharged by connecting it to an external DC voltage source. Bubbles of hydrogen and oxygen, a potentially explosive mixture, form above the electrodes. Recharging should thus be carried out in a well ventilated place.

An electric fish *(Gnathonemus petersi),* sometimes called an elephant fish because of its long snout, has a pair of electric organs astride its tail that act like biochemical batteries, emitting rapid pulses of electricity. The fish uses this power to attack prey or defend itself.

tive charge and making it the anode of the cell. A hydrogen-oxygen fuel cell essentially reverses the electrolysis of water (in which an electric current splits it into hydrogen and oxygen gases).

Secondary cells

With certain combinations of electrodes and electrolyte, it is possible to rejuvenate the cell when it is exhausted and return it to its original condition by forcing a current through it in the reverse direction. An external voltage source, such as a battery charger, is used to pass current in at the positive terminal and out at the negative one.

A cell that can be recharged in this way is called a secondary cell, or accumulator. Two particularly important types are the lead-acid accumulator and the nickel-iron alkali accumulator.

The lead-acid type of accumulator has electrodes consisting of lead oxide and lead plates immersed in sulfuric acid; the plates are large and close together, so as to give a very low internal resistance. The EMF is about 2 volts. A set of 6 cells arranged in series comprises the familiar 12-volt car battery, which is capable of delivering a current of up to 200 amps for a short period. The nickel-iron alkali type uses a nickel hydroxide cathode, an anode of iron, and an electrolyte of potassium hydroxide solution. There are also small nickel-cadmium rechargable cells, with an alkaline electrolyte. Alkali cells have EMFs of only 1.2 to 1.5 volts, but they have a longer lifetime than do lead-acid cells. They are also lighter and more robust.

Efficiency and cost

Cells (and batteries) convert chemical energy to electrical energy with an efficiency of about 90 per cent—much higher than any other method of electricity production, including the burning of fossil fuels (such as coal and oil) in conventional power stations. But the cost of electricity from cells is much higher because the chemicals are expensive. Research continues into ways of using fossil fuels or their derivatives to generate electricity chemically in fuel cells.

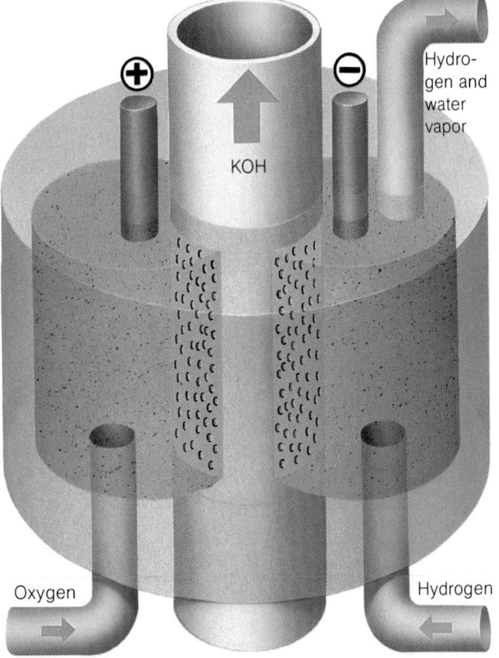

A schematic fuel cell *(left)* uses gaseous hydrogen as a chemical fuel. Hydrogen and oxygen are pumped into electrodes made from a porous nickel "sponge" (shown green), which are in contact with a circulating electrolyte of potassium hydroxide (KOH) solution. Hydrogen and oxygen combine to form water, which passes out of the cell as vapor with excess hydrogen. The American Apollo Moon lander and its electric "car" *(below)* had to derive all its electric power from solar panels, batteries, or fuel cells.

Sparks fly around a welder as he makes use of the heat of an electric arc to join two pieces of steel. Some of the electrical energy is also converted to light. The production of heat and light are among the most common ways in which electricity is put to use.

Electricity at work

Electricity is a convenient and flexible form of energy. It can be generated, stored, and distributed with comparative ease and—more importantly—it can readily be converted into other forms of energy, such as heat, light, sound, and mechanical energy. By means of electrolysis, electricity can be made to bring about chemical reactions. It is also clean and, with adequate safety precautions, safe to use.

Electrical heating and lighting

When an electric current (DC or AC) flows through a resistor, heat is produced—known as Joule heating. A large current flowing through a high resistance produces more heat than does a small current through a low one. The heat passes to the surroundings by a combination of conduction, convection, and radiation. In a domestic electric stove, radiation is the most important of the three; an immersion heater relies mostly on convection; and a cooker or electric blanket heats principally by conduction.

In an electric stove, the heat element (a resistor) becomes sufficiently hot to glow red. In an electric lamp, the tungsten filament glows white hot and becomes incandescent, reaching a temperature of about 4,500° F. (2,482° C). But in an ordinary filament lamp, only about 10 per cent of the electrical energy is emitted as light (the rest is lost as heat). Fluorescent strip

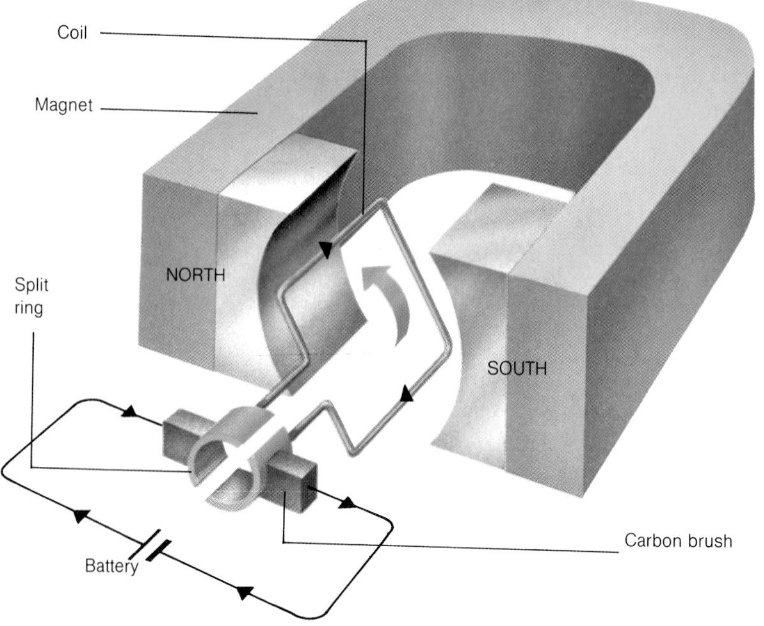

In a simple DC motor current is fed from a battery to brushes pressing on a split ring (commutator), which are in turn connected to a coil in the field between the poles of a magnet. The split ring reverses the direction of the current every half-turn, so that the coil continues to turn in the same direction.

lights are more efficient. The current is carried by electrons and positive ions that collide with gas atoms in the tube and cause them to emit ultraviolet light, which in turn activates the fluorescent coating on the inside of the tube.

Electrolysis

Electrolysis is the chemical decomposition that some substances undergo when an electric current passes through them. The substance to be decomposed—the electrolyte—is often an aqueous solution. The current enters or leaves it through metal or carbon plates called electrodes, entering through the anode (positive plate) and leaving via the cathode (negative plate). In aqueous solution, most acids, bases, and salts are dissociated into ions, and it is these that carry the current. Fused (molten) salts and alkalis are also ionized and can be employed as electrolytes.

A typical example is the electrolysis of copper sulfate solution using platinum electrodes. Both the copper sulfate and the water are dissociated into ions. When a voltage is applied between the electrodes, positive ions are attracted to the negatively charged plate (the cathode) and the negative ions to the positively charged plate (the anode). This flow of ions constitutes the electric current. At the anode negative hydroxyl ions (OH^-) from the water lose an electron each and then combine in pairs to form water and oxygen atoms; the oxygen atoms combine in pairs to form oxygen molecules, which are emitted as bubbles of gas. At the cathode copper ions (Cu^{2+}) gain two electrons each, producing copper atoms that are deposited as a thin layer of very pure copper.

The electrolysis of copper sulfate solution is employed on a commercial scale to purify copper. Large plates of impure copper are used as anodes, and thin sheets of pure copper form the cathodes. When a current passes, copper is removed from the anodes and deposited in pure form on the cathodes. Impurities are left behind as a sludge in the plating tank.

Electrolysis is the only commercial method for extracting aluminum from its oxide, alumina. The alumina ore (bauxite) is melted with two other minerals, cryolite and fluorspar, and electrolyzed, using copper electrodes. If the bottom of the furnace is made to form the cathode, a pool of aluminum is formed and can be run off at regular intervals. Because the process needs a lot of electricity, aluminum smelters are often sited where cheap hydroelectricity is available.

Electroplating

Steel or other cheap metals may be coated with a thin layer of another one, either to improve their corrosion resistance or their appearance. Steel car bumpers may be coated with chromium; cutlery may be coated with silver or (more cheaply) nickel or chromium.

The coating may be carried out by an application of electrolysis called electroplating. The plated article forms the cathode of the plating bath, with usually a piece of the coating metal as the anode. With the correct techniques, the electroplated layer can be made extremely thin but very even. Alloys such as brass can also be electroplated onto other metals.

A giant electric motor for a petrochemical plant has this stator, 16 feet (nearly 5 meters) across. The complete motor weighs more than 88 short tons (80 metric tons) and provides 7,800 brake horsepower of mechanical power (5.8 megawatts) to drive a huge compressor.

The motor effect

If an electric current flows along a wire placed across a magnetic field, a force is exerted on the wire, which, if not rigidly fixed, moves. This is the principle of the electric motor, often called the motor effect. The size of the force depends on the strength of the field and the current, and on the length of wire and its orientation in the field. The maximum force occurs when the wire is at right angles to the field.

Nickel plating is used on steel outer cases for electric toasters to improve their appearance and resistance to corrosion. Electroplating is one application of electrolysis; others include the extraction of reactive or precious metals, such as aluminum or gold, and the refining of metals, such as copper, to a high state of purity for use as electrical conductors.

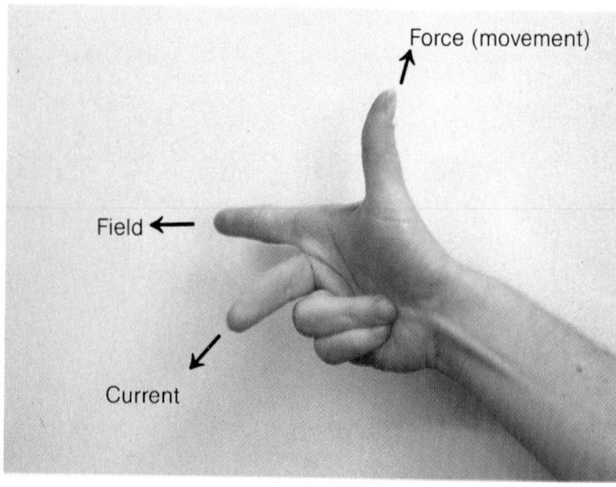

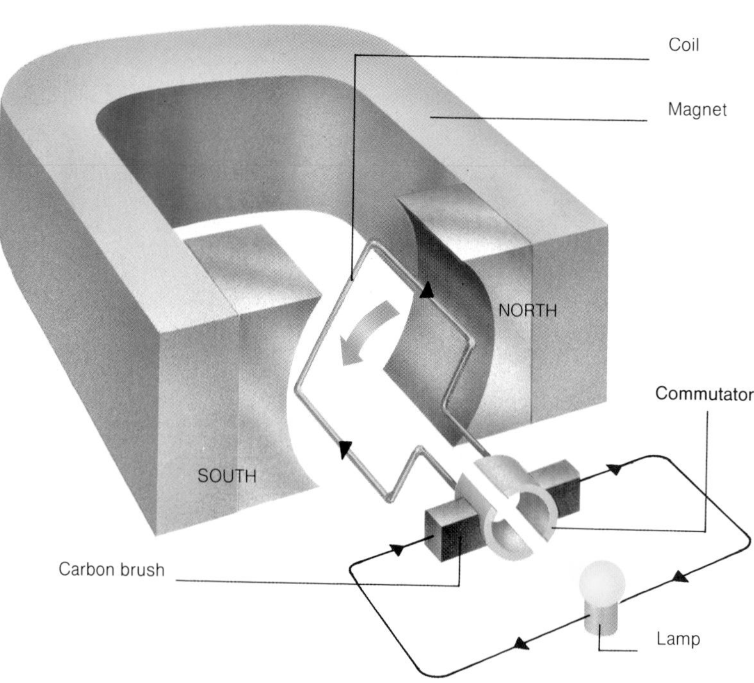

Fleming's right-hand rule relates the direction of the current in a conductor moving in a magnetic field to the directions of movement and the field. In a simple DC generator or dynamo *(right)*, current induced in a coil rotated in the field between the poles of a magnet is passed via a split ring (commutator) and brushes so that it always flows in the same direction, which can be determined by applying Fleming's right-hand rule.

In an electric motor, current flows through coils of wire that are free to rotate on an axle in a magnetic field. The motor effect gives rise to forces on the wire, which create a couple or torque about the axle that causes the coils and axle to rotate. In small motors the magnetic field is provided by permanent magnets; in large ones, electromagnets are used.

Galvanometers, which form the basis of most instruments for measuring current and voltage, employ the same principle, but the rotation of the coils and axle is opposed by a spring or torsion wire. The rotation stops when the torque due to the motor effect is exactly balanced by the counter effect offered by the spring or torsion wire. The larger the current, the farther the axle rotates before stopping. The angle through which it does rotate is a measure of the current flowing through the coils.

In a moving-coil loudspeaker, a varying current passes through a coil of wire placed in a magnetic field. The motor effect causes movement of the coil, which is attached to a cone of paper or plastic. The resulting movements of the cone vibrate the surrounding air, creating sound waves.

Electromagnetic induction

If a conductor is made to move across a magnetic field, an electric current flows in the conductor. This phenomenon of electromagnetic induction is in many ways the reverse of the motor effect. It is a completely different way of producing electricity to cells and batteries (in which chemical reactions are the source of energy). In electromagnetic induction, the source of the energy is the mechanical work that must be done to move the conductor. It is also the only practical way at present of producing electricity in large quantities.

The voltage induced depends on the strength of the field, the length of the conductor in the field, and the speed at which it moves. If any of these is increased, the induced voltage increases. The voltage also depends on the orientation of the conductor within the field and the direction in which it is moved. Maximum voltage occurs when the

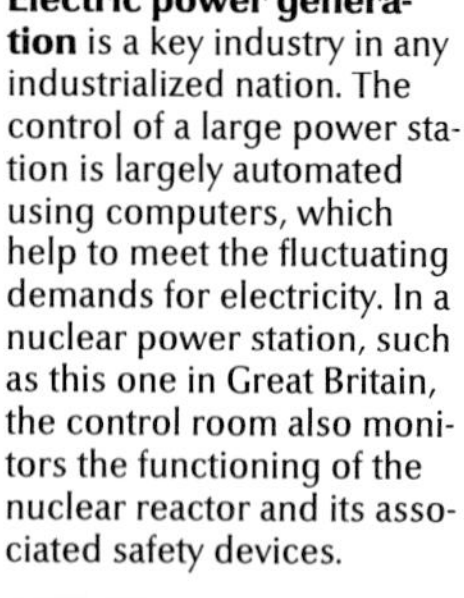

Electric power generation is a key industry in any industrialized nation. The control of a large power station is largely automated using computers, which help to meet the fluctuating demands for electricity. In a nuclear power station, such as this one in Great Britain, the control room also monitors the functioning of the nuclear reactor and its associated safety devices.

conductor moves at right angles to the field. The direction of the induced current, relative to the directions of movement and field, is given by Fleming's right-hand rule.

The dynamo

The most important application of electromagnetic induction is the dynamo. This is a device in which mechanical energy (rotational energy) is converted into electrical energy by inducing currents in coils of wire made to rotate in a magnetic field. As the coils rotate, they cut through the field lines first in one direction and then in the opposite direction. Each time the direction changes, so does the direction of the induced current; an alternating current (AC) is therefore produced. The current is led out of the coils using two slip rings and brushes. The whole device is called an AC dynamo, or alternator.

In a DC dynamo, the AC produced by the coils is converted to DC using a different system to "collect" the current. A single ring with two gaps in it, called a split ring, is employed. As the current in the coils reverses, the gaps pass the brushes, effectively reversing the connections to the wires and allowing the current to continue to flow in the same direction.

A moving-coil microphone also works on the principle of electromagnetic induction. Sound waves hitting a diaphragm cause it to vibrate, and the diaphragm is fixed to a small coil in the field of a magnet. As the diaphragm moves, so too does the coil and a varying voltage is induced in it. The small varying current that results is led away and amplified.

The transformer

A transformer is a device for changing low voltages to high voltages (a step-up transformer) or high voltages to low ones (a step-down transformer). It works only with alternating current. It consists of an iron core, wound with two sets of insulated coils. The input current, in the primary coil, sets up a varying magnetic field in the core, which induces a varying voltage in the output, or secondary, coil. The ratio of the output voltage to the input voltage is equal to the ratio of the number of turns of wire on the secondary coil to the number of turns on the primary.

Alternators for generating alternating current (AC) can be driven directly by large diesel engines. There is usually a bank of five or six generating sets, with one or two kept as standby machines in case of failure or during maintenance of one of the others.

In the transmission of electric power from one place to another, losses due to heating are smaller using a large voltage and a small current. High-voltage power transmission is, therefore, much more efficient than transmission at low voltage. For this reason, electricity from power stations is stepped up to high voltages before being fed into transmission lines. At its destination it is usually stepped down again before being used. The voltage changes are achieved using transformers.

The induction coil

This is essentially a step-up transformer redesigned to work with DC. The primary circuit, usually containing a battery, is switched on and off repeatedly using a "make-and-break" mechanism (as in an electric bell). At "break," the current—and thus the magnetic field—decays very rapidly. The voltage induced in the secondary circuit is correspondingly large, particularly if the secondary is wound with a large number of turns.

The piezoelectric property of quartz crystals is the basis for the operation of quartz watches and clocks. Voltage applied to a quartz crystal plate causes the plate to expand and contract, producing vibrations at a regular rate. These vibrations can then be translated into seconds, minutes, and hours.

Electricity from stress, heat, and light

The most familiar ways in which electricity is generated involve chemical or electromagnetic processes. But it can also be produced—although generally in only small amounts—as a result of the effects of physical stress (piezoelectricity), heat (thermoelectricity), or light (photoelectricity).

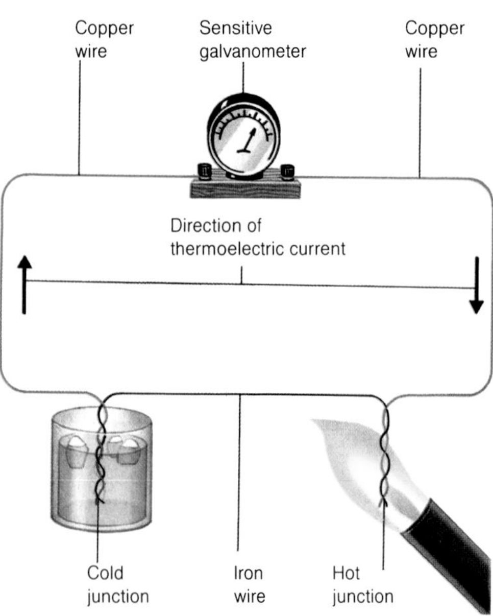

Thermoelectricity can be demonstrated using this simple apparatus, which consists of two pieces of copper wire joined to a length of iron wire to form two junctions; the other ends of the copper wires are connected to a sensitive galvanometer. Creating a temperature difference between the junctions by placing one in iced water and the other in a flame causes a small thermoelectric current to flow—measured by the galvanometer.

Piezoelectricity

When certain types of assymetrical crystals (notably quartz and Rochelle salt—sodium potassium tartrate) are subjected to pressure and deformed, opposing surfaces of the crystals become oppositely electrically charged. This phenomenon is called the piezoelectric effect. If the pressure is replaced by tension, the sign of the charges on each face changes—the face that was, say, positively charged when the crystal was compressed, becomes negatively charged when the crystal is under tension. As a result of the separation of charges, a stressed piezoelectric crystal generates a small voltage, which can drive a small current around a circuit. Crystals that generate piezoelectricity also exhibit the reverse effect, becoming deformed when a voltage is applied across them.

The piezoelectric effect is put to practical use in ceramic record player pickups and dynamic microphones. In a dynamic microphone, sound waves cause movements of a diaphragm, which deform a piezoelectric crystal and thereby produce a varying voltage.

Piezoelectricity is also used to control and sustain vibrations in crystals. When a piezoelectric crystal vibrates, the varying voltage it produces can be used to drive a current and this, in turn, can be used to sustain the vibration of the crystal. The frequency of vibration is exceptionally constant for some piezoelectric crystals, such as quartz; this is why quartz crystals are used in extremely accurate clocks and watches.

Thermoelectricity

When two wires of different metals are connected to form a loop and the junctions are kept at different temperatures, a small current flows. This phenomenon is called the Seebeck effect (after Thomas Seebeck, the German physicist who discovered it in 1821), and it is an example of a thermoelectric effect—that is, one in which there is a conversion of heat to electricity. The current flow is caused by a thermoelectric *e*lectro*m*otive *f*orce (EMF) generated between the junctions, which itself arises because electrons (the charge carriers) exist in different energy levels in the two metals. The thermoelectric EMF is very small, as is the resulting current.

As the temperature difference between the junctions increases, the thermoelectric EMF first increases, then reaches a maximum at a temperature difference (called the neutral temperature) specific to the metals used, and finally decreases with further temperature increase. Because of this, the Seebeck effect can be used as the basis for a thermometer. By keeping one of the junctions at a constant known temperature, usually 32° F. (0° C), and placing the other junction in the environment whose temperature is required, the magnitude of the thermoelectric EMF can be used to determine the unknown temperature.

If the two junctions of the basic thermoelectric circuit are at the same temperature

The five solar panels projecting like wings from the spacecraft consist of arrays of hundreds of selenium photocells, each of which generates a small photoelectric current when exposed to light. Although there are large numbers of solar cells, the total amount of electricity generated by them is insufficient for the spacecraft's requirements, and other power supplies, such as fuel cells, are needed.

and a battery is added to the circuit, one of the junctions heats up and the other becomes cooler. This phenomenon—called the Peltier effect (after Jean Peltier, the French physicist who discovered it in 1834)—is a reversal of the Seebeck effect. Moreover, if the direction of the current is reversed, the effects at the two junctions are also reversed, so that the junction that was initially heated becomes cooled, and vice versa. Thus, the Peltier effect is different from the normal Joule heating that occurs when a current passes through a wire, because the latter effect always generates heat in all parts of the circuit, irrespective of the direction of the current.

Photoelectricity

When some types of electromagnetic radiation—notably visible light, ultraviolet radiation, and X rays—strike certain substances (principally metals such as zinc), electrons are emitted from the surface of the substance; this phenomenon is called the photoelectric effect. The emitted electrons can be collected by a piece of metal at a higher electrical potential than the emitting substance, and if the two are then connected to form a circuit, an electric current (termed a photoelectric current, although it is no different from an ordinary current except in its method of generation) flows around the circuit. Hence the photoelectric effect exemplifies the conversion of light to electrical energy.

The interactions involved in the photoelectric effect are complex. Stated simply, electron emission occurs as a direct result of a collision between a photon of electromagnetic radiation and an electron in the substance's surface.

The photoelectric effect is put to practical use in photocells (also called photoelectric cells), which detect the presence of light or measure its intensity in such devices as photographic exposure meters. One of the main types is the selenium cell, which consists, in its basic form, of an outer layer of transparently thin metal, an intermediate thin barrier layer (also transparent), and an inner layer of selenium (an element that emits electrons when exposed to visible light). Light passes through the two outer layers and, on striking the selenium, causes it to emit electrons, which then pass across the barrier layer to the outer metal layer. As a result, a small potential difference is set up between the selenium and metal, and this can be used to drive a small electric current, whose size depends on the intensity of the incident light.

The other main type of photocell is the photoresistor. These devices are often made out of semiconducting materials (such as silicon) that emit electrons when struck by light, with the result that their electrical resistance decreases.

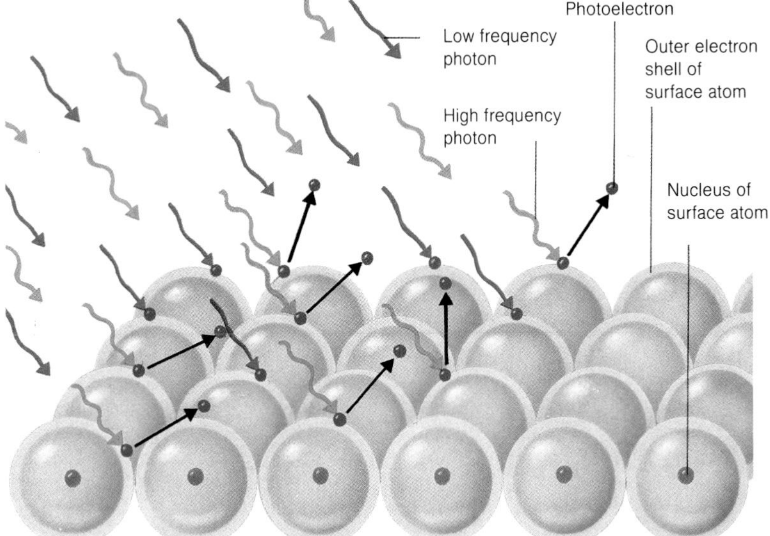

Electron emission from a photoelectric substance results from collisions between photons of electromagnetic radiation and outer electrons in the substance's surface. To provoke electron emission, the incident photons must be above a certain minimum threshold frequency (the green photons in the diagram), the specific value of which varies according to the substance involved. If the photons are below the threshold (the red photons), electrons are not emitted, irrespective of the intensity of the incident radiation.

Electronics

Electronics is a general term for the production and processing of electric signals (consisting of moving electrons, hence the name electronics) that convey information—for example, the sound reproduction by a radio or CD player and the data output of a computer. Electronics also includes the production and processing of beams of electrons, used in such devices as cathode-ray oscilloscopes, television sets, and electron microscopes.

Electric signals

Electric signals are produced by devices that convert the primary information source (which is in the form of another type of energy—sound or light, for instance) into small electric currents. For example, a microphone converts sound into an electric current, and an input unit of a computer converts data. These currents, which constitute a signal, are then transmitted along wires or, after conversion to radio waves, via radio links. On reaching a suitable receiver, they may be electronically manipulated—to reduce distortion in the signal, for instance—and are then amplified so that they can drive an output device, such as a loudspeaker, television set, or computer printer.

There are two types of electric signal: analog and digital. Analog signals vary continuously in voltage or current, corresponding to variations in the primary information source; thus the electric signals produced by a microphone are electrical "copies" of the original sound waves. Digital signals are not continuous but consist of numerous bursts of electric current between two voltage levels (one of which may be zero). In digital sound recording, for example, the amplitude of the sound wave is measured thousands of times every second and each instant converted into a binary code signal made up of rapid on-off bursts of current. On playback, the resulting train of digital amplitude code signals is converted back to analog signals that can drive a loudspeaker.

A computer uses digital signals in another way. They consist of binary codes that represent the numbers, letters, and symbols in the input data and in the various program instructions. In the computer's central processor, the data codes are subjected to processes of binary arithmetic or are compared with each other to obtain the results required by the program.

In modern electronics, electric signals may be processed in two main ways—by passing

An oscilloscope is an electronic instrument that displays changing electrical signals. Light, mechanical motion, and sound are all studied with oscilloscopes. The screen of an oscilloscope is the front of a *cathode-ray tube,* a special type of vacuum tube. Inside the tube, a device called an *electron gun* projects a beam of electrons onto the fluorescent screen. A circuit is used to make the beam move repeatedly from left to right. At the same time, the signal being studied is fed into the oscilloscope. The signal causes the beam of electrons to move up and down. This movement corresponds to *oscillations* (vibrations) in the signal. A pattern is traced as the movement of the beam of electrons leaves a glowing line on the screen. Oscilloscopes are used in industry, medicine, and scientific research. Electronics engineers used them to test computers, radios and other electronic equipment. Doctors use them to study electrical impulses from the brain or heart.

One of the earliest computers with memory storage—built in the late 1940's at Manchester University, England—contained 1,300 vacuum tubes, 8 cathode-ray tubes, and thousands of other electronic components. The device occupied an entire room yet could perform fewer functions than a modern pocket calculator *(above)*.

them through semiconductor devices (such as transistors) or by converting them into a beam of electrons, as in the formation of a picture by a television set (which also uses semiconductors).

Semiconductors

Semiconductors are substances (such as silicon and germanium) whose electrical resistance lies between that of conductors and insulators. They have this intermediate resistance because they have a few free electrons that can drift from atom to atom—unlike conductors, which have many free electrons, and unlike insulators, which have none.

The semiconductors used in electronic devices are "doped" to change their electrical properties. Doping involves introducing minute traces of other elements into the semiconductor's crystal lattice. Silicon and germanium each have four outer electrons per atom; doping them with an element with five outer electrons, such as phosphorus, frees the fifth electron so that the semiconductor has an electron excess. It is then known as an *n*-type (negative-type), because electrons have a negative charge.

Doping with an element that has only three outer electrons, such as boron, produces a crystal lattice with spaces, known as holes, which free electrons readily fill. This type of semiconductor has a lack of free electrons, which is equivalent to an excess of positive charges, and it is therefore known as *p*-type.

If a piece of *n*-type semiconductor is joined to a piece of *p*-type, the resulting device is called a *p*/*n* junction diode. If the diode is then connected to a battery so that the negative terminal is joined to the *n*-type semiconductor and the positive terminal to the *p*-type, current flows through the junction. The electromotive force (EMF) of the battery forces free electrons from the *n*-type material across the junction and into the *p*-type, where they fill the holes. Such a junction is said to be forward-biased.

But if the battery is connected the other way around, so that its positive terminal is connected to the *n*-type semiconductor and its negative terminal to the *p*-type, then free electrons in the *n*-type material are forced away from the junction. Because there are no free electrons in the *p*-type semiconductor to replace them, no current should flow across the junction, which is described as being reverse biased. In practice, a very small current does flow because a few electrons are freed as a result of the battery heating the junction.

By offering a high resistance to current flowing in one direction but little resistance to current flowing in the opposite direction, diodes act as rectifiers. They are therefore used to rectify the alternating current main supply to produce the direct current supply that devices such as radios and televisions require.

Transistors

Transistors are arrangements of semiconductor diodes that act as amplifiers or switches. There are two main types: junction and field-effect transistors. A junction transistor is essentially a current amplifier. It consists of two semiconductor diodes joined back to back in either an *n-p-n* or a *p-n-p* configuration. A main supply current is connected across the two outer pieces of similar semiconductor—one of which is called the emitter because it emits electrons (in an *n-p-n* transistor) or holes (in a *p-n-p* transistor), the other being the collector. The central piece is called the base and is supplied with a small varying current, which affects the flow of current across the transistor as a whole. The overall result is that the main

A photomicrograph of part of a single microchip illustrates the remarkable complexity of such devices. The inset photograph gives an idea of the actual size of a microchip (the small square between the forceps) sufficiently "powerful" to perform all the mathematical calculations in a sophisticated pocket calculator.

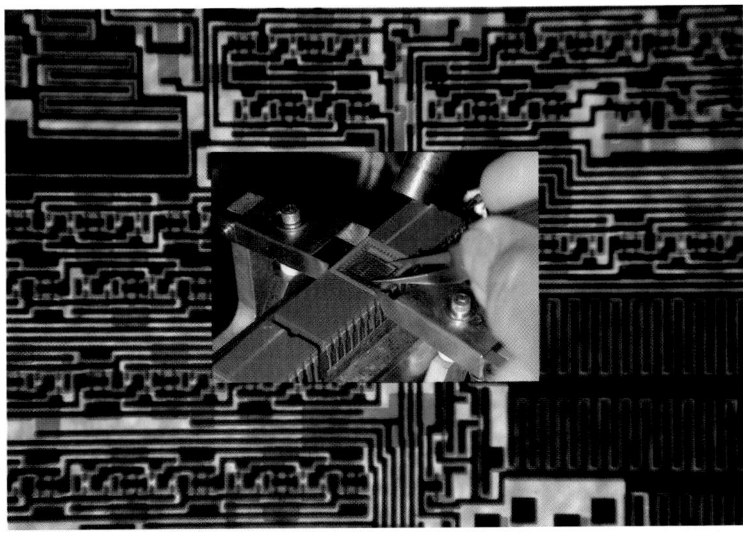

A television camera utilizes lenses and complex electronic devices to convert optical images into electric signals. The lenses focus an image onto a light-sensitive mosaic, which converts the image into a pattern of electrically charged particles. On being scanned by an electron beam, the particles discharge, thereby causing a small electric current to flow from the mosaic. This current—which is the output signal—is then converted into a modulated radio wave for transmission.

supply current is amplified (typically between 20 and 500 times) as it passes across the transistor, the variations in the base current being reproduced on a larger scale by the main supply current.

Field-effect transistors (FETs) work on a similar principle but consist of a bar of one type of semiconductor material; semiconductor of the opposite type is diffused into its sides to form what is called a gate (which performs the same function as the base in a junction transistor). The main supply current is connected across the ends of the bar, one of which is called the source (the counterpart of the emitter in junction transistors), the other being the drain (the counterpart of the collector). If the bar consists of *n*-type material, electrons move from the source to the drain; if it consists of *p*-type, holes move in the same direction. A varying current is connected to the gate, producing a varying electric field that affects the current flowing across the bar in the same way that the base affects the current flowing across a junction transistor. FETs are widely used because they produce less distortion and electrical noise than do junction transistors. They also require less power to operate. But they are much slower than the junction transistors.

Computer applications

A transistor can be made to act as a fast electronic switch by causing the current fed to the base or gate to vary widely, so that the main supply current effectively changes from on to off or from off to on. The microchips used in microprocessors and computers contain many thousands of miniature transistors formed by diffusing doping substances in complex patterns in several layers of silicon or, more rarely, some other semiconductor material. In microprocessors, the transistors form a series of switches that either pass or stop pulses of current. The operations of the transistors are controlled by the sequence of program instructions fed to the processor, and in this way the binary codes of data signals are processed.

In memory devices, data codes or program instructions may be stored temporarily or permanently. Dynamic memories have rows of transistors that switch pulses to capacitors in order to store the codes as on-off patterns of electric charge. Static memories have transistors that remain in on-off configurations to produce a particular code when required. Both types of memory need a supply of current to keep the transistors in their particular configurations, and so the memory contents are lost when the current is switched off. For this reason, data for permanent storage are recorded on magnetic tapes or disks.

Electron beams

Beams of fast-moving electrons can be generated using a diode tube, which, in its basic form, consists of an evacuated glass envelope containing a heated metal filament (the cathode) and a positively-charged plate (the anode). The hot cathode emits electrons that, being negatively charged, stream toward the anode, thereby forming an electron beam. Such beams have many uses—in television sets, cathode-ray oscilloscopes, and electron microscopes, for example. A characteristic of electron beams—and of almost all other beams of small particles—is that they can exist only in a vacuum.

The cathode-ray oscilloscope

A *c*athode-*r*ay *o*scilloscope (CRO) is one of the most widely used instruments in experimental physics. It utilizes a type of diode valve system with several anodes for focusing and controlling the intensity of the electron beam. The beam passes between two series of plates—X-plates and Y-plates—before hitting a fluorescent screen, which shows the position as a small dot of light. Potential differences applied to the X-plates move the beam and, therefore, the light dot, sideways. Potential differences applied to the Y-plates move it up or down. The amount of movement is proportional to the potential difference applied, which means that a CRO can be used as a voltmeter.

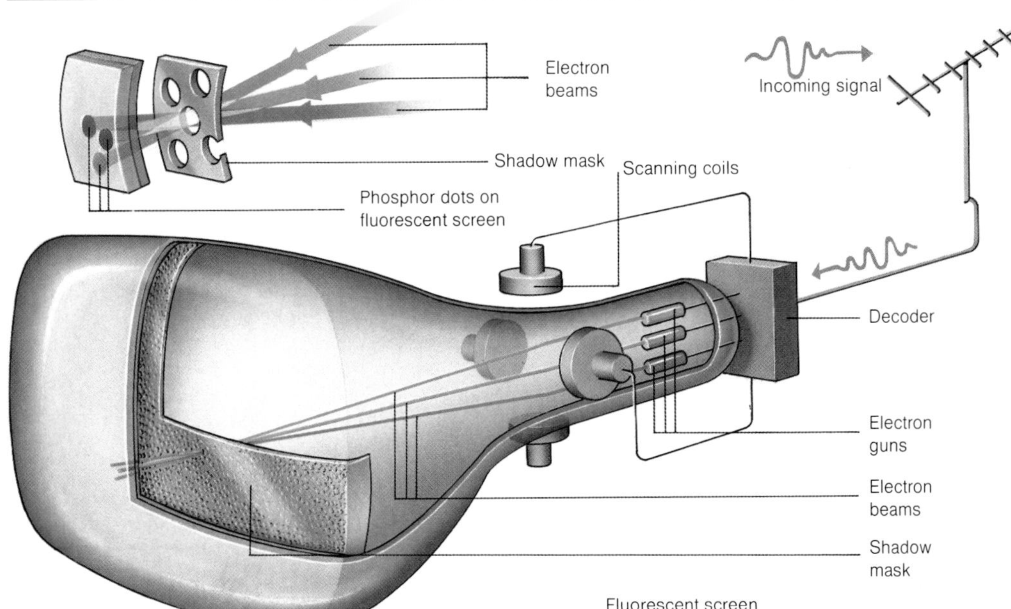

In a color television, the incoming signal is processed by a decoder, the output from which controls the beams from the three electron guns and also the ring coils. These coils deflect the beams so that they scan over the fluorescent screen, which is covered with tiny phosphor dots that emit red, blue, or green light. Before hitting the screen, the beams pass through a metal shadow mask, which has thousands of holes arranged so that the beam from each of the three guns strikes only those phosphor dots that emit the appropriate color.

Another important use of the CRO is to display waveforms. This requires using the CRO's timebase. When the timebase is switched on, a special circuit connected to the X-plates causes the dot to move across the screen; when it reaches the edge, it flicks back very quickly and begins again. Hence if a varying input is applied to the X-plates, the timebase "spreads out" this input horizontally so that the way the input varies with time is displayed. For example, using the main electricity supply as the input gives a vertical line on the screen when the timebase is not used, but a regular sine wave when the timebase is used.

Television

A black-and-white television set works in a similar way to a CRO, except that magnetic fields produced by currents in coils are used to deflect the electron beam. Moreover, instead of forming only a single line, the electron beam in a television set repeatedly moves across the screen forming a complete picture consisting of hundreds of adjacent lines in about $\frac{1}{30}$ second.

Each complete scan produces a slightly different picture, thereby enabling movement to be portrayed in the same way as in a cine film, in which the rapid succession of a series of still images gives the impression of continuous movement.

A color television set works on the same basic principles as a black-and-white set, except that it uses three electron beams, one for each of three colors (red, blue, and green). The beams themselves are not colored, but each beam affects only one of the sets of red, blue, and green phosphors on the screen. The amalgamation of the three single-colored images produces a full color picture.

An electron microscope is similar in principle to an optical microscope, but uses a beam of electrons instead of light. Also, the beam is focused by magnets instead of glass lenses. The main advantage of the electron microscope is its extremely fine resolving power—a result of the very short wavelengths of electrons—which, in turn, enables much higher magnifications than are practicable with optical microscopes. For example, the photomicrograph of a virus *(below)* is magnified about 530,000 times. The image has also been computer processed to yield colors; normal electron micrographs are black-and-white only.

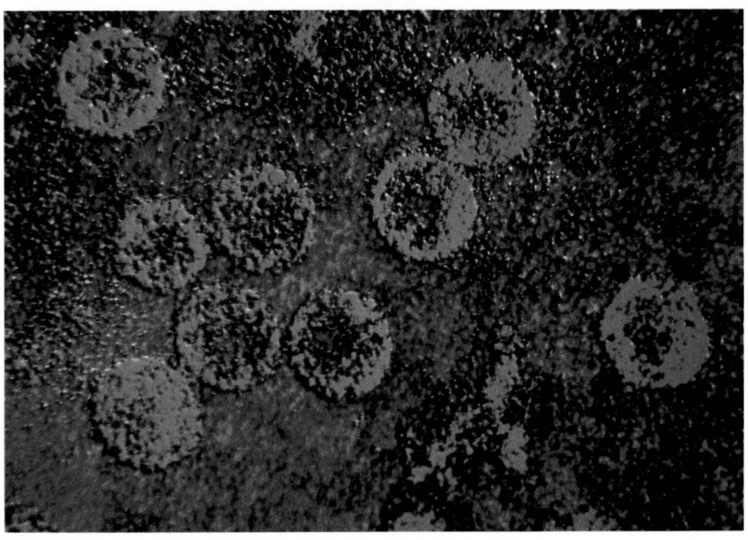

LIGHT THROUGH A WIRE: *fiber optics*

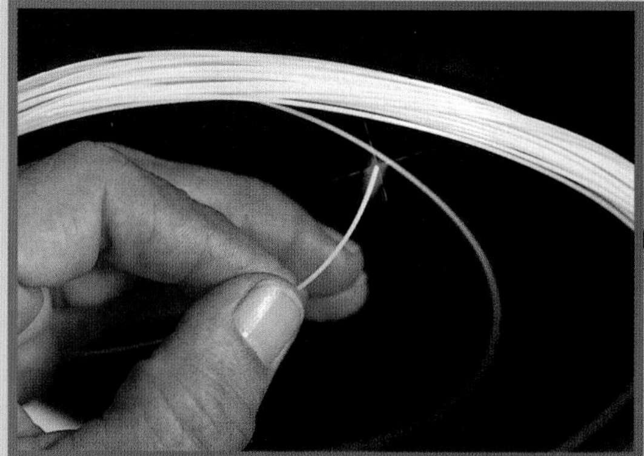

Optical fibers, such as the fiber optic telephone cable shown above, can carry thousands of times as much information as traditional copper cables.

Talking through a beam of light may sound like something from a science fiction movie, but in fact this is exactly how millions upon millions of phone conversations occur throughout the world every day, thanks to the branch of physics known as fiber optics.

Fiber optics is based on the transmission of light through strands of transparent glass or plastic called optical fibers. These fibers, about as big around as a human hair, can carry light over distances of more than 100 miles (160 kilometers).

Optical fibers consist of a highly transparent core of glass or plastic surrounded by a covering called a cladding. Light impulses from a laser, light bulb, or some other source enter one end of the optical fiber. As this light travels through the core, it is kept inside it by the cladding. Light rays that strike the inside surface of the cladding are bent or reflected inward. At the other end of the fiber, a detector, such as a photosensitive device or the human eye, receives the light.

Optical fibers have many uses. They are especially well-suited for medical use. They can be made into extremely thin, flexible strands that can be inserted into blood vessels, lungs, and other hollow parts of the body. These optical fibers are used in a number of techniques that enable physicians to look and work inside the body through tiny incisions.

But perhaps the most widespread application of optical fibers is in communications networks such as phone and cable TV systems. In a fiber-optic telephone system, a laser at one end of a fiber transforms electric signals

This cancer patient's tumor has been coated with a drug that is activated only by certain wavelengths of light. This photograph shows an optical fiber guiding light from a laser down the patient's throat. The light triggers chemical reactions that attack the tumor.

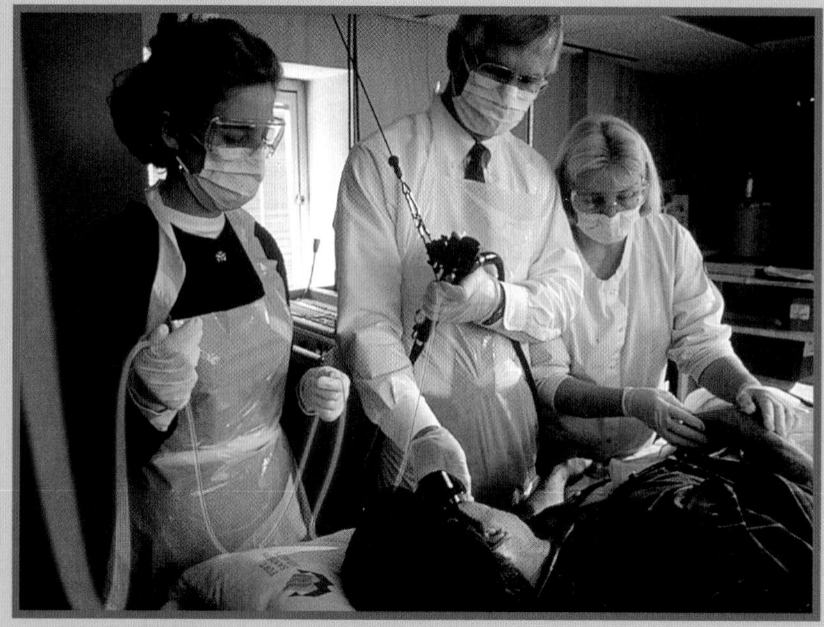

The U.S. government intends to complete the "information superhighway" early in the 21st century. This means individual homes and businesses must be wired with fiber-optic cables. The photograph above shows a truck laying fiber-optic cable in Idaho.

corresponding to a telephone message into a sequence of high-speed flashes of light representing "0's" and "1's." At the other end of the fiber, a special device transforms the emerging flashes back into electrical signals.

Fiber-optic communications systems have a number of features that make them superior to systems that use traditional copper cables. Because they use a laser beam to transmit signals, they have a much larger information-carrying capacity, they are not subject to electrical interference, and they need less amplification than do signals sent over copper cables of equal length. Thus, a strand of optical fiber can carry thousands of times as much information as a pair of copper wires.

Laser light is ideal for fiber-optic communication because it is highly coherent. This means that unlike regular light, laser light spreads little, even over long distances. The waves of a laser beam move in one direction along a narrow path. Therefore, laser light can be focused precisely, and all of its energy can be introduced into the fiber. In this way, tremendous amounts of telephone, television, and other data can be communicated relatively cheaply.

Because of their superior data-carrying capacity, fiber-optic cables are expected to become the main avenue on the "information superhighway," which the United States government has targeted for completion by 2015. Some of the services expected to be delivered via the superhighway include interactive video, sophisticated home shopping services, movies on demand, and perhaps even long-distance medical diagnoses.

Despite the many advantages of fiber optic communication, technical difficulties have delayed the wiring of individual homes and businesses with fiber-optic cables. In the mid-1990's, most lines that connected directly to homes and businesses were still low-capacity copper cable. One obstacle has been the high cost of the devices needed to align optical beams from the main fiber-optic lines to local ones. However, in 1994, researchers in Japan announced the development of a plastic fiber-optic cable that could be wired cheaply and easily and could handle the local delivery of the large amounts of data that would travel the information superhighway.

Magnetism

A magnetic compass points to the earth's magnetic north pole rather than to the true, geographical North Pole.

Magnetism is one of the fundamental properties of matter. All substances—even those such as glass, plastics, and wood, for example—exhibit magnetic properties to some extent, although most materials do so to such a minute degree that they are generally regarded as nonmagnetic. Only the metallic elements iron, nickel, and cobalt, some of their alloys (such as steel) and some rare earth compounds (such as neodyneum) are strongly magnetic; these are described as being ferromagnetic.

Magnetism is not only an important property of matter, it is also associated with electricity. A conductor carrying an electric current has a magnetic field around it as a result of the movement of the electric charges—electrons—that constitute the current.

In fact, all magnetic effects are caused by movements or properties of electric charges. This is true even of substances that behave as magnets without being connected to an external electricity supply. Their magnetic fields originate from electrons within the atoms or molecules of the substance itself. Thus, magnetism and electricity are not independent phenomena. Indeed, many of the basic concepts of magnetism are similar to those of electrostatics (the study of electric charges).

The magnetic field around a magnet is revealed using iron filings, which act like minute compasses and align themselves along the magnetic field lines *(see also the diagram below)*. As can be seen in this photograph of one pole of a bar magnet, the magnetic field is in three dimensions.

Basic concepts

An important concept in magnetism is that of the magnetic pole. There are two types of magnetic poles: north-seeking (or north) poles, and south-seeking (or south) poles. Like poles repel each other, whereas unlike poles attract (just as like electric charges repel each other and unlike charges are mutually attracted). The stronger the poles, the larger is the repulsive or attractive force between them. Also, the closer together they are, the greater is the force.

An important difference between magnetic poles and electric charges is that, unlike electric charges that can exist independently of other charges, an isolated magnetic pole has never been found to exist. That is, a north magnetic pole is always associated with a south magnetic pole, and vice versa. Although some speculative theories have proposed the existence of these magnetic monopoles, extensive observations have failed to uncover their exist-

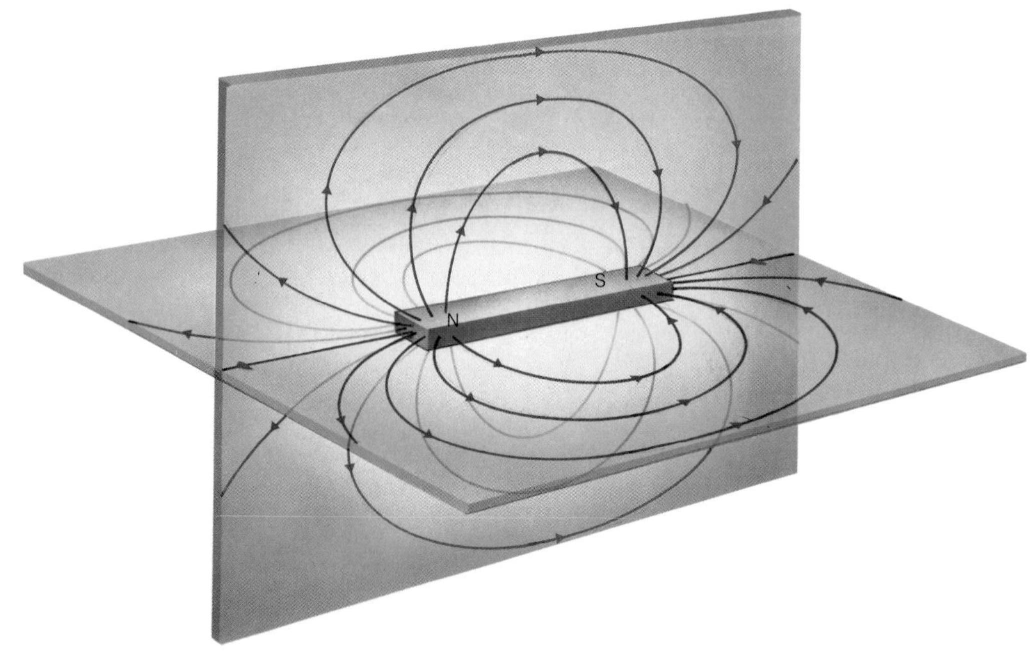

A bar magnet and its associated magnetic field (shown for clarity in only two perpendicular planes, although in fact the field completely surrounds the magnet) are illustrated in the diagram on the right. The arrows on the field lines indicate the direction in which a free magnetic north pole would move. The distance apart of the lines indicates the strength of the field, which is therefore strongest near the poles, where the field lines converge.

ence, and they certainly do not occur in ordinary matter. Thus, all magnets have both north and south poles. Cutting a bar magnet in half simply produces two smaller magnets, each with north and south poles.

A similarity between magnetism and electrostatics is the concept of a field. A magnet produces its strongest effects near its poles, but its influence also extends throughout the space surrounding it. The region in which this influence can be detected is known as a magnetic field (analogous to an electric field in electrostatics).

A magnetic field can be graphically represented by field lines, which indicate its strength and direction. Where the lines are close together, the field is strong; where the lines are widely separated, the field is weak. Arrows on the field lines show the direction in which a free north pole (if it could exist) would move at that point in the field; a free south pole would move in the opposite direction.

Theory of magnetism in materials

Iron, steel, and other ferromagnetic substances are magnetic as a result of their atomic structures. An atom consists of a central, positively-charged nucleus surrounded by negatively-charged electrons, which orbit rapidly around the nucleus while also spinning on their own axes. Because electrons are electrically charged, each one through its movements may give rise to a minute magnetic field. This effect occurs in all substances, but in most of them the electrons are arranged in such a way that their magnetic fields almost or completely cancel out each other.

In magnetized ferromagnetic materials, however, the magnetic fields of the electrons do not cancel, and each atom acts as a minute, weak bar magnet.

Furthermore, even when a ferromagnetic substance is unmagnetized, a certain amount of order exists among its "atomic magnets" (known scientifically as magnetic dipoles, because there is a minute separation of poles within each atom). The dipoles line up with each other in extremely small—typically less than a hundred-thousandth of a cubic inch (10^{-10} m^3)—regions called domains. Within each domain all the dipoles point in the same direction, but the domains themselves are randomly oriented in an unmagnetized material and, overall, the magnetic effects of the domains cancel out each other.

But when a ferromagnetic substance is magnetized by an external field, those domains that happen to be aligned with the field grow at the expense of others pointing against the field; yet others rotate so as to align themselves with the external field. As a result of the common alignment of the dipoles, their magnetic effects combine to give an overall induced magnetic field in the material. A substance is magnetically saturated when all its dipoles point in the same direction.

Magnetic effect of an electric current

A movement of electric charge is an underlying cause of magnetism. Hence an electric current, being a flow of charge, produces a magnetic field. The larger the current, the stronger is the magnetic field. If the current is flowing in a wire, the shape of the magnetic field depends on the configuration of the wire. A single straight wire carrying a current, for example, produces a cylindrical magnetic field.

Two straight current-carrying wires side by side attract each other when their currents flow in the same direction, and repel each other when their currents flow in opposite directions. The force of attraction or repulsion varies according to the size of the current, and so by measuring this force, the current can be determined. This is the principle of the current balance, the fundamental instrument of current measurement.

One of the most useful applications of this effect is a solenoid, which consists of a length of insulated wire coiled into a cylinder. It produces a magnetic field that, inside the coil, is highly uniform in both strength and direction. Overall, a solenoid's magnetic field resembles that of a bar magnet, and it behaves in a similar way—as if it had a north pole at one end and a south pole at the other. For example, a

Electromagnets, such as this one suspended from a crane and lifting scrap iron and steel, rely on the fact that an electric current produces a magnetic field. In most large electromagnets, a wire is wound many times around a pure iron core, which produces a stronger magnetic field than is possible using only the current-carrying wire. And, because pure iron is a temporary magnetic material, it loses its magnetism when the current is switched off.

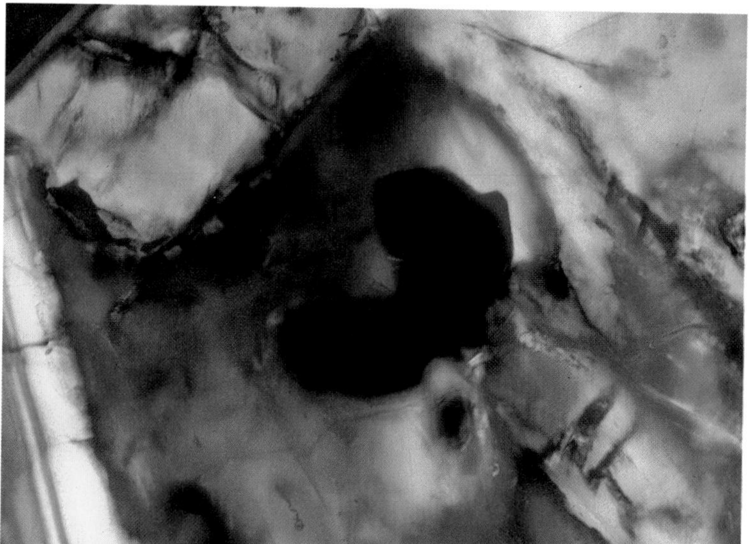

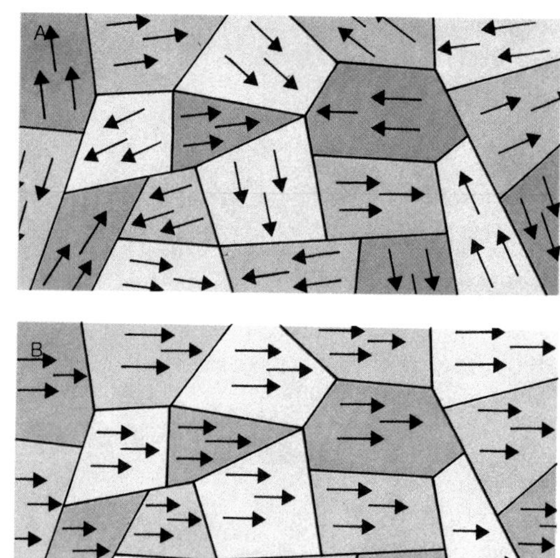

Magnetite, a form of iron ore, is the principal naturally occurring ferromagnetic material and is frequently found as grains in other rocks—for example, in the photomicrograph *(above)* of a sample of solidified lava (magnified 380 times), the central black mass is a piece of magnetite. More detailed examination reveals that in an unmagnetized ferromagnetic substance the magnetic domains are randomly arranged (A, *above right*), but in a magnetized ferromagnetic material they are aligned in the same direction (B, *above right*).

freely suspended current-carrying solenoid aligns itself with Earth's magnetic field.

Types of magnetism

All materials can be classed as diamagnetic, paramagnetic, or ferromagnetic, depending on the electronic structure of the atoms and their arrangement within the material. Each behaves in a characteristic way when placed inside a current-carrying solenoid (that is, in a uniform magnetic field).

In a piece of diamagnetic material (such as bismuth or copper), the electrons in each atom are arranged so that their magnetic effects cancel out each other; hence a diamagnetic substance has no overall magnetic field associated with it. If a bar of the material is placed in an inducing field, however, the electron paths are slightly distorted. As a result, the substance becomes very weakly magnetized, but in the opposite direction to that of the inducing field. Hence the susceptibility (the ratio of the induced to the inducing magnetic field) of diamagnetic materials is very small and negative (typically about -10^{-5}). If a bar of diamagnetic material is freely suspended in a strong uniform magnetic field, it aligns itself across the field.

In a paramagnetic material (such as platinum or aluminum) the magnetic effects of the electrons do not cancel completely and each atom acts like a minute, weak bar magnet. Under normal circumstances, these magnetic dipoles are oriented at random and there is no overall magnetic effect. Under the influence of an inducing field, however, the dipoles tend to align with the field. As a result, the magnetic field induced in the substance is in the same direction as that of the inducing field, and a freely-suspended bar of paramagnetic material therefore aligns itself with the inducing field.

However, the dipole alignment is not complete and so the induced magnetism tends to be weak, although it is still strong enough to overwhelm the diamagnetic effect. The susceptibility of paramagnetic substances is small and positive (in the order of $+10^{-3}$). When the inducing magnetic field is removed, the magnetic dipoles within the substance again become randomly oriented and the magnetization goes away.

Many substances are paramagnetic, but very few exhibit the third and most important form of magnetism—ferromagnetism. Ferromagnetic materials, such as iron and cobalt, display paramagnetic behavior, but much more strongly than most paramagnetic substances because many more atomic dipoles align when in an inducing field. Thus ferromagnetic materials have large, positive susceptibilities (often greater than $+10^{3}$). These magnetic fields can persist even after the inducing field is taken away.

Magnetization and demagnetization

The type of iron ore called magnetite is the chief naturally-occurring magnetic substance (that is, one that can be magnetized). Lodestone is the only natural substance that behaves as a magnet.

An ordinary piece of ferromagnetic material such as steel does not behave as a magnet but it can be made into one in any of several ways. One method involves simply placing the steel close to a strong magnet. The part of the steel nearest the magnet's north pole becomes a south pole (and vice versa) and the steel is therefore attracted to the magnet. It is as a result of this phenomenon of induced magnetism that a magnet attracts iron and other ferromagnetic materials. This method of magnetization usually induces only a weak magnetic field in the steel, because relatively few domains become aligned. Greater domain alignment—and therefore a stronger magnetic field—can be achieved by stroking the steel in the same direction with a bar magnet.

But the most effective way to magnetize a piece of steel is to place it inside a solenoid carrying a large direct current. The current creates a strong magnetic field inside the solenoid and this field, in turn, induces a magnetic field in the steel. When steel has been magnetized it tends to retain its magnetism, becoming a permanent magnet. In some other ferromagnetic materials, however, the induced magnetism is only temporary; pure ("soft") iron, for example, loses its magnetism when it is removed from the magnetizing field.

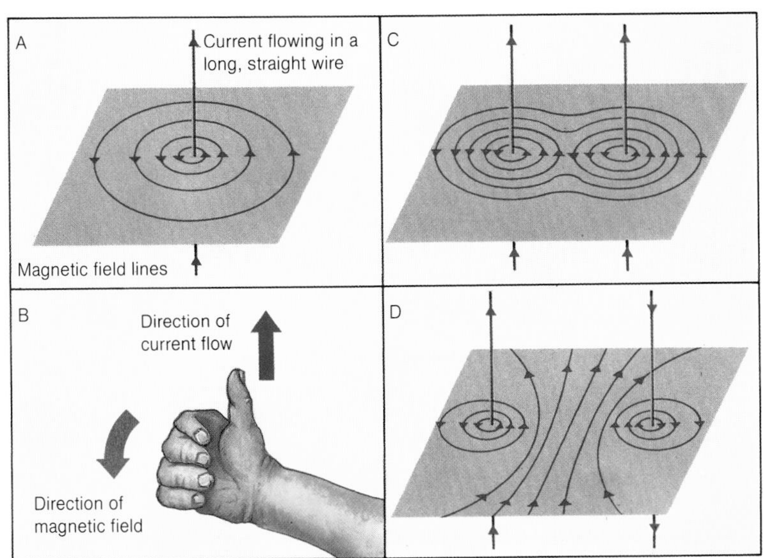

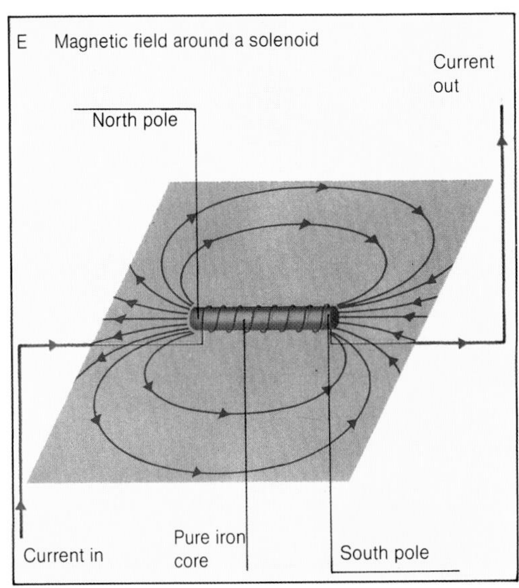

An electric current flowing though a conductor generates a magnetic field. Diagram A shows that a current flowing through a long, straight wire produces concentric magnetic field lines. The direction of the field can be found using the right-hand grip rule (B); if your right hand is held as if it is grasping the wire, with your thumb pointing in the direction of current flow, then the direction of your fingers gives the direction of the magnetic field. C illustrates the magnetic field around two straight wires placed close together and carrying currents in the same direction; D illustrates the field when the currents flow in opposite directions. The field around a solenoid (E) resembles that around a bar magnet.

Although permanent magnets tend to retain their magnetism, they can be demagnetized by various means. Rough treatment, by hammering for example, diminishes the strength of a permanent magnet by physically jolting the domains out of alignment with each other. This method rarely completely demagnetizes a permanent magnet, however. More effective methods of demagnetization include heating the magnet to near its melting point—about 2,550° F. (1,400° C) for steel—or placing it in a solenoid through which an alternating current is flowing, then gradually reducing the current to zero. Heating demagnetizes the material by increasing the random vibrations of its atoms and thus destroying their organized alignment. Similarly, passing a reducing alternating current through a solenoid reverses the magnetic field many times every second and leaves the ferromagnet with no overall field.

Hysteresis

When an unmagnetized ferromagnetic substance, such as iron, is placed in a continually reversing magnetic field (usually produced by passing an alternating current through a solenoid), the substance goes through many magnetic cycles every second. It becomes magnetized (by induction) first in one direction, then in the opposite direction. In this opposition, the magnetic field induced in the substance lags behind the changing magnetizing field produced by the solenoid. This is the phenomenon called hysteresis. A graph of the magnetizing field against the induced magnetic field for a complete cycle takes the form of a characteristic closed loop which is called a hysteresis loop.

A hysteresis loop enables various magnetic properties of a ferromagnetic substance to be determined, of which remanence and coercive force are two of the most important. After the material has been magnetically saturated and the magnetizing field is reduced to zero, the material is still magnetized. The strength of its magnetic field at that point on its hysteresis loop is the remanence of the substance. The higher a material's remanence, the stronger is its residual magnetism.

The coercive force (or coercivity) of a substance is the strength of the magnetizing field necessary to reduce its remanent magnetism to zero. Hence it is a measure of how well a material can retain its magnetism.

Iron has a high remanence but a low coer-

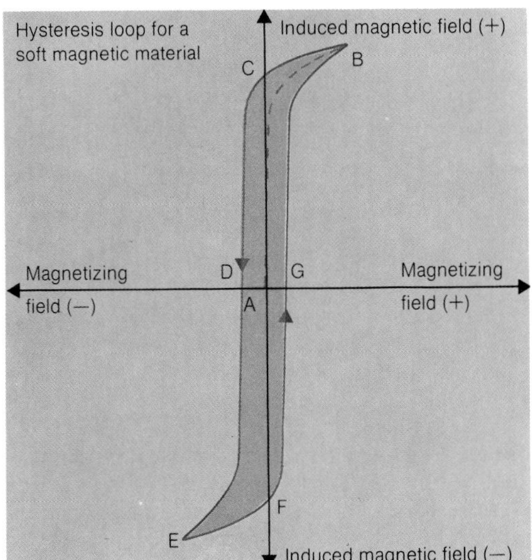

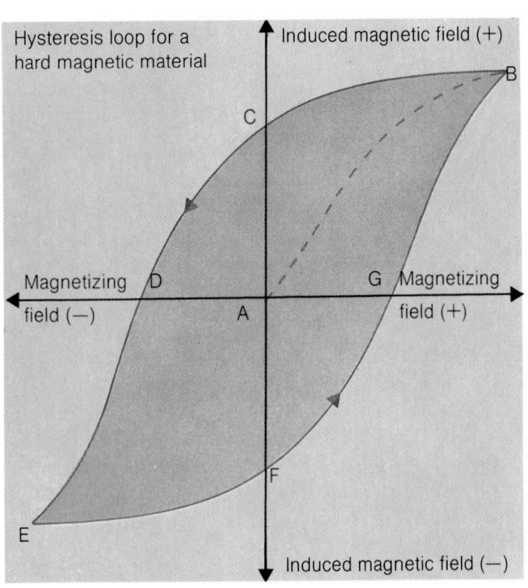

Hysteresis loops for a soft magnetic material such as iron *(far left)* and for a hard magnetic material such as steel *(left)* reveal characteristic differences between the two materials. On the graphs, the dotted line AB represents initial magnetization to saturation (at B and E the substances are saturated in opposite directions). AC (equal but opposite to AF) is a measure of the substances' remanences. And AD (equal but opposite to AG) indicates their coercivities.

civity, so it can be magnetized strongly but is easily demagnetized. Substances such as iron are called "soft" magnets. Steel, on the other hand, has a slightly lower remanence than iron (so makes slightly weaker magnets) but a much higher coercivity, so it retains its magnetism well. Steel is an example of a "hard" magnetic material.

Ordinary steel (an alloy of iron and a small proportion of carbon) is not the only hard magnetic substance; other, more complex alloys have been made, some of which—alnico, for example—have a higher remanence and coercivity than steel.

Uses of hard and soft magnetic materials

Because of their high coercivities, hard magnetic materials are most suitable for making permanent magnets, used for magnetic catches on doors, for example. Soft magnets are generally used in devices where it is essential that the magnetic field can be easily "switched off" or reversed. Hence soft magnets are used for making the cores of electromagnets. A simple electromagnet consists of a solenoid wound around a bar of pure iron. When the electric current is switched on, the bar becomes strongly magnetized; when the current is switched off, the bar loses its magnetism.

Soft magnetic materials are also the most suitable for transformer cores, in which the direction of the magnetic field reverses many times every second. The continual reversing of the magnetic field leads to the generation of heat within the transformer core as a result of friction between its magnetic domains as they constantly change direction. Hence, there is an energy loss within the transformer and a reduction in its efficiency. The amount of energy lost is directly proportional to the area of the hysteresis loop for the material used in the transformer core. For minimum energy loss and maximum efficiency, the core must therefore be made of a substance with a low coercivity and a small hysteresis-loop area. Pure iron has both these properties, and several alloys have been produced with even smaller hysteresis loops; a typical example is mumetal.

But even when materials with small hysteresis loops are used for transformer cores, there is still some energy loss (again, largely dissipated as heat) as a result of electric currents called eddy currents induced within the core. Eddy currents can be eliminated only by using magnetic materials with high electrical resistances, and by constructing the core in laminar form—using alternate layers of magnetic material and an insulator. The most important materials with high resistances are ferrites (magnadur, for example), which are ceramic iron oxides that act as ferromagnets but do not conduct electricity.

Earth's magnetic field

A freely-suspended piece of magnetic material—the needle of a magnetic compass, for example—points North-South because earth itself has a magnetic field. The shape of this field is similar to that which would be produced by an enormous bar magnet with its south pole in the Northern Hemisphere and its north pole in the Southern Hemisphere, and with its axis slightly inclined to the earth's axis of rotation.

The earth's magnetic poles are geographical poles. The magnetic north pole is situated about 1,000 miles (1,600 kilometers) south of the geographical North Pole, on one of the islands north of mainland Canada. The magnetic south pole lies about 1,600 miles (2,570 kilometers) north of the geographical South Pole, on the ice shelf about 1,550 miles (2,500 kilometers) southwest of New Zealand.

The angle between true (geographical) North and magnetic north is called the magnetic declination (or deviation). It varies at dif-

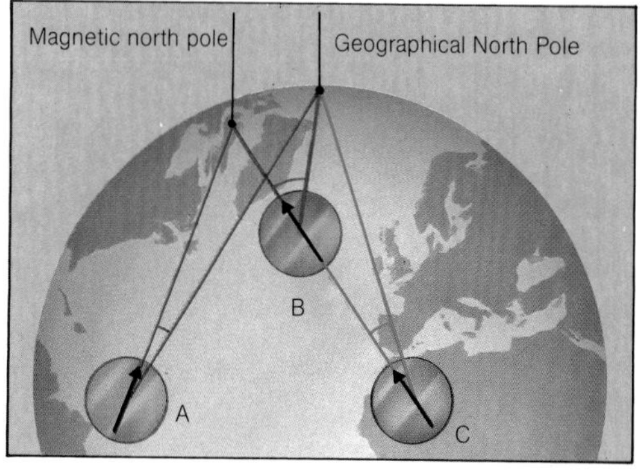

The angle of declination is between true, geographical North and magnetic north. As can be seen by comparing A, B, and C, this angle varies according to both latitude and longitude.

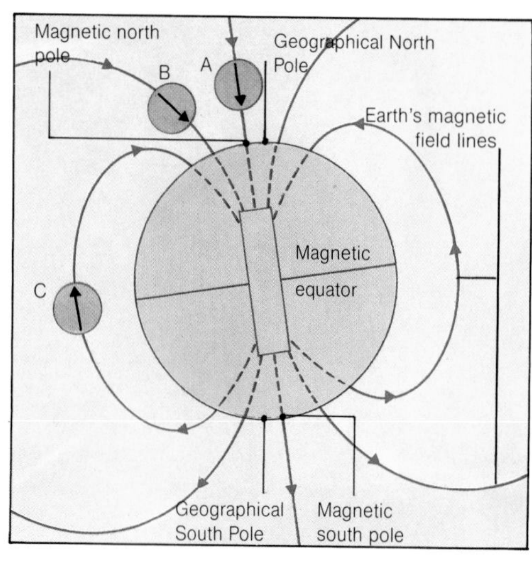

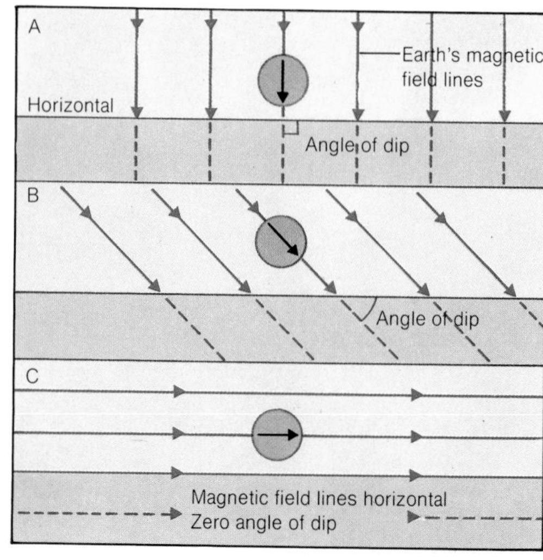

The angle of dip is between the horizontal and earth's magnetic field, as shown in the diagrams *(right and far right).* It varies according to location: at the magnetic north pole (A) the angle of dip is 90°; it decreases southwards (B), eventually becoming zero at the magnetic equator (C).

Many birds use Earth's magnetic field to help them navigate when flying long distances. Scientists have not yet discovered the precise mechanisms involved, but it seems that birds can in some way detect changes in the angle of dip and in the magnetic declination (*see* the diagrams at the bottom of the previous page), which together give an indication of latitude and longitude.

ferent points on the earth's surface, according to the observer's position relative to the two poles and his distance from them. It also varies at different times, corresponding to the "wandering" of the magnetic poles. Not only do the magnetic north and south poles move slightly from year to year, but at certain periods in the earth's past they have been completely reversed, so that the present magnetic north pole was at one time the magnetic south pole. Despite considerable scientific investigation, the reason for the seemingly arbitrary movement of the magnetic poles is not completely understood. Furthermore, the earth's magnetic field is not parallel to the surface (except at the magnetic equator) but is inclined. The angle between the magnetic field at any point and the horizontal is called the angle of dip. It varies according to the distance from the magnetic pole.

The cause of the earth's magnetism has long been a matter of conjecture and is still not completely understood. It is certainly not due to the existence of a bar magnet deep within our planet. Earth's core is extremely hot, and although it contains a large amount of iron, the iron is molten and therefore not magnetic (being molten, the iron atoms are moving too much to allow alignment of magnetic dipoles).

One widely accepted explanation of the earth's magnetism is based on the theory of a self-exciting dynamo. As applied to our planet's magnetism, the theory postulates that the earth, rotating in its own magnetic field, generates (by electromagnetic induction) a potential difference within its core. This potential difference drives an electric current which, in turn, produces the magnetic field. Hence the system is self-sustaining.

Earth is not the only planet with a magnetic field. The moon, Mercury, Jupiter, and Saturn also have magnetic fields, as do some other celestial objects—the sun, for example. Satellite and space-probe investigations have enabled these fields to be measured, which has provided much valuable information about the composition of the sun and planets.

Time and relativity

Sundials indicate the time by the position of the shadow cast by a vertical pole. The modern sundial—at the Royal Greenwich Observatory, Herstmonceux, England—shows 09:30 hours.

The first chronometers were made in the eighteenth century by the British horologist John Harrison (1693-1776). This one is the second of a series of four he made in an attempt to win a large prize offered by the British government for an accurate marine chronometer. In 1762 the last of the series met all the standards required for the prize, but Harrison did not receive the full award until 1773.

Over the ages philosophers have argued about the basic nature of time, whereas scientists have been involved in the study of phenomena that seem to demonstrate the "flow" of time from the past to the future, and how phenomena take place within this flow.

As a result, much effort has been expended on measuring intervals of time, so that the flow of time can be used as a backdrop against which to examine natural phenomena. Such phenomena have become increasingly esoteric over the years, and our understanding of the nature of time has had to be modified accordingly. In particular, developments in physics during the past 100 years or so have overthrown one of the longest-standing concepts of time: that it flows at the same rate for everyone in the universe. The realization that time is not, as Isaac Newton stated, absolute, but is in fact relative, led Albert Einstein to develop the physics of relativity.

Time and its measurement

Instruments that measure the passage of time have been in existence for about 4,000 years. The earliest were probably sundials, which used the movement of the shadow cast by a pole to show the passage of time. Other early forms of clocks used the change in the level of water or sand as it flowed from a reservoir. Such clocks had numerous disadvantages and limited applicability, but it was not until the late 1200's that mechanical clocks became available in Europe. These clocks—probably based on Chinese timepieces—used falling weights or springs to power them but, because the whole clock took part in the timekeeping process, their accuracy was poor, and they had to be corrected frequently against the more accurate sundial readings.

Not until the mid-seventeenth century and the invention of the pendulum clock by the Dutch scientist Christiaan Huygens did reliable clocks become available. In this, the timekeeping pendulum was separated from the rest of the timepiece by a mechanism that supplied energy to the pendulum sufficient to overcome air and other resistances.

A century later, the need for precise timekeeping to enable ships to navigate accurately led to the invention (by the British horologist John Harrison) of a chronometer with an accuracy of better than one second a week.

Refinements to mechanical chronometers gradually improved their accuracy, but it was not until the twentieth century and the advent of clocks based on natural physical processes that dramatic improvements in accuracy were possible. The most familiar type of such clocks is the quartz clock, which relies on the fact that quartz crystals exhibit the piezoelectric effect—they vibrate with a specific frequency when subjected to a resonant alternating electric field. These clocks are capable of accuracies of better than one minute a year, and they

The cesium clock is a complex but extremely accurate device used to define the second. It relies on the fact that certain subatomic changes in atoms of cesium-133 cause the emission of radiation of a specific frequency (that is, a specific number of oscillations per unit time). Hence a second can be defined as the duration of a certain number of oscillations—9,192,631,770 for the standard second.

are also small enough to be used in wristwatches.

The other principal type of natural clock—the atomic clock—is the most accurate of modern timekeepers. When certain changes occur within atoms of some elements, such as cesium, radiation of a specific frequency is emitted. Because frequency is the number of oscillations per unit time, it is possible to define the second in terms of the number of oscillations of a particular type of radiation.

Currently the second is defined as the duration of 9,192,631,770 oscillations of a specific radiation emitted by an atom of cesium-133.

The cesium clock is extremely accurate—to within several seconds in 100,000 years. It is possible to achieve even greater accuracy by using a hydrogen maser, but this device is less reliably consistent than the cesium clock and so is not suitable for defining a standard second.

Time standards

Before the invention of the very accurate atomic clocks the standard of time was based on astronomical phenomena, such as the time taken for the earth to complete one rotation about its axis, used to define one day. This definition is, however, too imprecise for most scientific purposes because the rotation, as well as being erratic, also differs (by about four minutes) according to whether it is measured relative to the sun or the stars.

Similarly, the time taken for earth to orbit once around the sun has been used to define a year; but again, several definitions of an orbit are possible.

Since 1964, with the adoption of the cesium clock second as the standard time unit for scientific purposes, other time standards have generally fallen into disuse. Nevertheless, standards of time based on the earth's orbit are still used occasionally—the ephemeris second, for example, which is defined as 1/31556925.9747 of the tropical year of A.D. 1900. The tropical year is defined as the time taken for the sun to pass between successive spring equinoxes.

The adoption of the cesium clock standard has had a particularly important consequence for astronomers and physicists. Because it enables time to be measured independently of the earth's motion, the cesium clock standard has made it possible for scientists to determine the rate of change of the orbital motion of the earth and planets and to discover if the orbits are shrinking or expanding with time.

Observations carried out using atomic timekeepers have also enabled astronomers to investigate the possibility that the strength of gravity (that is, the magnitude of the universal gravitational constant) is changing. Such changes—which have yet to be detected—would reveal themselves in slow variations in the orbital periods of planets and other objects within the solar system.

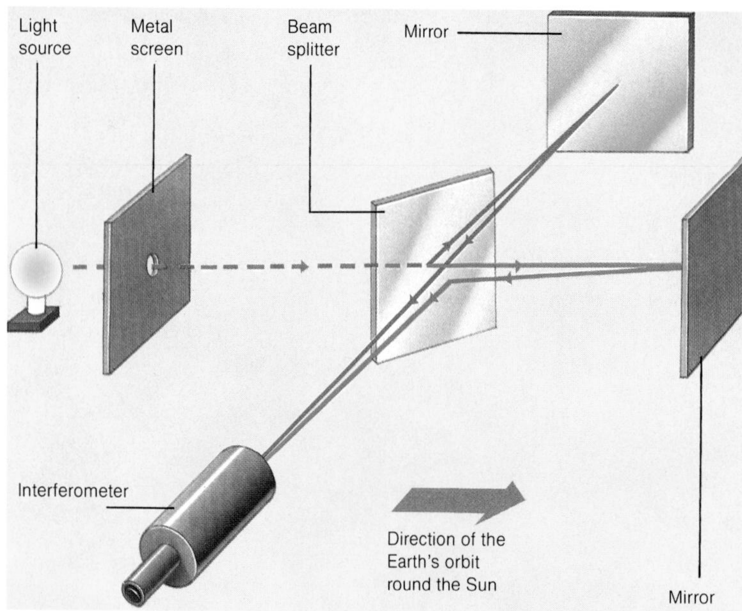

Michelson and Morley's 1887 experiment attempted to show that the speed of light moving with the earth's orbit (blue beam) was greater than that of light traveling across this path (red beam). By making the beams interfere, any velocity difference would show as a shift in the interference fringes seen through the interferometer when it was turned through 90°. But no shift occurred, indicating no velocity difference.

Special Relativity

The special theory of relativity immediately reminds us of Albert Einstein, the German-born American physicist who developed the theory (and published it in 1905). Some of the essential background work was, however, done in the previous century.

In 1887, the American physicists Albert Michelson and Edward Morley carried out one of the most famous experiments in the history of science. They attempted to show that the speed of light depended on whether the light was traveling in the same direction as earth's motion about the sun or at right angles to this path. They expected to find a difference in speed just as the speed of a man walking inside a moving train is greater than that of a man walking along the track, when seen by somebody standing by the track. In fact they found that the speed of light is the same in both cases. They then went on to measure the speed of light to a high degree of accuracy.

That such an unlikely result might be obtained was indicated some years before the Michelson-Morley experiment, in the laws of electromagnetism as described by James Clerk Maxwell in the 1860's. In these laws the speed of light emerges from the equations as a constant, the value of which does not depend on the velocity of the observer who is attempting to measure it.

In response to Michelson and Morley's findings, attempts were made within the framework of conventional Newtonian physics to account for the apparently anomalous behavior of light, by devising mathematical transformations relating the behavior of moving objects to their motion. But the correct explanation did not emerge until Einstein's publication of his special theory.

What Einstein did was to take the Michelson-Morley result at face value and use it to show that the Newtonian idea that time and space measurement are independent of the observer is incorrect. In Einstein's new conception of space and time, the two are not separated but are part of the more general entity, space-time. In using this concept, instead of specifying only the position of an object, it is necessary also to include time, thereby obtaining not a point in three dimensions of space, but an event in four dimensions of space-time.

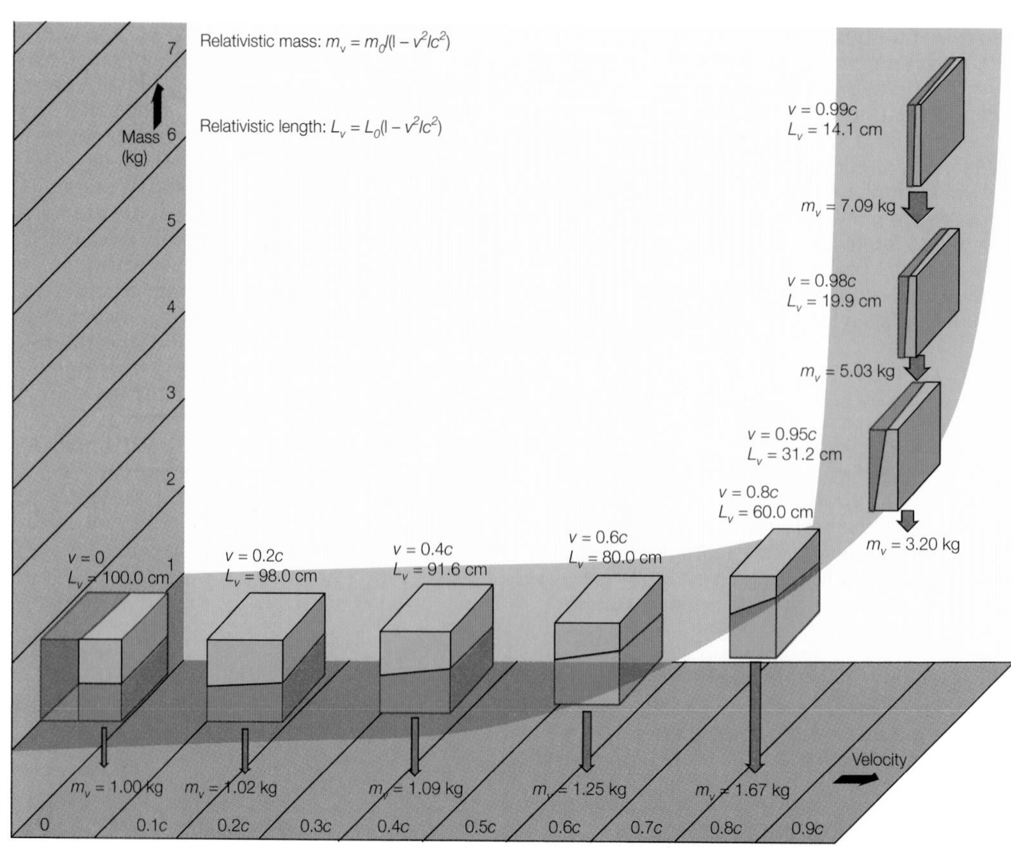

Two of the consequences of special relativity are that the mass of an object increases and its length decreases (compared with their values when the object is at rest) as the velocity increases. The diagram illustrates these phenomena, showing how a cube with sides of 100 centimeters and a mass of 1 kilogram when at rest changes in mass and dimension on acceleration to near-light speed. As can be seen from the diagram, there is relatively little change until the speed approaches that of light: on acceleration from rest (velocity, $v = 0$) to 0.6 times the speed of light ($v = 0.6c$) the length (l) of one side decreases by only a quarter and the mass (m) increases by the same proportion; from $0.6c$ to $0.99c$, however, l decreases more than fivefold and m increases more than fivefold.

Having lost the notion that space and time intervals are the same for all observers, we now have two new "invariant" quantities. The first is the speed of light; the second is the space-time interval. This latter is comprised of the square of the time and the space interval combined in a special way, in which space and time are treated similarly.

Using these two invariants, we can deduce the new transformation laws that give the coordinates of one body that is moving relative to another. And from these transformations (obtained mathematically by the Dutch physicist Hendrik Lorentz), it is possible to see that the "commonsense" transformations (derived several centuries before by Galileo) in fact give good agreement with observations only when dealing with speeds much slower than light. Although this is acceptable for the vast majority of circumstances, it is too inaccurate when dealing with near-light speeds. At such high speeds, the Lorentz transformations predict that some unusual phenomena start to become observable.

Consequences of the special theory

The Lorentz transformations enable calculation of the space and time intervals observed when traveling at any speed up to that of light. The length of a rod, for example, is a case of a space interval. Using Lorentz transformations, we find that as a moving rod approaches the speed of light, the length of the rod appears to decrease (if the rod is parallel to the direction of motion) as compared to an identical stationary rod. (In fact, the length L_v at a speed V is given by $L_o(1-V^2/c^2)^{\frac{1}{2}}$, where c is the speed of light and L_o is the rest length. The velocity V can be either that of the rod relative to the observer, or vice versa.)

As one expects from a theory that puts space and time on the same footing, there is a similar Lorentz effect on moving clocks: time measured by a clock moving at a speed V relative to a stationary clock runs slower by a factor of $(1 - V^2/c^2)^{\frac{1}{2}}$. An example of a moving clock is an unstable elementary particle that decays after a certain "lifetime" (as measured in its own reference frame). Experiments have shown that unstable particles moving at near-light speeds continue to exist for several lifetimes longer than their stationary counterparts by exactly the amount predicted from special relativity.

The postulates of special relativity also lead to the conclusion that the mass of a moving object increases relative to a stationary counterpart: at velocity V the mass m_v is $m_o(1 - V^2/c^2)^{-\frac{1}{2}}$, where m_o is the rest mass. This phenomenon leads, in turn, to one of the most important conclusions of special relativity: no material body can travel at speeds equal to the speed of light. Only entities that have zero rest mass—photons, for example—can travel at such speeds. In all other cases the mass tends toward infinity as the speed of light is approached. Thus an infinite amount of energy would be needed to reach light-speed.

Finally, we can deduce from the postulates that even when a particle is not moving, it possesses a certain amount of energy, given by the famous equation $E = m_oc^2$, where m_o is the rest mass of the particle, and c is the speed of light. The importance of the equation lies in its implication that matter and energy are equivalent. In addition, the appearance in the equation of the square of the speed of light leads us to expect that a vast amount of energy could be released if all the matter of even a very small object could be transformed into energy. In theory, the total conversion of just three tons of material could supply the whole world's energy requirements for a year. In fact, however, practical converters are far less efficient. In the sun, for example, hydrogen atoms fuse to form helium, a process that involves the direct conversion of matter to energy. This conversion is only about 1 per cent efficient, but countless billions of atoms are fusing every second and therefore the total amount of energy generated is vast. The steady, controlled output of a nuclear power station and the uncontrolled nuclear explosion are also produced by the partial conversion of mass into energy. When a particle of matter encounters a particle of antimatter (such as when an electron and an anti-electron (positron) collide, the masses of both particles are completely converted into energy in the form of electromagnetic radiation.

The aerial view above of the CERN research establishment (situated on the Swiss-French border near Geneva) shows the large size of the underground Super Proton Synchrotron (indicated by the dotted white line), a device that can accelerate protons to near-light speed. A proton synchrotron consists basically of a ring of electromagnets surrounding a circular vacuum chamber; the protons are continually accelerated by an electromagnetic field as they repeatedly circle around inside the vacuum chamber. The CERN synchrotron also has storage rings (part of one of which is shown *left*), which make it possible to produce two particle beams traveling in opposite directions. This arrangement permits proton collisions at much higher energies than can be achieved by other means, and such very high energy collisions are essential for research into fundamental particles such as quarks.

Space-time curvature can be visualized using the rubber sheet analogy. Large masses with strong gravitational fields cause space-time (the blue "rubber sheet") to curve around them; the greater the mass, the greater the curvature. Thus the geodesics (the blue grid lines)—the paths that small objects follow when near a large mass—are also curved.

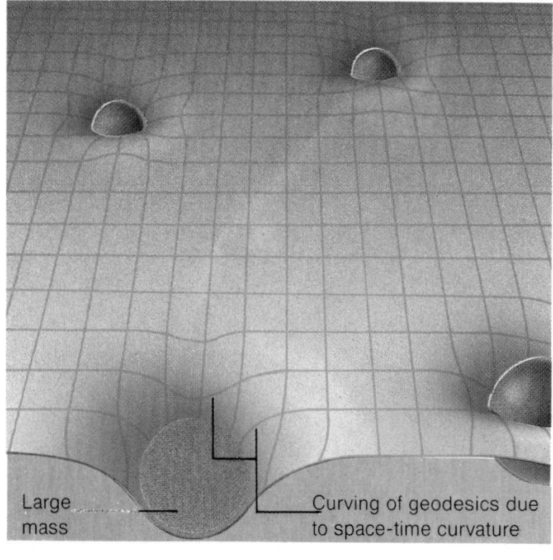

General Relativity

After putting forward his special theory of relativity in 1905, Einstein went on to develop his general theory of relativity (published in 1916), which includes the effects of accelerations and gravity in addition to those of uniform velocities. (The special theory applies only to the latter.) Mathematically, it is far more complex than the special theory, and its implications are even more far-reaching.

General relativity and gravity

General relativity is primarily a theory of gravity. To arrive at it Einstein extended his principle of relativity (according to which all observers are equivalent, irrespective of their velocities) to accelerations. Including accelerations in this way naturally brings in gravity. Consider an elevator in deep space, far from any gravitational field. If the elevator is made to accelerate upward at a rate equal to the acceleration due to gravity on earth, an object released within the elevator appears to accelerate toward the floor at the same rate as it would if released on earth. On our planet, however, the acceleration would be the result of the earth's gravitational field. Suppose now that two different masses are held by the observer in the elevator. As the floor accelerates upward to meet them, the two masses strike the floor at the same time—exactly what happens when two different masses are released from rest above the earth's surface.

Thus, if the region of space is small (so that the convergence of the paths of falling objects toward the center of gravity of the gravitating body is negligible), it is impossible to distinguish between an accelerated system and one that is at rest because, as our hypothetical comparison shows, an accelerated system is equivalent to a system at rest in a gravitational field. This is the principle of equivalence that, in its general form, asserts that the laws of physics must be the same for all observers, irrespective of how they move.

To obtain a theory of gravity from this principle, we consider Newton's first law of motion: a moving object continues to move in a straight line unless acted upon by a force. Returning to our hypothetical elevator in space, if a bullet were shot through from one side to the other while the elevator accelerated upward, the bullet's point of entry would be higher than its exit point; no force acted on the bullet but it appeared to have followed a curved path. Because the force of gravity was absent, the bullet, according to Newton's law, should have followed a straight path; that it did not means that Newton's law seems not to be universally applicable. Nevertheless, the essential features of this important law can be retained simply by dropping all reference to forces and merely stating that all bodies travel in a straight line when left to themselves (after suitably redefining the term "straight").

This leads to the view of gravity as being curvature of space-time. We know that when a ball is thrown on earth, its path is not a straight line, but a parabola. Thus in the modified version of Newton's law we need the idea of a generalized straight line—a geodesic. A geodesic is the shortest distance between two points in a given type of geometry. When the geometry is flat (Euclidean), we get the familiar straight line. But in the earth's gravitational field objects do not move along these straight line geodesics but instead follow curved geodesics. In other words, gravitational fields manifest themselves as curved, non-Euclidean geometries, giving rise to curved geodesics.

The easiest way to visualize Einstein's conception of gravity, not as a force but as curvature of space and time, is to use the rubber-sheet analogy. In this, a massive weight (representing a body with a strong gravitational field—a planet or star, for example) is placed on a rubber sheet (representing space-time), which curves under the mass (representing

The "double quasar" (indicated by the two arrows on the image-enhanced radio map, *right*) is, in fact, only a single quasar. The second *(lower)* image can be seen because light from the real *(upper)* quasar is deflected toward earth by a massive galaxy—as shown in the diagram below. The discovery of this double quasar in 1979 is consistent with the prediction of Einstein's general theory that gravity can bend light.

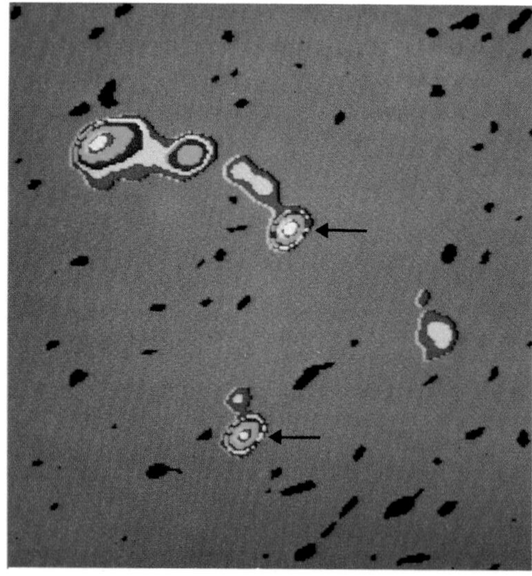

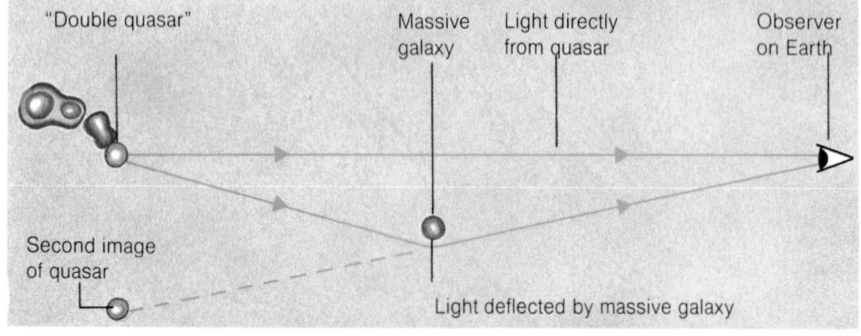

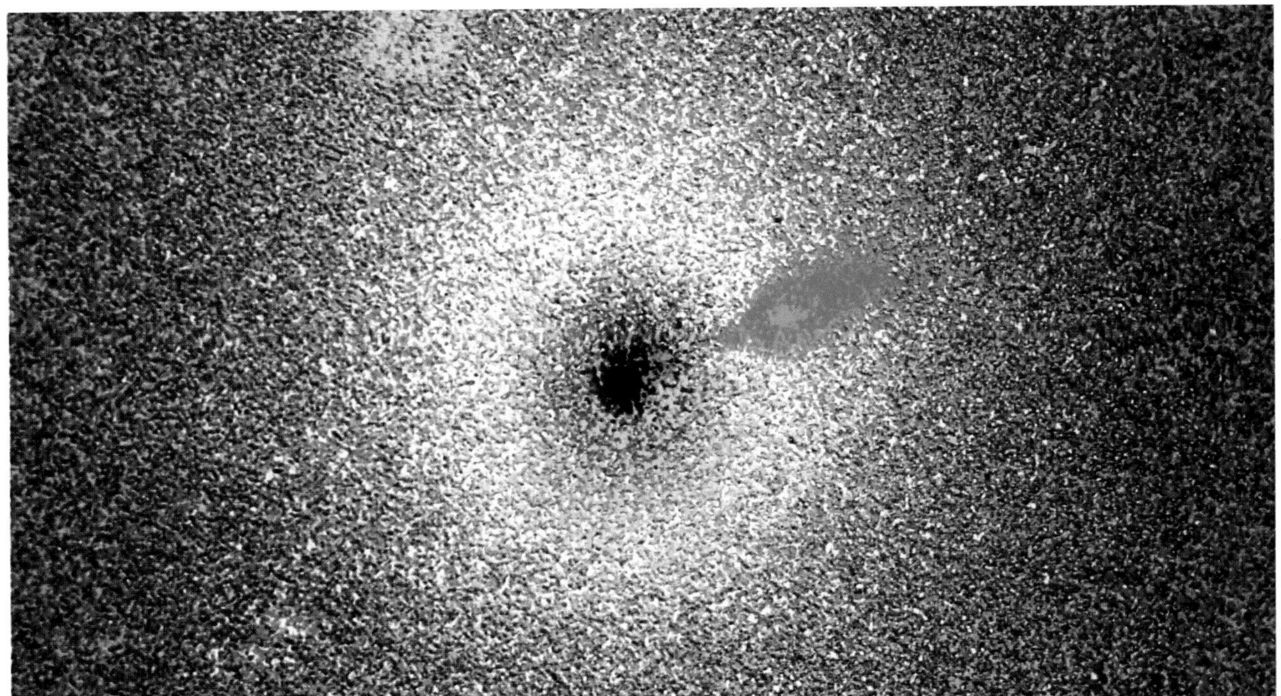

The giant galaxy M87—which lies about 50 million light-years from Earth in the constellation of Virgo—contains at its core a small, extremely dense object that some scientists believe is a black hole. Black holes are so compact and massive (the object in M87 has a mass about five thousand million times that of our sun) that their gravitational fields are unimaginably strong—so strong that, according to general relativity, even light cannot escape. In contrast, other regions of M87 are powerful radiation emitters—the blue jet in the photograph, for example, is a strong X-ray and radio source.

the curvature of space-time caused by gravitational fields). Small masses moving near the large mass follow the paths that approximate most closely to the shortest-distance straight lines, the geodesics, which are inevitably curved except at distances far from the large mass.

Einstein then went on to produce an equation that showed how much curvature was produced by various amounts of mass. For the weak gravitational fields of small masses, the curvature is also small and Einstein's equation reduces to Newton's law of gravity. But for stronger fields significant differences show up. For example, the elliptical orbit of Mercury (a relatively fast-moving planet, strongly influenced by the sun's gravitational field) slowly rotates in a manner explained precisely by general relativity; Newton's law of gravity cannot account for the rotation. Other tests also support Einstein's conception of gravity.

Cosmology

When the nature and evolution of the universe is studied, it is again found that Newton's law of gravity is inapplicable and that Einstein's theory of general relativity has to be used. By assuming that, on a large-scale view, the universe looks the same in all directions from all points within it, it is possible to derive a set of equations that describe the behavior of the universe. Unfortunately, a wide variety of model universes are possible, despite the fact that there is, in reality, only one universe; general relativity by itself seems too "weak" to arrive at one unique model without invoking additional concepts and theories. Another problem with general relativity is that the equations lead to the conclusion that the universe began in a state of infinite density and curvature (called a singularity), to which existing laws of physics do not apply. Again, the indications are that general relativity is not the whole story, but perhaps its integration with quantum theory—as yet unachieved—may lead to a fuller understanding of the origin of the universe.

In the meantime, the observational data suggests that the expansion of the universe predicted by Einstein's equations may continue forever, with the universe "dying" by slow cooling rather than recollapsing to a big bang from which another universe might originate in the same way that the present universe is thought to have come into existence.

Glossary

In the following glossary, small capital letters (for example, IMAGE) indicate terms that have their own entries in the glossary.

A

aberration The degration of an IMAGE produced by a lens or mirror. It is caused by light rays coming to a FOCUS at different positions.

absolute temperature A temperature measured in KELVINS (K) on the kelvin or absolute scale of temperature.

absolute zero The lowest temperature that is theoretically possible, at which a material would have zero ENERGY. It is equal to 0 K (−459.67° F. or −273.15° C).

absorption 1. The take-up of ENERGY, particularly RADIANT ENERGY and SOUND, by a body or system. The absorbed energy is converted to other forms of energy, often heat. 2. The movement of MOLECULES of gas or liquid into the whole of a liquid or solid object.

absorption spectrum A spectrum of discrete dark lines or bands produced by passing a CONTINUOUS SPECTRUM through a medium that selectively absorbs specific WAVELENGTHS.

adhesion The INTERATOMIC or INTERMOLECULAR FORCES that bind the ATOMS or MOLECULES of two different materials placed in contact at the boundary between them.

adiabatic process A process in which HEAT does not enter or leave the system. Any change in temperature is produced by ENERGY changes within the system.

adsorption The movement of MOLECULES from a surrounding gas or liquid into the surface of a solid, where they are held by INTERMOLECULAR FORCES to the solid molecules.

alpha particle A positively charged particle consisting of two PROTONS and two NEUTRONS emitted from the nucleus of a radioactive ELEMENT. Alpha particles are helium nuclei.

alternating current (AC) Electric current in which the flow of ELECTRONS continually reverses at a particular FREQUENCY.

ampere (amp) The SI UNIT of electric CURRENT. It is defined in terms of the force that exists between two wires, each carrying a current of 1 ampere.

amplitude In any body that oscillates or vibrates about a central position, the maximum amount of displacement from this position. In a WAVE MOTION, the square of the amplitude is equivalent to the amount of ENERGY transmitted by the wave.

amplitude modulation (AM) A method of transmitting information in which the AMPLITUDE of a CARRIER WAVE is modulated (varied) at the FREQUENCY of the information signal.

analog system A measuring or computing system in which the magnitude of a quantity is represented by a change in a physical body, such as the movement of a pointer over a scale to represent voltage.

angular momentum *See* MOMENTUM.

angular velocity *See* VELOCITY.

anode An ELECTRODE that has a positive POTENTIAL or CHARGE.

antinode A point of maximum oscillation in a STANDING WAVE. Each antinode in a standing wave is half a WAVELENGTH apart.

antiparticle A SUBATOMIC PARTICLE that is of equal mass but opposite charge to another particle; the POSITRON is the antiparticle of the ELECTRON, for example.

atom The smallest part of a chemical ELEMENT that retains the identity of the element.

atomic number The number of PROTONS in the NUCLEUS of any ATOM of an ELEMENT, or the number of ELECTRONS in an uncharged atom.

atomic weight The average mass of an ATOM of an ELEMENT or ISOTOPE measured relative to the mass of the isotope carbon-12, which has an atomic weight of 12.

B

band spectrum A SPECTRUM consisting of bands made up of groups of lines.

baryons A class of SUBATOMIC PARTICLES consisting of PROTONS, NEUTRONS, and HYPERONS.

beats A WAVE MOTION produced by the interference of two other waves. The beat FREQUENCY is the difference in frequency of the two interfering waves.

beta particle An ELECTRON or POSITRON emitted by an atom of a RADIOACTIVE ELEMENT.

black body An object that absorbs all FREQUENCIES of RADIANT ENERGY incident upon it and which is capable of emitting all frequencies. A perfect black body cannot be made, but a cavity in the wall of a totally black container approximates to one.

black body radiation RADIANT ENERGY of the complete range of FREQUENCIES.

boundary layer The layer of FLUID that lies next to the surface of a body in RELATIVE MOTION to the fluid. The proximity of the surface reduces the rate of flow in the boundary layer.

Brownian motion Microscopic random movements of solid particles suspended in a liquid or gas. It is caused by collisions with the liquid or gas MOLECULES.

bubble chamber A device in which the motions of SUBATOMIC PARTICLES produce trails of bubbles in a liquid in the chamber.

C

calorie A unit of HEAT equal to 4.1855 JOULES. The Calorie used to measure the energy content of foods is equal to 1,000 calories (1 kilocalorie).

candela The SI UNIT of LUMINOUS INTENSITY. It is defined as the luminous intensity of a certain BLACK BODY at a particular temperature and pressure.

capacitance (or capacity) The ability of an electric circuit to store CHARGE.

capacitor An electrical device that stores an electric CHARGE when a POTENTIAL DIFFERENCE is placed across it.

capillary action The movement of the surface of a liquid up or down a narrow tube placed in the liquid. It is caused by the SURFACE TENSION of the liquid.

Carnot cycle An ideal cycle of temperature and pressure changes in a HEAT ENGINE; the cycle is reversible.

carrier wave A beam of RADIANT ENERGY at radio or other frequencies that is modulated to carry information, either by AMPLITUDE MODULATION or FREQUENCY MODULATION.

cathode An ELECTRODE with a negative POTENTIAL or CHARGE.

cathode rays A beam of ELECTRONS emitted by a CATHODE, for example in a discharge tube.

Celsius temperature A temperature measured on the Celsius scale, in which the freezing point of water is 0° C and the boiling point is 100° C. Also known as centigrade temperature.

center of gravity The fixed point within a body through which the resultant force of gravity always passes for any position of the body. In a uniform gravitational field, this point is the CENTER OF MASS.

center of mass The fixed point within a body or system at which the entire mass of the body or system can be considered to be concentrated.

centrifugal force A fictitious (that is non-physical) force invented to describe the feeling one experiences when moving in a curved path.

centripetal force The force pushing on a body to make it go in a circular path instead of a straight line. This force is directed towards the center of the circle.

chain reaction Any chemical or nuclear reaction that is self-sustaining once it has started. In particular, a process of nuclear FISSION in which the products of nuclear reactions between the nuclei taking part initiate further reactions.

charge, electric An electrical property possessed by most SUBATOMIC PARTICLES, ELECTRONS having unit negative charge and PROTONS unit positive charge, for example. The motion of electrons in conductors and insulators produces differences in charge. Electric fields exist around charges and interact such that like charges attract and unlike charges repel one another. A force also exists between a moving charge and a magnetic field.

chemical energy ENERGY that is either consumed or liberated by substances undergoing a chemical reaction.

cloud chamber A device for detecting the motion of SUBATOMIC PARTICLES in which the particles produce trails of droplets in a supersaturated vapor.

coefficient A constant value for a given material that expresses how a particular property of the material changes under certain conditions. Expansivity (the coefficient of expansion), for example, indicates the amount of expansion that occurs with increasing temperature.

coherent light or radiation Light or radiation in which all the waves are exactly in PHASE.

cohesion The INTERATOMIC or INTERMOLECULAR FORCES that hold together the ATOMS or MOLECULES within a solid or liquid.

colloid A system in which one substance is distributed within another in the form of very fine particles. The particles are larger than MOLECULES, but too small to group together and settle or coagulate.

compound A substance in which the atoms of two or more chemical ELEMENTS are bound together in a definite proportion by INTERATOMIC or INTERMOLECULAR FORCES or both.

condenser 1. Electrical condenser—a CAPACITOR. 2. Optical condenser—a system of lenses that concentrates light rays from a source into a beam, generally to illuminate objects. 3. Chemical condenser—apparatus used to condense vapors into liquids during distillation.

conduction The flow through a substance of HEAT, or electric CURRENT or CHARGE.

conductor A substance that readily permits the passage of HEAT, or electric CURRENT or CHARGE.

constant A quantity that does not charge. Examples include the velocity of light and the charge on the electron. Constants are factors in equations and formulas that enable calculations of the magnitudes of variable quantities to be made. Accurate measurement of constants is therefore important.

continuous spectrum A SPECTRUM of LIGHT or RADIANT ENERGY in which all FREQUENCIES exist so that no distinct lines or bands are produced.

converging lens A lens that causes rays of light passing through it to converge toward a single point, the FOCUS. Converging lenses are also known as convex lenses.

coulomb The SI UNIT of electric CHARGE. It is defined as the charge produced by the flow of 1 AMPERE of current in 1 second.

couple Two forces of equal magnitude that act in parallel but opposite directions at different points on a body or system. A couple generally acts to rotate a body or system.

critical angle The angle of INCIDENCE at which a ray of light is refracted parallel to the surface of the medium that it enters. At angles of incidence greater than the critical angle the ray is reflected from the surface, not refracted.

critical mass The minimum amount of a fissile material required to produce a nuclear CHAIN REACTION. It depends on the shape of the material as well as the identity and purity of the fissile elements.

critical temperature The temperature above which a gas cannot be liquefied, regardless of the pressure applied.

current, electric A movement of ELECTRONS through a CONDUCTOR in the same direction (direct current) or reversing at a particular FREQUENCY (alternating current).

cycle In a system that undergoes oscillations or repeated patterns of changes, a single complete oscillation or series of these changes.

D

decay The change of one ELEMENT into another that occurs in RADIOACTIVITY.

degree 1. The general unit of temperature. Its value is the same in the absolute, CELSIUS, and centigrade scales and is equal to the KELVIN. The Fahrenheit degree is equal to 5/9 kelvin. 2. The unit of plane angle, a complete circle having 360 degrees.

density 1. The ratio of the mass of an object to its volume, expressed either in units such as kilograms per cubic meter or as a numerical ratio known as the RELATIVE DENSITY or specific gravity. 2. Optical density—the relationship between the REFRACTIVE INDICES of two mediums. If one medium is optically denser than another, then it has a greater refractive index.

diamagnetism The small magnetic field created inside all substances when they are in the vicinity of a magnetic field created by another object. The diamagnetic field always opposes this other field.

dielectric An INSULATOR or non-conductor of electricity, particularly that used in a CAPACITOR to store CHARGE.

diffraction 1. The bending of rays of ELECTROMAGNETIC RADIATION that occurs as they pass through a narrow opening or pass the edge of an obstacle. 2. X-ray or electron diffraction—the bending of beams of X RAYS or ELECTRONS as they pass through matter.

digital system A measuring or computing system that represents the magnitudes of quantities as numbers composed of digits, either in decimal or binary form. A digital watch, for example, displays the time as a decimal number, and digital computers process all data in the form of binary codes.

diode An electronic device, generally a transistor or a component of a microchip, consisting of an ANODE and CATHODE such that a current flows only from the cathode to the anode and not in the reverse direction.

dipole 1. A radio aerial or antenna made up of two rods. 2. Two equal and opposite electric CHARGES or magnetic POLES that are separated, for example in the distribution of charge to opposite ends of a MOLECULE.

direct current (DC) A continuous flow of ELECTRONS through a CONDUCTOR in the same direction.

dispersion The splitting of light comprised of a mixture of WAVELENGTHS into its component wavelengths or colors, for example by a prism or by DIFFRACTION.

diverging lens A lens that causes rays of light passing through it to diverge. Diverging lenses are also known as concave lenses.

Doppler effect The change in FREQUENCY of RADIANT ENERGY or SOUND that occurs as a result of the RELATIVE MOTION of the source and observer.

dynamics The study of bodies in motion.

E

efficiency The ratio of a machine's ENERGY or WORK output to the energy or work input; efficiency is usually expressed as a percentage.

electrode The part of an electric or electronic device at which an electric CURRENT or a beam or flow of ELECTRONS either enters (at the electrode called the CATHODE) or leaves (at the ANODE).

electromagnet A device that produces a magnetic field when a current flows through it. It basically consists of a coil of wire wound around an iron bar.

electromagnetic radiation Energy that is propagated in the form of rays or waves consisting of simultaneous WAVE MOTIONS in an electric field and a magnetic field. The group of electromagnetic radiations comprise radio waves, microwaves, infrared rays, visible light, ultraviolet rays, X rays, and gamma rays.

electromotive force (EMF) The electric "force" possessed by a source of electric energy that causes a current to flow through a circuit connected to the source. It is measured in VOLTS.

electron A stable SUBATOMIC PARTICLE bearing the fundamental (indivisible) amount of negative electric CHARGE. This charge is equal to 1.602×10^{-19} coulombs. The electron seems to be a point-like particle (without size) and a rest-mass of 9.110×10^{-31} kilogram.

electron-volt (eV) A unit of energy equal to the energy required to drive an ELECTRON through a POTENTIAL DIFFERENCE of 1 volt; 1 eV is equal to 1.602×10^{-19} joules.

electroscope A device that detects electric CHARGE.

element (chemical) A substance in which all the ATOMS have the same ATOMIC NUMBER.

emission spectrum A SPECTRUM produced by a self-luminous source of RADIANT ENERGY from which the radiation is received without being passed through an intervening medium as with an ABSORPTION SPECTRUM.

energy The capacity of a body or system to do WORK. The greater a body's energy, the more work it can perform.

energy level The energy state of an ATOM or NUCLEUS. An atom or nucleus is excited by the gain of energy to achieve one of a series of energy states, each at a particular level of energy. The lowest stable energy level is called the GROUND STATE.

entropy A measure of how randomly the objects in a system are arranged.

equilibrium A state of rest or of no change in a system produced by a balance of forces or energy.

F

Fahrenheit scale A scale of temperature in which the freezing point of water is set at 32° and the boiling point at 212°. The degree Fahrenheit is equal to $\frac{5}{9}$ of the degree CELSIUS or KELVIN.

ferromagnetism The ability of some substances to create and sustain their own magnetic fields even after all other magnetic fields have been removed.

fission A nuclear reaction in which the nuclei of a heavy fissile element break apart, each producing two smaller nuclei, NEUTRONS, and a large amount of energy.

fluid A substance that can flow, i.e. a liquid or a gas.

flux The ratio of the rate of flow of energy, mass, or volume to the area through which it is flowing, measured perpendicular to the direction of flow.

focus Real focus—the point at which light rays from the same part of an object are made to meet by a CONVERGING LENS or concave mirror. Virtual focus—the point from which light rays appear to diverge after reflection by a mirror or refraction by a DIVERGING LENS.

force An external effect capable of changing the state of rest or motion of a body. All forces are basically in the form of a push or a pull.

free electron An ELECTRON that is not bound to a single ATOM but which is free to move among atoms or from one atom to another.

frequency The rate at which an oscillating or vibrating system such as a WAVE MOTION undergoes CYCLES.

frequency modulation (FM) A method of transmitting information in which the FREQUENCY of a CARRIER WAVE is varied (modulated) at the frequency of the information signal.

friction A FORCE that acts to oppose RELATIVE MOTION between two bodies and which is caused by INTERACTION of the surfaces of the bodies.

fusion 1. Melting. 2. Nuclear fusion—nuclear reaction in which the NUCLEI of light ELEMENTS are forced together to produce the nucleus of a heavier element.

G

g The acceleration due to GRAVITY. It is the acceleration that occurs in any body that falls freely in a vacuum due to gravity. The standard value for g at earth's surface is 32.1740 feet (9.80665 meters) per second per second.

galvanometer An instrument that can detect or measure weak electric CURRENTS.

gamma rays ELECTROMAGNETIC RADIATION emitted by the nuclei of RADIOACTIVE ELEMENTS.

gas laws The basic laws that relate the effects of pressure, volume, and temperature on a quantity of gas. Only hypothetical ideal gases exactly obey the gas laws, which therefore have to be modified slightly to account for the behavior of real gases.

gravitational constant (G) The constant that relates the force of GRAVITY produced between two bodies to their masses and distance apart. It is equal to 6.664×10^{-11} N m^2 kg^{-2}.

gravity 1. The force of mutual attraction that exists between any two bodies. 2. A measure of the magnitude of force. A force of one gravity (1 g) is equal to the force of gravity at the earth's surface.

greenhouse effect A heating effect caused by the penetration and subsequent trapping of INFRARED RADIATION beneath or behind a transparent surface, such as the glass walls of a greenhouse. Earth is warmed by the greenhouse effect, caused by the atmosphere trapping the sun's radiation.

ground state The lowest and most stable ENERGY LEVEL that an ATOM or NUCLEUS may possess.

H

hadron A SUBATOMIC PARTICLE that may be involved in a strong INTERACTION within the NUCLEUS. *See also* LEPTON.

half-life The time required for the RADIOACTIVITY of a sample of a radioactive ISOTOPE to decrease to half its value.

harmonic A secondary WAVE MOTION associated with a primary wave motion and having a FREQUENCY that is an integral multiple of the frequency of the primary.

heat The flow of thermal energy from one substance into another.

heat capacity The amount of heat required to raise the temperature of a body by one DEGREE.

heat engine Any engine in which heat energy is converted to motion (kinetic energy), such as an internal combustion or steam engine.

hydrodynamics The study of the physical behavior of liquids in motion.

hydrostatics The study of the physical behavior of liquids at rest.

hyperons A class of SUBATOMIC PARTICLES that are very short-lived and of comparatively large mass.

hysteresis The elastic or magnetic properties of bodies that result from a variation in stress or magnetization.

Hz (hertz) The unit of FREQUENCY; the number of cycles of an oscillating or vibrating system (such as a WAVE MOTION) that occur in 1 second.

I

ideal (or perfect) gas A theoretical substance that obeys the GAS LAWS exactly. Ideal gases do not exist in practice.

image A representation of an object. A real image is formed by converging light rays and can be formed on a screen. A virtual image is formed by parallel or diverging light rays and cannot be formed on a screen.

impedance The property of a circuit carrying an ALTERNATING CURRENT that determines the AMPLITUDE of the current at a particular voltage. For a DC circuit, impedance is equal to resistance.

incidence, angle of The angle between a ray of light striking a surface and the normal (a line perpendicular to the surface).

induction The production of an ELECTROMOTIVE FORCE in a conductor by RELATIVE MOTION in a magnetic field, or the production of a magnetism in a body by the presence of an external magnetic field.

inertia The resistance of a body to acceleration or deceleration. The inertia of a body depends only on its MASS, and therefore the two properties can be considered to be equivalent.

infrared radiation ELECTROMAGNETIC RADIATION with WAVELENGTHS longer than those of light but shorter than microwaves.

insulator Any substance that does not conduct either heat or electricity.

intensity 1. In a WAVE MOTION, the ratio of the POWER transmitted to the area through which the wave passes. In light, intensity is perceived as brightness, and in sound as loudness. 2. Electric intensity—the strength of an electric field. 3. Magnetic intensity—the strength of a magnetic field.

interaction 1. Any action or influence that takes place between bodies, particles, or systems. 2. One of the four basic forces that constitute all interaction as defined above. These are: the strong (or nuclear) interaction, which holds nucleons in the NUCLEUS; the electromagnetic interaction, which holds ATOMS and MOLECULES together; the weak interaction, which produces BETA DECAY; and the gravitational interaction or GRAVITY.

interatomic forces The forces that produce the bonds that hold ATOMS together in molecules or lattices. Attractive forces are electrostatic, caused by a transfer of ELECTRONS from one atom to another in ionic bonds, or by a sharing of electrons between atoms in covalent bonds. Metal bonds and pi-bonds result from extended orbits of electrons that encompass several atoms. The attractive forces in the bonds are balanced by repulsive forces produced by INTERACTION between the electron clouds around the atoms.

interference The effects produced by the changes in AMPLITUDE that occur when wave motions combine.

intermolecular forces The forces that hold MOLECULES together. Attractive forces include VAN DER WAALS' FORCES and hydrogen bonds, which both result from electrostatic attraction caused by a distribution of electric CHARGE over the molecules so that one part becomes negative and another positive. These attractive forces are balanced by repulsive forces produced by INTERACTION between the ELECTRON clouds of the ATOMS in the molecules.

inverse-square law Any law that states that a quantity varies inversely with the square of distance. GRAVITY obeys an inverse-square law, its force decreasing by four if the distance between the gravitating bodies is doubled, for example.

ion An ATOM that has either gained or lost one or more ELECTRONS from the electron shells surrounding the NUCLEUS, or a group of atoms containing one or more such atoms. A cation is an ion that has lost electrons and has a positive charge. An anion has gained electrons and has a negative charge.

isothermal Changes taking place in a system at constant temperature.

isotope The isotopes of an ELEMENT differ only in the number of NEUTRONS in the NUCLEI of their ATOMS.

J

joule The SI UNIT of ENERGY and WORK, defined as the energy consumed and work produced when a force of 1 NEWTON is made to move a distance of 1 meter in the direction of application.

K

kelvin (K) The degree of temperature in the kelvin or absolute scale. 0 K is ABSOLUTE ZERO. The magnitude of the kelvin degree is equal to the degree CELSIUS or centigrade.

kilogram The SI UNIT of MASS, defined as the mass of the standard kilogram (a particular piece of platinum).

kinetic energy The ENERGY possessed by a moving body, particle, or system that is due solely to the motion. Kinetic energy may be translational (motion from one point to another), rotational, or vibrational.

kinetic theory The kinetic theory of gases, which explains the GAS LAWS by relating the ENERGY of gas MOLECULES to their motion.

L

latent heat The amount of HEAT that is either absorbed or emitted as a substance changes state and remains at its freezing, melting, or boiling point.

leptons A family of six fundamental particles. Included in this family are electrons.

line spectrum A SPECTRUM in which single, well-defined lines are observable.

load The FORCE that has to be overcome to produce WORK in a system or machine; for example, the weight of an object to be lifted by pulleys. The force produced by the system or machine acts in the opposite direction to the load.

longitudinal wave A WAVE MOTION in which the vibration takes place in the same direction as the wave is moving. SOUND waves are longitudinal waves.

luminous intensity The amount of light produced by a point source in 1 steradian (unit solid angle) in 1 second.

M

mass The amount of matter in a body, which determines its INERTIA and also its WEIGHT. The SI UNIT of mass is the KILOGRAM.

mass number The total number of PROTONS and NEUTRONS in each ATOM of a particular ISOTOPE of an ELEMENT.

mass spectrometer An instrument used to determine the masses of IONS by measuring their deflection in magnetic and electric fields.

mechanical equivalent of heat A CONSTANT that expresses the amount of HEAT that is totally converted into a certain amount of WORK. In SI UNITS, both heat and work are measured in the same unit (JOULES), so the mechanical equivalent of heat is equal to 1.

mechanics The study of the action of FORCES on matter. The behavior of bodies in motion is studied in DYNAMICS, and that of bodies at rest in STATICS. The behavior of matter on an atomic scale is described by quantum mechanics (*see* QUANTUM THEORY) and WAVE MECHANICS.

meniscus 1. The curved surface of a liquid adjacent to the sides of its container. 2. A lens having one convex side and one concave side.

mesons A class of SUBATOMIC PARTICLES that are found in cosmic rays and are produced by the bombardment of nuclei. Mesons are involved in the formation of the strong nuclear forces that hold nuclei together.

meter The SI UNIT of length, defined as 1,650,763.73 WAVELENGTHS of light produced by krypton-86.

mho The SI UNIT of electrical conductivity or conductance, also known as the siemens. It is equal to a reciprocal OHM, that is, a circuit has a conductivity of 10 mhos (or siemens) if its resistance is 0.1 ohm.

modulus A factor specific to a given substance that expresses its behavior under certain conditions. The bulk modulus and elastic modulus, for example, define changes in volume and length of a material under stress.

molecule Made of atoms, the smallest part of a substance that can exist and retain the identity of the substance.

moment of a force The product of a FORCE and the distance (measured perpendicular to its direction) of the force from the point at which it is applied.

moment of inertia The moment of inertia of a body rotating about an axis is the sum of the product of the mass of each particle of the body and the square of the distance of that particle from the axis of rotation. It is a measure of how resistant an object is to changing the way it is rotating.

momentum 1. Linear momentum—the product of the MASS and VELOCITY of a moving body. 2. Angular momentum—the product of the moment of inertia of a rotating body and its angular velocity.

muon An unstable SUBATOMIC PARTICLE with a negative or positive CHARGE and mass 207 times that of the ELECTRON.

N

N *See* NEWTON.

neutrino A stable SUBATOMIC PARTICLE that has no electric CHARGE and no (or possibly a very small) mass when at rest. Neutrinos or their antiparticles, antineutrinos, are produced in beta decay, together with POSITRONS or ELECTRONS respectively.

neutron A SUBATOMIC PARTICLE found in the nuclei of all ISOTOPES except hydrogen-1. The neutron has no electric charge, and its mass is 1.675×10^{-27} kilogram. Outside the nucleus, a neutron is unstable and decays (with a HALF-LIFE of 12 minutes) to give a PROTON, ELECTRON, and antineutrino.

newton (N) The SI UNIT OF FORCE, and also the correct scientific unit of WEIGHT. A force of 1 N produces an acceleration of 1 m s^{-2} in a mass of 1 kilogram.

node In a STANDING WAVE, the points of zero oscillation or AMPLITUDE. The nodes are half a WAVELENGTH apart.

normal An imaginary line that is perpendicular to a surface. In the reflection and refraction of light, the angles of the rays are measured between the rays and the normal at the point where the rays strike the surface of the reflecting or refracting medium.

nuclear energy (or atomic energy) Energy released when changes take place in the nuclei of atoms. It appears as RADIOACTIVITY, and heat produced by the increased motion of the atoms involved.

nuclear reaction A reaction in which the NUCLEUS of an ELEMENT changes, thereby producing a new element, usually with the release of energy. Nuclear reactions include RADIOACTIVITY, the production of artificial radioactive ISOTOPES, FISSION, and FUSION.

nucleon A stable SUBATOMIC PARTICLE that makes up the nucleus, i.e., a PROTON or a NEUTRON.

nucleus The central core of an ATOM. It consists of a certain number of PROTONS (the atomic number) and NEUTRONS, with the exception of the ISOTOPE hydrogen-1, which has a nucleus consisting of only one proton.

O

ohm The SI UNIT of electrical RESISTANCE. A conductor has a resistance of 1 ohm if a POTENTIAL DIFFERENCE of 1 VOLT applied across it produces a current of 1 AMPERE.

orbit The path that a satellite, such as the moon, follows in its motion around its parent body. The path of an ELECTRON around the NUCLEUS is also called an orbit.

P

parallel In electric circuits a group of CONDUCTORS are described as being in parallel when they are joined in such a way that at the junction point the current divides to flow through all of them simultaneously, rather than flowing through each element in turn—as happens when elements are joined in SERIES.

paramagnetism The property of a substance that causes it to align itself in the direction of a magnetic field if freely suspended in the field. *See also* DIAMAGNETISM.

perfect gas *See* IDEAL GAS.

period The time taken for one CYCLE to occur in an oscillating or vibrating system, or the time interval between successive crests or troughs in a WAVE MOTION.

phase 1. One of the three states of matter—solid, liquid, or gas. 2. In a WAVE MOTION two points are in phase or of equal phase if their AMPLITUDES or displacements from a central position of rest or zero energy are the same.

photoelectric effect The property exhibited by certain metals and SEMICONDUCTORS of generating an electric current when illuminated.

photon A QUANTUM or particle of ELECTROMAGNETIC RADIATION, particularly of visible light and radiations at higher FREQUENCIES.

piezoelectric effect The property exhibited by certain crystals of generating an electric current when they are distorted. The reverse effect may also take place, i.e., the crystal distorts if subjected to a current, and vibrates if subjected to alternating voltage.

plasma A state of matter achieved at very high temperatures in which all the ATOMS in a gas form positive IONS and FREE ELECTRONS.

point source An ideal source of light that is infinitely small. A point source cannot exist in practice but is considered to exist when making optical calculations.

polarized light Light in which all the magnetic vibrations are in the same plane, called the plane of polarization. The electric vibrations take place in the plane at right angles to the plane of polarization, called the plane of vibration.

pole 1. Magnetic pole—the point at which a magnetic field appears to be concentrated and from which the magnetic field lines diverge. Magnetic poles can exist only in pairs of north poles and south poles, the field lines extending from one pole to the other. 2. Optical pole—the center of a curved mirror.

positron A unit positively charged SUBATOMIC PARTICLE that has the same mass as an ELECTRON. Positrons are the antimatter equivalents of electrons.

potential difference A measure of the difference in energy between any two positions.

potential energy The ENERGY that a body, particle, or system possesses by virtue of its position.

power The rate of doing WORK, measured as the ENERGY expended or gained per unit time.

precession The motion of the axis of a rotating body that occurs if a force is applied to the axis to cause it to turn. When precession occurs, the axis turns in a direction at right angles to the applied force.

primary cell An electric cell or battery that cannot be recharged, unlike a SECONDARY CELL or accumulator.

primary coil The coil of a TRANSFORMER or induction coil that receives the input current.

primary color One of the three basic colors that cannot be made by mixing other colors but which can be mixed in various proportions to give all other colors. In light, the three primary colors are red, green, and blue. In pigments, the three primaries are yellow, cyan, and magenta.

proton A very stable SUBATOMIC PARTICLE present in the nuclei of all atoms. The proton has a positive electric charge equal in magnitude to the negative charge of the electron. Its mass is 1.673×10^{-27} kilogram.

pyrometer An instrument that measures extremely high temperatures.

Q

quantum The fundamental (indivisible) amount of ENERGY, only integral multiples of which can exist. In RADIANT ENERGY (electromagnetic radiation) the actual value of the quantum varies with the FREQUENCY.

quantum mechanics A branch of physics dealing with microscopic particles and their interactions. It is based on the concept that particles also have wave-like properties.

quantum numbers A set of numbers that specify certain properties or the state of a system on an atomic scale. Each ELECTRON has a unique set of quantum numbers that specifies such properties as its energy and spin.

quarks A family of six fundamental particles. Protons and neutrons are made out of "up" and "down" quarks.

R

radiant energy ENERGY that is transmitted in the form of rays from one body or system to another. It comprises ELECTROMAGNETIC RADIATION.

radiation Any rays, WAVE MOTIONS, or particles that are emitted by a source, but more specifically: 1. RADIANT ENERGY or electromagnetic waves; 2. subatomic or atomic particles.

radioactivity The emission of particles or electromagnetic radiation when unstable particles disintegrate.

ray The straight path along which RADIANT ENERGY travels from one point to another.

reaction 1. The FORCE that always accompanies the production of another force (the action); the reaction is equal in magnitude to the action but opposite in direction. 2. Chemical reaction—a process in which ELEMENTS or COMPOUNDS interact to produce other elements or compounds. All chemical reactions involve exchange of outer ELECTRONS between the ATOMS taking part in the reaction. 3. Nuclear reaction—a process in which the nuclei of atoms taking part in the reaction change.

rectifier An electrical device that converts an ALTERNATING CURRENT into a DIRECT CURRENT. A DIODE that prevents current from flowing in one direction is an example of a simple rectifier.

reflection, angle of The angle between the light ray reflected from a surface and the normal (perpendicular line) at the point on the surface at which reflection occurs.

refraction The change in direction of a WAVE MOTION that may occur if the wave motion changes velocity as it passes from one medium into another.

refraction, angle of The angle between a ray that is refracted on entering a medium and the NORMAL (perpendicular line) at the point of surface at which refraction occurs.

refractive index The ratio of the velocity of light in a vacuum to the velocity in a medium is the refractive index of that medium. In practice, it is measured as the ratio of the sine of the angle of INCIDENCE to the sine of the angle of REFRACTION of a wave passing from one medium to another.

relative density (or specific gravity) The ratio of the DENSITY of a liquid or solid to the density of water.

relative motion Two bodies are in relative motion if either one or both are moving so that the distance between them changes. If both are moving but remain the same distance apart, then their relative motion is zero.

relative velocity The difference in VELOCITY between two moving bodies.

relativistic effect Any effect, such as a mass increase, which becomes apparent at speeds approaching that of light. *See also* RELATIVITY.

relativity Two important theories formulated by Albert Einstein that were developed from the postulates that all motion is relative and that the velocity of light is constant and independent of the motion of the observer. The special theory explains how MASS, length, and time change at velocities approaching that of light, and how mass and ENERGY are interconvertible. The general theory expresses GRAVITY as a property of space.

resistance The property of a substance that causes it to resist the flow of electric CURRENT. Electrical resistance is measured in OHMS.

resonance An increase in the AMPLITUDE of the oscillation of a vibrating system by the effect of another system oscillating at the same or a similar FREQUENCY.

rest-mass The MASS of a body when it is not in motion relative to the observer. RELATIVISTIC EFFECTS cause the mass of a fast moving object, such as a SUBATOMIC PARTICLE, to be greater than its rest-mass.

S

saturated vapor pressure The maximum pressure that a VAPOR can exert on its liquid at a given temperature.

scalar Any quantity that is defined solely by its magnitude. *See also* VECTOR.

scattering The deflection of beams of SUBATOMIC PARTICLES or rays of RADIANT ENERGY by their interaction with the nuclei or electrons of atoms or with whole atoms or molecules.

second The SI UNIT of time, defined as the duration of a particular number of CYCLES of an ELECTROMAGNETIC RADIATION of a specific FREQUENCY.

secondary cell An electric cell or battery, such as an accumulator, which can be recharged.

secondary coil The coil in a TRANSFORMER or induction coil that produces the output voltage.

secondary color Any of three basic colors that are produced by mixing two PRIMARY COLORS. In light, the secondary colors are yellow, cyan, and magenta. In pigments, they are red, blue, and green.

semiconductor A substance, such as silicon, whose electrical RESISTANCE decreases as its temperature increases, unlike normal conductors. These are widely used in electronics as their electrical properties can be finely controlled by including small amounts of impurities.

series A group of electrical elements is said to be in series when they are connected so that the current flows through them in sequence. *See also* PARALLEL.

SHM *See* SIMPLE HARMONIC MOTION.

siemens *See* MHO.

simple harmonic motion (SHM) The enforced oscillation of a body about a central point of equilibrium such that it accelerates toward the central point by an amount proportional to its distance from the point.

SI unit A unit in the *Système International d'Unités* (the international system of units) used in science throughout the world. The system is based on seven fundamental units—the meter, kilogram, second, ampere, kelvin, candela, and mole—from which all others are derived.

sound ENERGY radiated in the form of pressure waves into the medium surrounding a vibrating body.

space-time The view, formulated in the general theory of RELATIVITY, that the three dimensions of space and the one dimension of time are interrelated to form a four-dimensional space-time continuum.

specific gravity An alternative term for RELATIVE DENSITY, often used for liquids.

spectrum A display of RADIANT ENERGY, usually light, that has been split into its component WAVELENGTHS, by a prism or diffraction grating, for example.

standing wave A WAVE MOTION in which the oscillations or vibrations do not travel through space so that each point oscillates or vibrates by an unchanging amount.

statics The study of the behavior of bodies acted on by balanced forces.

strain The change in length or volume of a body subjected to a STRESS, divided by the original length or volume.

stress The force applied to a body. If the force applied is perpendicular or tangential to one side of a body, the stress is the magnitude of the force divided by the area of the side. If the force is applied to the whole body, the stress is simply the pressure exerted on the body.

subatomic particles Particles smaller than atoms. They are also known as elementary particles and fundamental particles. There are two main groups of particles. LEPTONS are subject to the electromagnetic or weak INTERACTIONS and comprise ELECTRONS, NEUTRINOS, and MUONS. HADRONS are subject to the strong interaction and comprise PROTONS, NEUTRONS, HYPERONS, and MESONS.

superconductivity The loss of electrical RESISTANCE that occurs in certain substances at very low temperatures approaching ABSOLUTE ZERO.

surface tension The property of the surface of a liquid that causes it to act like an elastic film. It is produced by the INTERMOLECULAR FORCES between the liquid molecules in the surface.

T

thermal energy The amount of energy stored in a substance due to the random motions of the atoms and molecules making up the substance.

thermocouple A device for measuring temperature. It consists of two junctions of wires of different metals, one junction kept at a known temperature and the other at the temperature to be measured. An electric current proportional to the temperature difference is produced.

thermodynamics The study of processes that involve the interconversion of HEAT and WORK.

transformer An electrical machine that alters the voltage of an ALTERNATING CURRENT, either increasing it (stepping-up) or decreasing it (stepping-down).

transistor An electronic device made of various kinds of SEMICONDUCTORS that can amplify a current passing through it or switch the current on or off in response to a small controlling signal.

transverse wave A WAVE MOTION in which the vibration takes place in a plane perpendicular to the direction in which the wave is traveling. Water waves and electromagnetic radiations are transverse waves.

U

ultrasonics The study and application of sound waves with frequencies too high to be audible.

ultraviolet light RADIANT ENERGY with WAVELENGTHS shorter than those of light but longer than X RAYS.

V

Van der Waals force An attractive force that is part of the INTERATOMIC and INTERMOLECULAR FORCES that exist between atoms and molecules.

vapor A gas that is below its CRITICAL TEMPERATURE so that it can be liquefied by pressure alone.

vector Any quantity that is defined by both the direction in which it acts and its magnitude. *See also* SCALAR.

velocity 1. The rate at which a body moves in a particular direction. 2. Relative velocity—the difference in velocity between two bodies. 3. Angular velocity—the rate of rotation of a body spinning about an axis. It is equal to the angle turned in unit time.

viscosity The resistance that a FLUID exhibits to its own motion or the motion of bodies through it.

volt The SI UNIT of ELECTROMOTIVE FORCE, electric potential, or POTENTIAL DIFFERENCE. A potential difference of 1 volt is required to drive a current of 1 AMPERE through a conductor of resistance 1 OHM.

W

watt The SI UNIT of POWER, equal to the expenditure of 1 JOULE of energy in 1 second.

wavelength The distance between two successive points in a WAVE MOTION that are of the same PHASE.

wave mechanics The theory that describes the behavior of SUBATOMIC PARTICLES in terms of WAVE MOTIONS. *See also* QUANTUM MECHANICS.

wave motion The movement of regular oscillations or vibrations in ENERGY outwards from a source of energy. The energy is transmitted by the wave motion from the source to a point through which it passes, where it produces the oscillations. Wave motions include sound waves, which are pressure oscillations, water waves and vibrating strings, which are oscillations in position; and electromagnetic waves, which are oscillations in electric and magnetic fields.

weight The FORCE with which GRAVITY acts on a body. Weight is correctly measured in pounds or NEWTONS and not in KILOGRAMS because the weight of a body may vary depending on the magnitude of the force of gravity, whereas the MASS of body, which is measured in kilograms, does not change. Weight is equal to mass multiplied by the acceleration due to gravity *(g)*. At earth's surface, in metric units, a mass of 1 kilogram has a weight of about 9.8 newtons.

work The ENERGY expended or gained when a force results in the motion of a body or system.

X

X rays RADIANT ENERGY shorter in WAVELENGTH than ultraviolet rays and originating from the motion of electrons.

Units and measurement

Basic Si Units

Quantity	Unit	Symbol
Length	meter	m
Mass	kilogram	kg
Time	second	s
Electric current	ampere	A
Temperature	kelvin	K
Luminous intensity	candela	cd
Amount of substance	mole	mol

Metric Prefixes

Prefix	Multiple	Symbol	Prefix	Multiple	Symbol
kilo	10^3	k	centi	10^{-2}	c
mega	10^6	M	milli	10^{-3}	m
giga	10^9	G	micro	10^{-6}	μ
tera	10^{12}	T	nano	10^{-9}	n
			pico	10^{-12}	p
			femto	10^{-15}	f
			atto	10^{-18}	a

Principal Physical Constants

Quantity	Symbol	Value	Quantity	Symbol	Value
Velocity of light			Gas constant (molar)	R	8.314510 J mol^{-1} K^{-1}
in vacuum	c	2.997925×10^8 m s^{-1}	Mass of neutron	m_n	1.67482×10^{-27} kg
Permeability of vacuum	μ_o	$4\pi \times 10^{-7}$ H m^{-1}	Avogadro constant	N	6.0221367×10^{23} mol^{-1}
Permittivity of vacuum	ϵ_o	8.85419×10^{-12} F m^{-1}	Boltzmann constant	k	1.38054×10^{-23} J K^{-1}
Mass of electron	m_e	9.10908×10^{-31} kg	Faraday constant	F	9.64870×10^4 C mol^{-1}
Radius of electron	r_e	2.817777×10^{-18} m	Gravitational constant	G	6.670×10^{-11} N m^2 kg^{-2}
Elementary charge	e	1.60210×10^{-19} C	Planck constant	h	$6.6260755 \times 10^{-34}$ J s
Charge/mass ratio for electron	e/m_e	1.758796×10^{11} C kg^{-1}	Rydberg constant	$R\infty$	1.097373×10^7 m^{-1}
Mass of proton	m_p	1.67252×10^{-27} kg	Stefan-Boltzmann constant	σ	5.6697×10^{-8} W m^{-2} K^{-4}

Customary units and conversion factors

Length		Mass		Weight/Force	
1 mile	1.6093 km	1 kg	0.06854 slugs	1 newton	0.2248 pounds
1 yard	0.9144 m	1 slug	14.59 kg	1 pound	4.448 newtons
1 foot	30.48 cm				
1 inch	25.4 mm			**Temperature**	
				1 degree Fahrenheit	5/9 kelvin
1 kilometer	0.62137 miles				
1 meter	39.37 in.				
1 centimeter	0.3937 in.			1 kelvin	9/5 degrees Fahrenheit
1 millimeter	0.03937 in.				

Index

Credits

The following have provided photographs for this book: Cover photo—Doug Johnson/SPL/Photo Researchers. Aga Thermovision/Science Photo Library 69; Air Products 30; Heather Angel 73, 99, 106, 121; Avery Denison 25; Dr Alan Beaumont 33; S.C. Bisserot/Nature Photographers Ltd 109; Paul Brierley 27, 35, 47, 48, 73, 89, 90, 93, 94, 95, 113, 120, 126, 129; Robert Brenner/PhotoEdit 130; Nicholas Brown/Nature Photographers Ltd 109; Dr J. Burgess/Science Photo Library 98; Michael Burgess/Science Photo Library 110; Central Electric Research Laboratories 110; CERN/Science Photo Library 19, 143; Chris Christodoulou 107, 108; Gill Clark/Owenbank 30, 43, 85, 87; Jean Collombet/Science Photo Library 17; Colour Library International 64, 97, 121; E. H. Cook/Science Photo Library 131; Peter David/Seaphot 93; Dick Durrance/Woodfin Camp & Assoc. 9; H. Edgerton/Science Photo Library 60; H. F. Epco 47; ESA/Science Photo Library 75; Mary Evans Picture Library 140; Fermilab 10, 17; Fire Research Station, Borehamwood 35; Vaughan Fleming/Science Photo Library 134; Geoscience Features 34; Roberto Germaine 43; Leon Golub/Science Photo Library 74; Dr Steve Gull & John Fielder/Science Photo Library 96; Simon Fraser/SPL/Photo Researchers 8; David Frazier Photolibrary 132, 133; Tony Freeman/PhotoEdit 128; Jeff Greenberg 11; Hawker Siddeley Group 123, 125; Health & Safety Executive 104; Dr J.G. Hills/John Innes Institute/Science Photo Library 15; Fritz Hoffman/JB Pictures 39; Alan Hutchinson Library 33; Illustrated London News 128; Institute of Geological Sciences 22, 92; Douglas W. Johnson/Science Photo Library 11; David Jones 23, 40, 44, 50, 51, 57, 83, 91; Keystone Press Agency 25; Gary Ladd/Science Photo Library 101; London Transport 42; Dr Jean Lorre/Science Photo Library 145; Jerry Mason/Science Photo Library 17; Tom McCarthy/PhotoEdit 18; CoCo McCoy/Rainbow 128; NASA 53; NASA/Bandphoto 53; NASA/Science Photo Library 37, 49, 53, 72, 108, 125; National Physical Laboratory 141; OMI 124; Martyn Page 56, 62, 67, 74, 80, 89, 105; David Parker/Science Photo Library 2, 79, 82, 84, 85, 136; RCA Records 107; Louis A. Raynor/Sportschrome East/West 55; Dr D. R. Roberts/Science Photo Library 144; Royal Collection Hampton Court/Department of Technology Courtauld Institute 100; Royal Greenwich Observatory 140; Graham Saxby/Advanced Holographics Ltd 95; Science Photo Library 131; Dr Gary Settie/Science Photo Library 9; Lee Snyder/Photo Researchers 10; Sinclair Stammers/Science Photo Library 27, 30; Tony Stone Images 14, 18, 21, 28, 41, 45, 54, 57, 63, 68, 70, 75, 78, 87, 88, 100, 115, 116, 119, 122, 125, 139; William Strode/Woodfin Cam & Assoc. 38; David Sutherland 117, 122, 129; Trepel 46; UKAEA 101; A. V. Crewe, M. Ohtsuki & M. Utlaut/University of Chicago 12; University of Tokyo 17; Vautier-De-Nauxe 130, 135; S & J Walker/Science Photo Library 36; John Walmsley 32; Thomas J. Watson Research Center/IBM 39; Ralph Wetmore/Science Photo Library 103.